Advances in OFFSHORE OIL & GAS PIPELINE TECHNOLOGY

Edited by

R F de la Mare

Chairman of the Postgraduate School of Technological Management
University of Bradford

GULF PUBLISHING COMPANY
Houston, London, Paris, Tokyo

First published in Great Britain by
Oyez Scientific and Technical Services Ltd
Bath House, (3rd Floor), 56 Holborn Viaduct, London EC1A 2EX, England

This edition published 1985 by
Gulf Publishing Company, Houston, Texas.

Library of Congress Catalogue card number: 84–82620

ISBN 0 87201 035 X

Printed and bound in Great Britain

To Alex my Champion

CONTENTS

PREFACE

In recent times, man's inexorable demand for petroleum products has intensified the search for oil and gas in regions of the world which hitherto were unexplored. This has led to the development of petroleum resources in offshore areas which are especially harsh due to deep water and/or the severity of prevailing climatic conditions imposed by high winds, stormy seas and low temperatures, which exist in areas such as the northern North Sea and polar regions. In many cases, the only means of economically exploiting these resources entails the construction of long and expensive submarine pipelines from offshore platforms to onshore terminals. Networks of interlinking pipelines have also materialised in several offshore regions to enhance the development of marginal fields and mitigate some of the risks arising from the possible failure of singular pipelines. Evidence suggests that the pace of recent developments will continue as onshore reserves of oil and gas diminish, with the result that submarine pipelines will become extremely important to our welfare and, effectively, the arteries in an increasingly energy-hungry world.

A review of the literature concerning pipeline technology and management shows that textbooks and journals dealing with topics such as the fluid flow through and around a submarine pipeline are literally legion, and others concerned with their material science properties abound. However, very few deal with offshore pipelines from a perspective which recognises clearly that pipelines need designing, fabricating, installing, operating, inspecting, maintaining, repairing and possibly replacing from a 'total system' viewpoint. This book endeavours to fill that 'gap'; an objective which determines its scope, presentation and content.

The overall *scope* of this book, therefore, is to demonstrate why it is important to appraise the fitness of purpose of a pipeline throughout its economic life – from its inception to selection, design and commissioning, operation and maintenance, through to its eventual replacement and retirement – and show how this life-cycle appraisal can be accomplished.

As regards *presentation*, this book initially formulates those design criteria which must be met if a pipeline is to fulfil its intended purpose, and then it proceeds with a sequential examination of each design factor, cognizant of the fact that they are closely interrelated and that the final design solution really involves an iterative process in the search for an 'optimised' design.

In selecting the *contents* of this book, we have taken into account the needs of the engineer wanting to practise the general principles of pipeline technology (interpreted in its widest context) and the manager seeking an appreciation of the 'total system' implications of pipeline design. To meet these ends, the book starts with matters concerning the environment, fabrication and installation (such as geotechnology, lay-stress analysis, buckling and connecting) and then proceeds to problems encountered with surveillance, maintenance and repair. Excepting our interest in the economic implications of transporting gases in the 'dense phase', and the diameter-dependent and cost effect of pressure drop, this book is not generally involved with the *internal* fluid flow problems of pipelines because this subject is properly dealt with elsewhere. However, the book is particularly concerned with matters which relate to the *external* flows of fluids around a pipeline, especially when they threaten its integrity. Indeed, a concern for the safety and reliability of pipelines pervades this book and culminates in a case history which demonstrates how a 'balanced design' was achieved in practice.

In the process of optimising any system it behoves the decision-maker to select between alternative designs involving competing technologies. Petroleum transportation is no exception to this rule. To conclude the book, therefore, we examine the merits of single point moorings in conjuction with shuttle tankers as an alternative means of petroleum transportation, and we formulate those general conditions under which single point mooring can be favourable.

Since the substance of this book is virtually boundless, the greatest difficulty with its editing lay in knowing what to omit. As the title suggests, this book is concerned with advances in pipeline technology. Effectively therefore, its contribution is an incremental one, in that it builds on an existing foundation of knowledge in pipeline technology, which would require many books to describe in detail. To counter any deficiences which might arise from the limitations of space, we therefore provide comprehensive reference lists, incorporating books and papers, at the ends of most chapters, so readers should experience little difficulty in pursuing each topic beyond the scope presented here.

This book is based on the proceedings of recent Oyez Scientific Technical Services' European Offshore Oil and Gas Pipeline Technology Seminars from which nineteen papers were selected, updated and rewritten in a unified form. Permission to reproduce this material here is given by the authors whose biographies appear at the back of this book.

Besides my great debt to these authors, I am especially grateful to Robin Colquhoun and Deborah Hooper for their unstinting efforts in checking the technical and editorial content of this book. I would also like to thank Tony Richardson for the design and production of the book and Barbara Pickard who diligently and enthusiastically typed the manuscript.

Last but not least, my thanks are due to friends and colleagues too numerous to mention, who encouraged, assessed and made possible my editing this book. Based on their collective experience, we expect this book to be of value to engineers and managers alike; those who are working (or expect to be working) in the offshore industry in general, and the field of submarine pipeline technology in particular. In this respect, the book should be useful for postgraduate and post-experience (extension) programmes in offshore technology and management. It should also be of direct relevance to practising pipeline engineers who wish to develop a systems approach to pipeline design, and to practising managers seeking to expand their knowledge of such matters.

R F de la Mare
University of Bradford
March 1985

1

GEOTECHNICAL CONSIDERATIONS IN SUBMARINE PIPELINE DESIGN

The principal objective of pipeline design is to select the pipeline dimensions and route, and its method of fabrication, installation and protection so that it can transport the specified production requirement at an acceptable level of risk, whilst incurring minimum life cycle costs. To this end, the pipeline engineering process, as depicted in Figure 1.1 (overleaf), must consider the constraints imposed on the pipeline design by the nature of the environment, the methods of construction, operation and maintenance and the changing state of pipeline technology. These constraints are described briefly in Table 1.1 (see over) and more fully elsewhere in this book. Quite often geotechnical considerations play only a minor part in the design process, when the pipeline is located in tranquil and shallow water. However in hostile environments, unusual soil conditions and deep water, geotechnical considerations can assume a major role in this process.

In view of these constraints, the pipeline designer has to gather data and information to define them at the outset. The design process then focuses on a logical balancing of these constraints to yield an optimum design. Of special significance in this respect is the use of hazard mitigation techniques, such as: backflow valves to prevent fluid escape; pumping and compression shutdown systems to prevent over-pressurisation and minimise leakage; breakaway couplings to control the points of failure; the selective use of pipe coatings to minimise the effects of external forces; failure detection controls and sensible maintenance and repair systems. The most important hazard mitigation technique, however, is perceptive siting and routing of the pipeline to avoid or minimise exposure to initial and subsequent hazards. It is this aspect of pipeline design which is covered in this chapter.

1.1 Route Reconnaissance and Hazard Identification

The routing of pipelines to avoid hazardous conditions starts with detailed surveys and geological studies of all the potential routes. These surveys define the seabed conditions including, for example, the locations

Table 1.1 A Brief Description of the Constraints Involved in Pipeline Design

CONSTRAINT TYPE	DESCRIPTION
Environmental	These require the definition of currents, mudslides, fault movements, soil profiles and bathymetry, which could affect the stability and integrity of the pipeline during its economic life.
Constructional	These include the equipment needed for fabrication and installation, the specification of the pipeline steels, welding and quality controls, and pipeline bedding, backfill and armouring.
Operational and Maintenance	These must consider the need for tie-in points, flowrates, pressure and temperature profiles and the corrosivity of the fluids to be transported, methods of pipeline surveillance and monitoring, the need for maintenance and possibly repair, means of controlling fluid escape and emergency procedures.
Design	These include the methods of analysis to be used, route guidelines, regulatory requirements and codes, allowable stresses and factors-of-safety. Economic considerations must include the costs of construction, operation, surveillance, maintenance failure and repair. Furthermore, the potential effects of the pipeline on other systems must be fully explored for their economic, political, environmental and social effects, especially when these relate to a pipeline failure.

of faults, slumps, slump channels, diapirs and other geological features including the soil conditions and the density and nature of local shipping. Side-scan sonar, high resolution profiling, sub-bottom profiling, soil coring and drop penetrometers are but some of the techniques used to provide this information. Geology-based geotechnical studies are an essential ingredient for analysing the route survey data to define relict conditions, initial hazards and the likely location and characteristics of future hazards. Knowledge of the underlying conditions that create the seafloor characteristics becomes a key factor in predicting the locations and characteristics of these or similar hazards during the lifetime of the pipeline.

In deep water, pipelines can be susceptible to a multitude of environmental hazards with variable degrees of severity. The designer may be required to demonstrate that a pipeline has adequate serviceability under wave and current hydrodynamic loadings, displacements induced by earthquakes and faulting, spanning across bottom irregularities, and sufficient strength and flexibility to accommodate the forces induced by slides. In effect, therefore, to devise a pipeline to withstand multiple hazards, the designer is confronted with conflicting requirements in his choice of methods for pipeline protection. For example, to protect against hydrodynamic forces, it is preferable to bury the line. On the other hand, to reduce the forces caused by slides, it may be advantageous to place the line on the seabed, thus reducing potential soil loading and allowing the

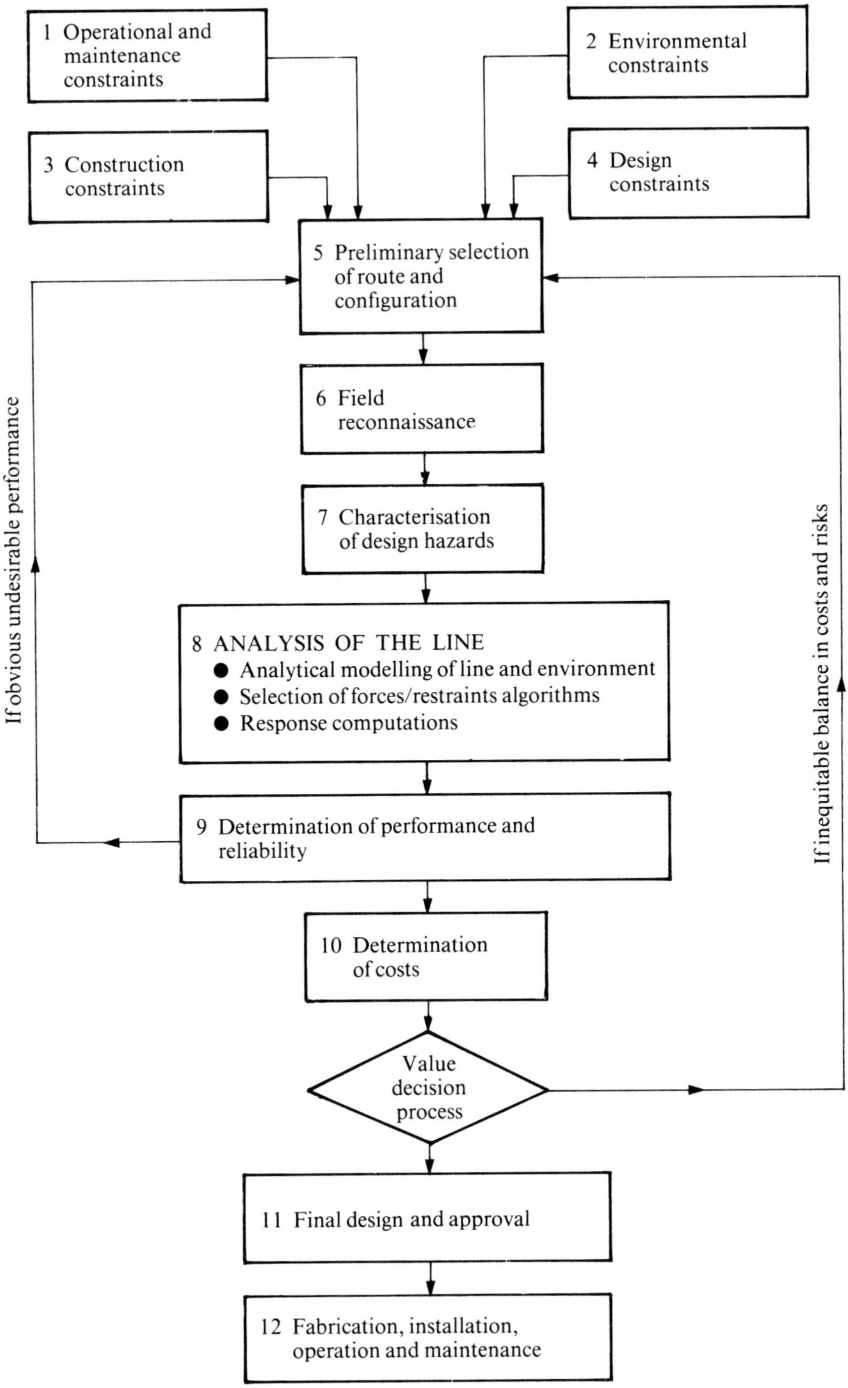

Figure 1.1 The Methodology of Pipeline Design

line to deform with lower induced stresses. An exposed pipeline is preferable in earthquake-prone areas, since the forces induced by an earthquake or fault movement are proportional to the amount of soil restraint around the pipeline. Furthermore, if the pipeline is unrestrained at the soil surface and crosses faults at advantageous angles, then the induced forces will be greatly minimised. Table. 1.1.1 presents a summary of balanced design considerations for pipelines. The positive and negative results of the various methods available for protection of the pipeline are noted relative to each of the primary environmental hazards. To reach an effective and efficient design for the pipeline, all conditions and constraints must be considered and their relative importance assessed so as to reach a balanced design that optimises the reliability and cost of the line.

Table 1.1.1 Balanced Design Considerations for Pipelines

ENVIRONMENTAL HAZARDS	METHOD FOR PROTECTION Bury line	Expose line	Weight coating	Pipe anchor	Stronger line
Hydrodynamic lift, drag and inertia	+	–	+	+	
Vortex shedding	+	–	+	+	+
Scour, pipe spanning	+	–	–		
Sand-waves	+	–		+	+
Mudslides, mudflows and turbidity currents	* – ** +	+ –	– –	– +	+ +
Flotation	+	–	+	+	+
Liquefaction	Design pipeline so that its unit weight be as close as possible to unit weight of fluidised soil				
Mudlumps, diapirs	–	+	–	–	+
Faulting, earthquakes	–	+	–	–	+

NOTE: (+) positive results produced from method of protection
(–) negative results produced from method of protection
* pipeline in mudslide area
** pipeline downstream of mudslide

The principal classes of geotechnical hazards in deep water include:

(i) earth-shifts and fault movements that induce lateral and vertical forces on a pipeline; and

(ii) bottom obstructions or irregularities, sand-waves, and seafloor depressions caused by degassing, sliding or solution (in calcareous materials) and the effects of scour, which can result in pipeline spans. Hydrodynamic effects can then induce pipeline overload or fatigue failures, and in some cases cause a loss of weight coating leading to pipeline flotation.

A wide variety of geologic and oceanographic processes combine to influence these hazards in deep water. These include: steep slopes; blankets of soft soils; outcrops of very old sediments (associated with the formation of the ocean basins involving calcareous or salt accumulations); the interactions of deep ocean and shallow water continental shelf processes (such as the breaking of internal waves and turbidity flows initiated by storms affecting shallow water sediments). They either induce loadings on pipelines (slides) or remove support from the pipeline (spans).

1.2 Characterisation Problems with Deep Seabeds

Currently, five basic techniques are used to characterise deepsea soil conditions. These comprise side-scan sonar surveys, high resolution geophysics, electric, nuclear and acoustic logging, sediment sampling and testing, and *in-situ* tests using a variety of penetrometer and pressuremeter devices. All these methods employ proven techniques and equipment, which are used regularly in shallow water, and with special adaptations they can also be used in deepwater service. The special difficulties with their application in deep water result primarily from equipment design problems associated with high pressure and the need for pressure relief upon retrieval. It should be noted that equipment problems have often been the main cause of misleading data and information on deepwater geotechnical surveys. Until these equipment problems are solved, the large uncertainties of identifying the seafloor conditions, hazard characteristics, and soil characteristics will prevail, and the geotechnical engineering of deepwater pipelines will remain more of an art than a science.

1.3 The Study of Soil-Pipeline Interaction

Four major categories of geotechnical problems are relevant to pipeline design, namely:

(i) the identification of seafloor movement zones;
(ii) the determination of soil loadings or deformations associated with these zones;
(iii) the determination of soil restraints active at the time of pipeline movement; and
(iv) a consideration of construction effects such as bottom tows, trenching and burial.

In this chapter we focus our attention on the second two categories of soil-related problems, hereafter referred to as 'soil-pipeline interaction'. The other two categories of problems receive proper attention elsewhere in this book.

The following considerations are of primary importance in the mechanics of soil-pipeline interaction:

(i) soils behave as nonlinear, inelastic, non-homogenous materials;
(ii) pipelines behave as nonlinear, inelastic elements;
(iii) most of the analytical models currently used to characterise soil-pipeline interaction fall far short of simulating reality, due principally to their deficiencies in dealing with the important elements of nonlinear behaviour;
(iv) engineering estimates involved with the characterisation of soils and pipelines that were originally intended to provide conservative designs can, under certain circumstances, provide unconservative results for some pipeline design problems, especially those associated with slides, faults and earth movements;
(v) construction effects are extremely important as they influence the characteristics of the soils surrounding the pipeline and the stresses and displacements induced into the line during the installation process.

Much experimental and analytical work has been done to characterise soil responses for a wide variety of pipeline-soil interaction problems. A detailed discussion of these matters is provided in Appendix A for those readers who wish to study this subject in detail. For the more casual reader, however, a basic understanding of the behaviour of a pipeline to earth movements can be obtained from the illustrations which follow in this chapter.

Figure 1.3.1 shows two cases of pipeline-soil interaction which are based on experimental results by Audibert *et al* (1979). They relate to a pipeline which is pulled laterally through soil when it is buried beneath the mudline. In the shallow burial case (a) there is a complex interaction of surface related failure surfaces. In the deep burial case (b) there is plastic flow around the line. In effect, therefore, the depth of burial of the pipeline has an important influence on the soil-pipeline interactive forces.

1.4 Pipeline Behaviour in Slides

One of the most common deepwater geotechnical hazards is that of slides which can be initiated by a wide variety of phenomena, including earthquakes, waves, currents, and simply by excessive deposition of weak soils on steep slopes. In this section we discuss the behaviour of pipelines in slides comprising soft, cohesive, clay soils which typify those found in some deepwater locations. The soil strength and water content profile chosen for the purpose of this illustration is given in Figure 1.4.1 (overleaf), which shows that, close to the mudline, the soil is weak with a shear strength approximately 40 to 80 pounds per square foot (1.9 to 3.9 kPA) in the top 3 to 4 feet. However, the shear strength increases significantly with depth. It also shows that the water content of the near surface soils usually exceeds 100 per cent, in which case they behave more like fluids than plastic solids.

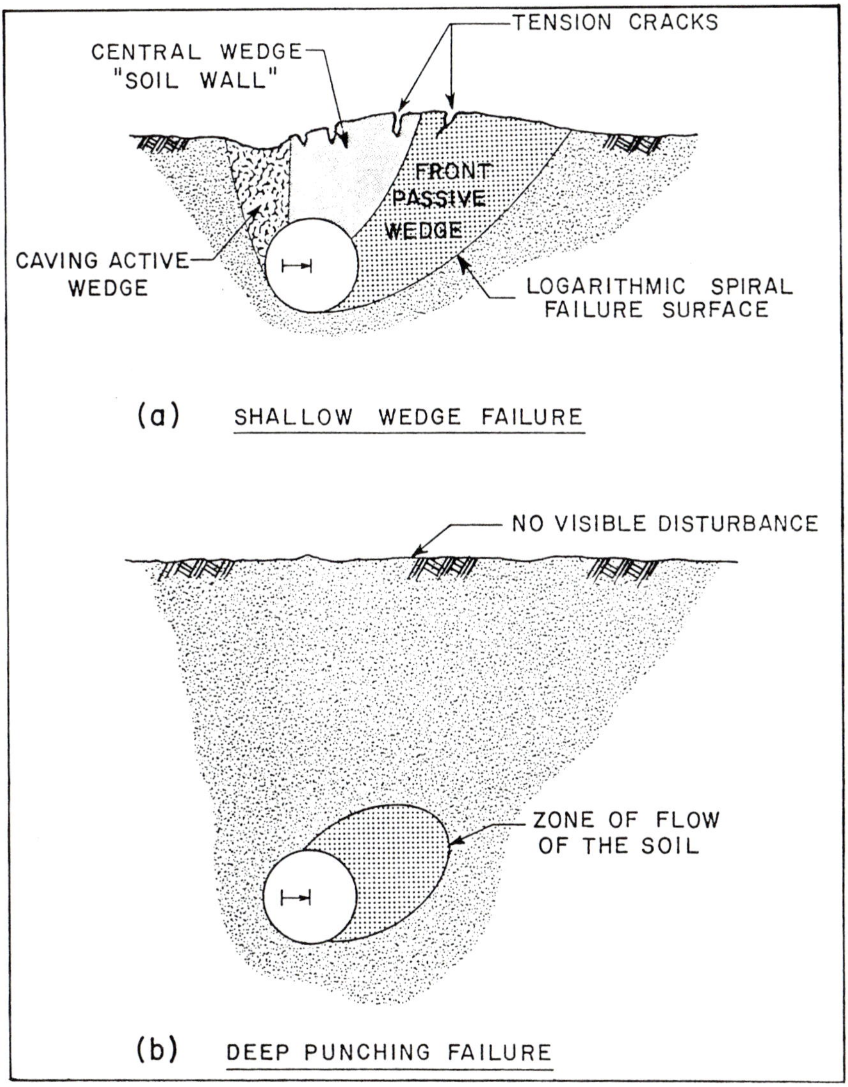

Figure 1.3.1 Soil-Pipeline Interaction

Figure 1.4.2 (overleaf) shows the stresses which are induced in a 12.75-inch diameter pipeline when it is subjected to a slide having a width of 500 feet moving perpendicular to the pipeline axis. The stress levels referenced are those of an equivalent elastic pipeline. The true non-linear yielding response of the pipeline was included in the analytical

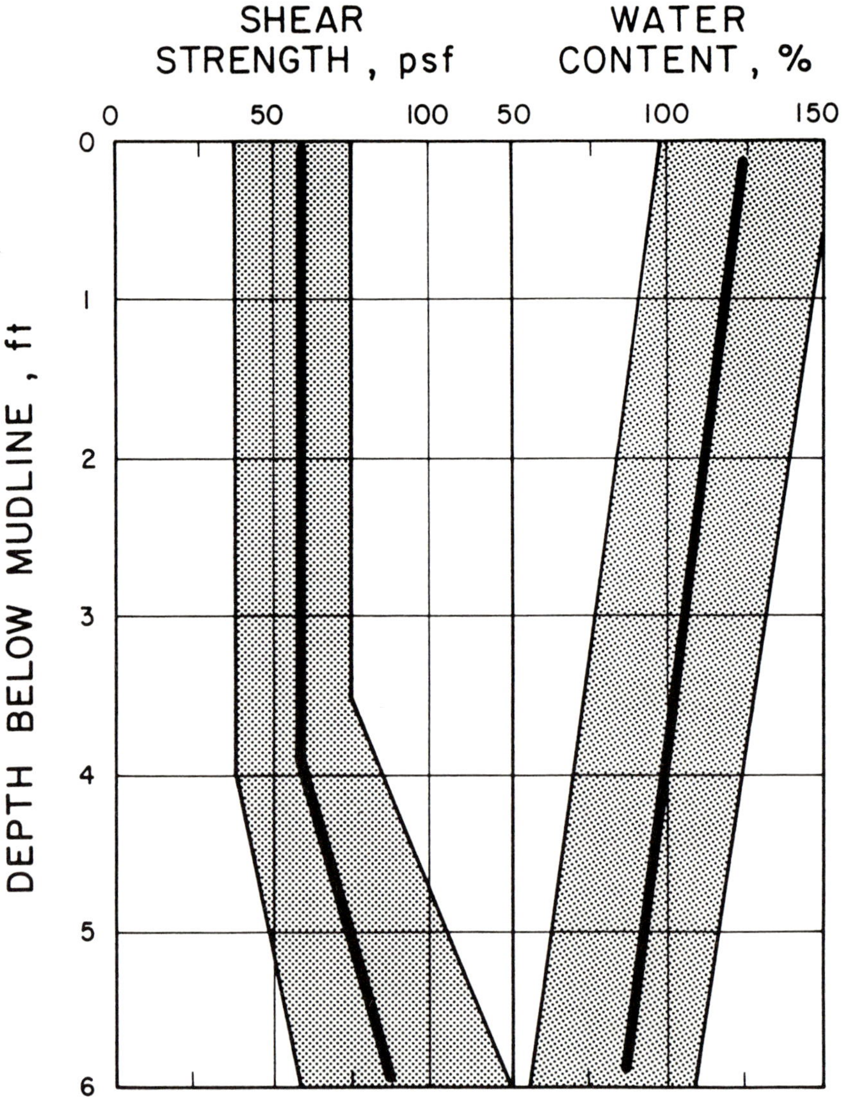

Figure 1.4.1 Soil Shear Strength and Water Content in Slide Area

model. These results show that if the pipeline can be kept or weighted to remain close to the mudline, then it would be able to withstand the range of potential slide forces. However, if the line were buried beyond a depth of 1 to 2 feet, then extensive yielding would result. Note that the concrete cover basically has the effect of increasing the diameter of the pipeline and increasing the forces induced by the slide, with no strength benefit.

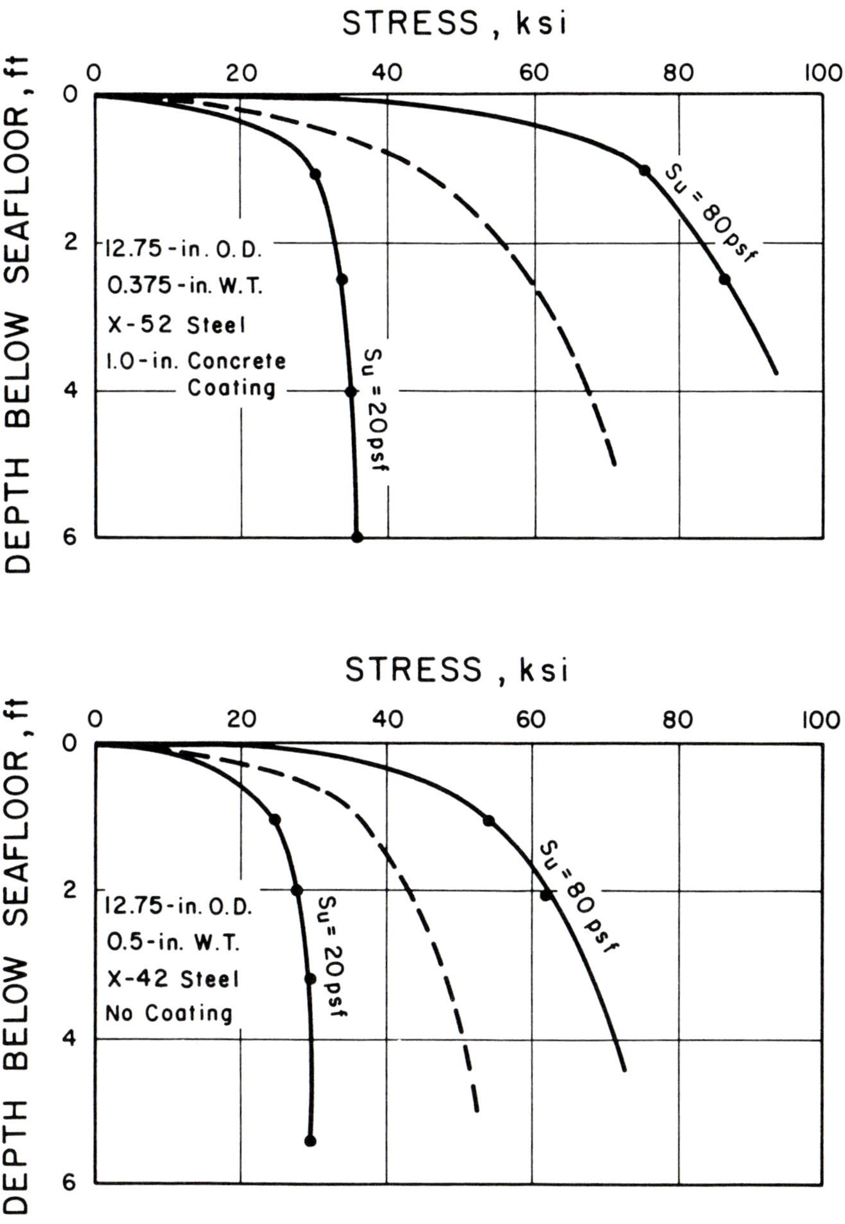

Figure 1.4.2 Combined Tensile and Flexural Stresses Developed in Pipeline Subjected to Transverse 500 Feet (150 Metres) Wide Slide

Figure 1.4.3 shows the stress and displacement effects on unweighted pipelines, of 8.625 and 10.75-inch diameter, resulting from mudslides of varying widths. In these cases a pipeline would sink to a depth at which the resisting soil strengths and soil-water buoyancy effects just balance the weight of the pipeline and its contents. In effect, this means that the heavier a pipeline is made or the deeper that it is forced, then the larger will be the soil forces imposed on the pipeline if the soils move. The results of Figure 1.4.3 show that, for the cases studied, the pipelines are capable of resisting slides with widths approximating 2,000 to 2,500 feet, in which case pipeline maximum downslope displacements of 150 to 300 feet are indicated.

Similar results to those illustrated in Figure 1.4.3 apply for a wide range of pipe diameters (D), wall-thicknesses (t), soils forces at the depth of self-

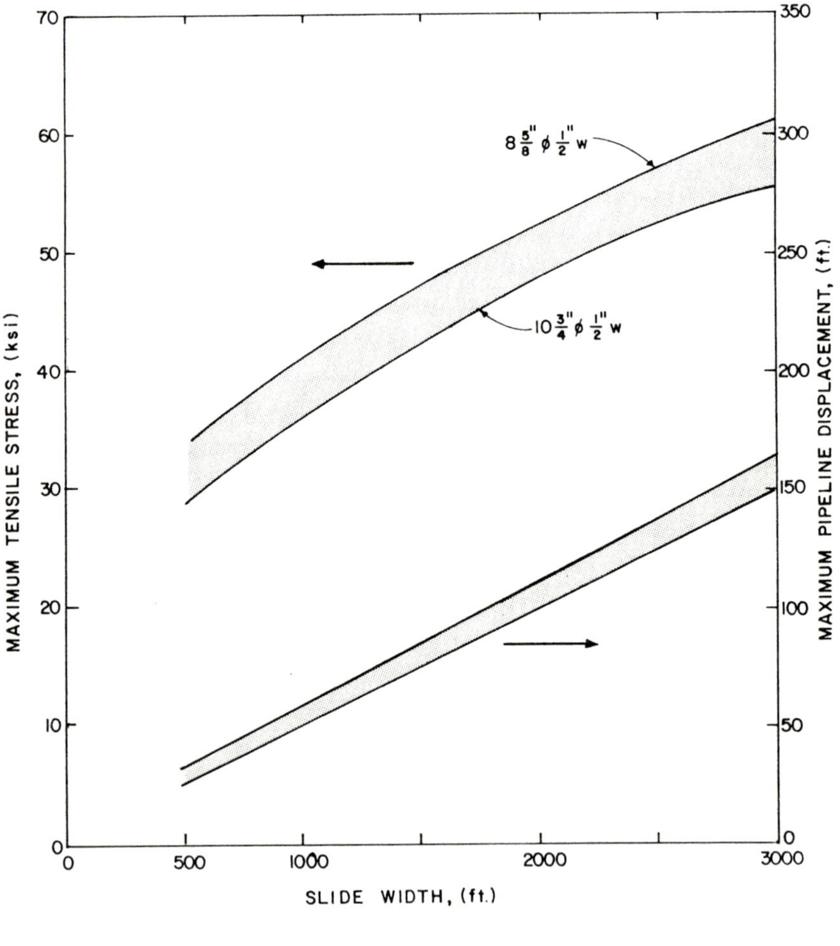

Figure 1.4.3 Predicted Pipeline Behaviour under Slide Forces

burial (S) and slide widths (W). These results are summarised in Figure 1.4.4, which plots the maximum tensile stress in the pipe, σ_t (measured in psi) against the pipeline slide force to size ratio, F/S (in psi) where:

$$F/S = \frac{S \times W}{D \times t}$$

Thus, given the slide force to size ratio, F/S, one can determine approximately the maximum tensile stress induced in a pipeline by a slide of a given width.

Similar results apply to the maximum displacements of the pipeline at the centre of a slide. These results are shown in Figure 1.4.5 (overleaf) in which case a good approximation is that the maximum downslope displacement of the pipeline is equal to about 1/20th of the slide width.

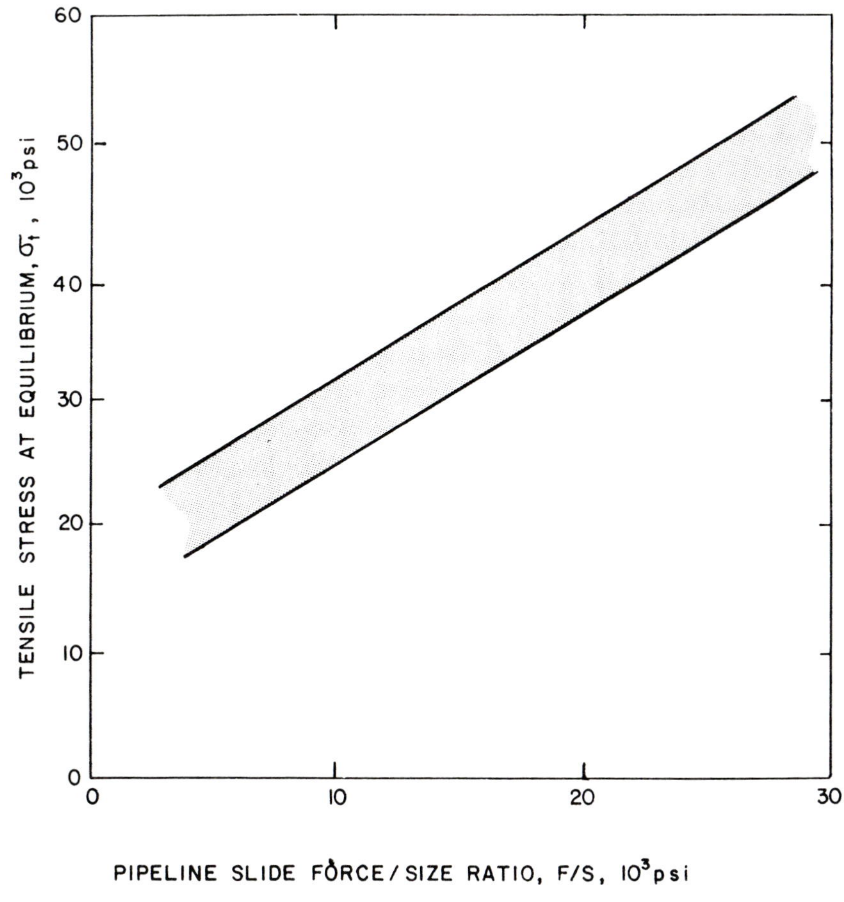

Figure 1.4.4 Design Guidelines for Maximum Tensile Stresses Induced in Pipeline by Slides

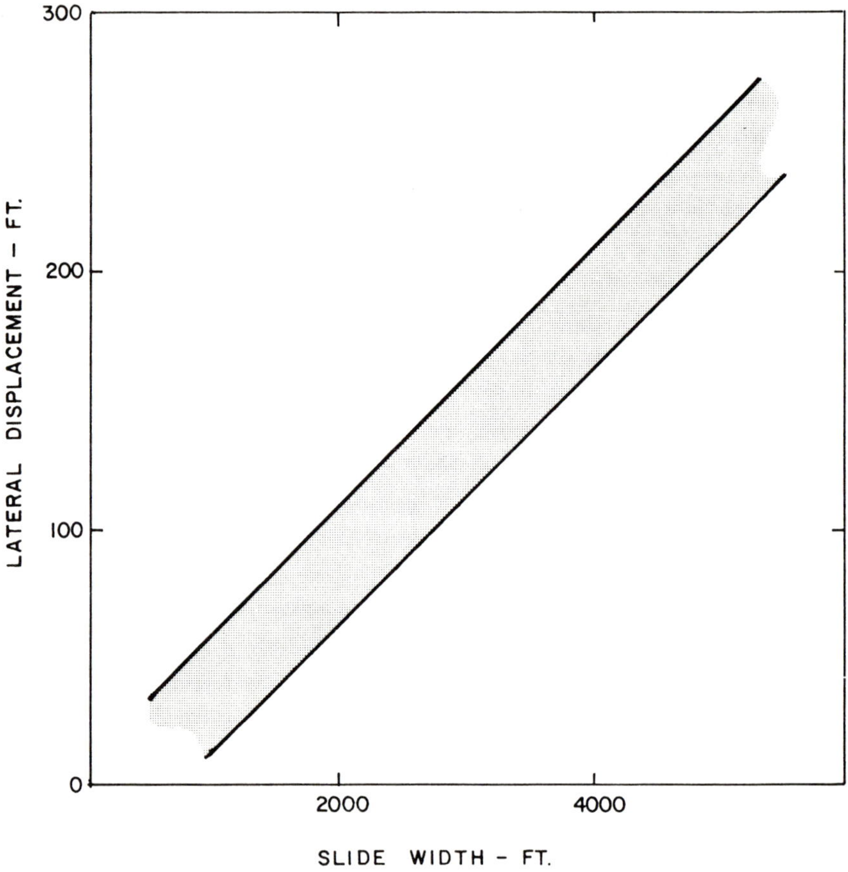

Figure 1.4.5 Design Guideline for Maxium Downslope Displacement of Pipeline Induced by Slides

1.5 Conclusions

For the majority of pipelines which will be founded in deep water, few unsolvable geotechnical problems exist. Most of their sites are expected to be relatively quiescent, with stable and reasonably understood soils. However, deepwater pipelines which intend to cross active marine slopes can expect a wide variety of important geotechnical problems, including slide forces/displacements and spanning forces/displacements. To this end, research and developement is needed to reduce existing uncertainties concerning soil loads and restraints on pipelines. The principal sources of these uncertainties lie in our lack of knowledge of the geologic-oceanographic processes and conditions that pose hazards and a lack of equipment, which is data-effective and reliable, to gather information on seafloor conditions and soils. Research and development is also needed to help identify and anticipate hazards during the lifetime of a pipeline. An adequate understanding of the basic phenomena and conditions, integrated into realistic analytical models, calibrated with prototype field data and experience would represent a viable approach to this challenge. Moreover, research and development is also needed to allow more reliable and realistic characterisation of the properties of the soils themselves, both as they exist *in-situ*, as they exist after construction, and as they exist during the hazard events of serviceability interest. Integrated, multi-sensor techniques using logging, sampling, laboratory testing, and *in-situ* testing should be developed to provide representative indices of soil behaviour which are needed for adequately designing and engineering deepwater pipelines.

1.6 Bibliography

Audibert, J. M. E., Lai, N. W. and Bea, R. G., 'Design of Pipelines – Sea Bottom Loads and Restraints', *ASCE Speciality Conference on Pipelines in Adverse Environments – A State of the Art*, (New Orleans, Louisiana, January 15–17, 1979).

R G BEA

2

PIPELINE LAY-STRESS ANALYSIS

Traditional marine pipeline design often assumes that the dynamic stresses which occur during installation will not exceed one third of the static stresses. As the industry moves into more exacting environments, this assumption becomes less tenable and careful analysis of the dynamic stresses becomes essential in order to define the limiting weather conditions which could result in fatigue failure and overloading of a pipeline.

In this chapter we describe a computer model which can simulate realistically the dynamic behaviour of pipelines under various lay conditions and several case histories are used to illustrate its effectiveness.

2.1 A Description of the Computer Model

The 'Dynamic Seapipe' computer program, belonging to the Danish Hydraulic Institute (DHI), was developed as a tool to determine stresses in pipelines during their installation and other post-installation dynamic situations. It is built on a mathematical concept formulated and coded originally by Abbott, a co-author of this chapter, and then developed by DHI to meet the practical requirements of the offshore industry. This program is versatile, simple to apply, and economical to use. It is designed to model the stresses which occur with S-method pipelaying, using fixed or articulated stingers, J-method and drill-ship laying, shore pulls with buoyancy, surface and sub-surface tows, deepwater anchor mooring lines, spans subjected to vortex shedding, and expansion devices subjected to environmental loading. A detailed description of the mathematical model forming the basis of this program is given by Bryndum *et al* (1982).

The basic model is derived using classical elastic rod theory in a small strain, large displacement formulation, with axial and torsional deformations neglected. Two separate coordinate systems are used, one fixed (x, y, z) and one local orthogonal to the pipe (L, V, H). The model comprises ten equations, two describing the equilibrium transverse to the pipe in the M and V directions, one describing axial equilibrium, two giving bending moments in the H and V planes and three reflecting kinematic compatibilities. Environmental conditions and reaction forces are expressed as equation parameters. As with most computer programs, the solution to these equations requires a specification of the initial situation and the boundary conditions at each instant in time.

During the initial time-steps of the computation, suitable boundary conditions involving motions and forces are applied to one or both ends of the pipe until the more complex real life initial situation is achieved. The time-steps used during this phase may be quite different from those used subsequently. At each phase in the computation it is necessary to prescribe ten independent boundary conditions. For static stress calculations these will be constant, whereas for dynamic stress calculations some will be time-varying. The program provides a range of facilities for the addition of regular/irregular motion and force function transients into the analyses.

The set of differential equations is solved numerically using finite difference approximations. To reduce the complexity of the computations, a linearised multi-level difference scheme is used. The force coefficients, and more particularly the reaction forces in the pipe, are dependent on the solution at all levels, so several iterations are required to define their correct values at each time-step. The code incorporates convergence tests and convergence rate options. At each level, the equations are solved with an efficient double-sweep algorithm which makes optimal use of the restricted band widths of the coefficient matrices. In the finite difference scheme, all boundary conditions are prescribed in a fully centred manner so as to introduce them with the same order of accuracy as that of the general difference scheme.

During the initial formulation of a modelling system it is difficult to conceive all possible configurations and loadings that may occur. To accommodate this, most common situations have been incorporated in the Dynamic Seapipe program as standards, whilst others can be integrated with the main routine by simple program additions. One of the basic assumptions of the model is that the seabed is elastic with a constant or variable reaction coefficient, which is demobilised when the pipe is lifted. Other assumptions can easily be accommodated into the code.

For laybarge applications, only the pipe dynamics are simulated by the Seapipe program, since it is *assumed* that the barge motion is not influenced by the pipe. This is probably true in moderate water depths even with stiff and heavy pipes, and with smaller diameter pipes even in deep water. However, the possibility exists for interfacing the Dynamic Seapipe program with a vessel dynamics program, should it be necessary to explore the interdependent effects due to the relative movements of the laybarge and the pipeline.

This program has been applied to a number of different types of pipeline installation, involving shallow and deepwater situations.

2.2 An Example of Shallow Water Pipelaying: The Danish Gasline System

During the design phase of the Danish 30-inch diameter natural gas pipelines, which are described in detail in Chapter 17, pipeline lay-stress analyses were performed in the static as well as the dynamic situation for water depths up to 53 metres. The preliminary analysis embraced two types of laybarge: a semisubmersible provided with a fixed stinger and a conventional barge with an articulated stinger.

For the semisubmersible, dynamic exitation comprising a ±0.5 metre heave motion in phase with a ±1 degree pitch motion resulted in a 60–70 per cent increase in the bending stresses relative to those in the static situation. Typical results for this part of the study are shown in Figure 2.2.1 (overleaf). These barge motions corresponded to a significant wave height of 4.5 metres and a peak period of 8 seconds, which represent collectively the worst sea state that can be tolerated without overstressing the pipelines. Heavier seas would necessitate abandoning the pipes to comply with the specified limit on static lay-stress of 60 per cent of SMYS.

Sensitivity analyses showed that an excitation in which maximum heave amplitude and maximum pitch amplitude were in phase was not necessarily the most critical for all points on the pipe. Since heave and pitch motions have their own separate phase relationships to each harmonic component of the incident irregular wave train, an accurate analysis must involve a detailed knowledge not only of the amplitudes of barge heave and pitch response to a given wave but also their phase relationship to it. This emphasised the need for reasonably accurate information on the motion characteristics of the laybarge: the quality of the computer input should match the quality of the program itself, especially as the motion of the pipe attachment point is an important boundary condition input. Unfortunately, it is often difficult to obtain adequate motion data from laybarge operators to relate the motion of the attachment point to the sea state. As a minimum, Response Amplitude Operators (RAO's) for heave and pitch, with phases, should be supplied. These data can be produced by model tests or strip-theory calculations. All relevant loading conditions, wave directions and water depths should be covered. As an example, typical RAO's for the semisubmersible *Viking Piper* are shown in Figure 2.2.2 (overleaf) and described in detail by Rasmussen *et al* (1982).

The importance of the phase information is illustrated in Figure 2.2.3, (see page 20). This shows how the pitch motion adds to the heave in the in-phase situation, whereas in the out-of-phase situation the pitch reduces the heave motion at the after end of the barge. In reality the situation will be somewhere between the two extremes. The phase angles will be different for the different wave frequencies which compose the sea spectrum. The combined vertical motion at the aft end of the laybarge in an

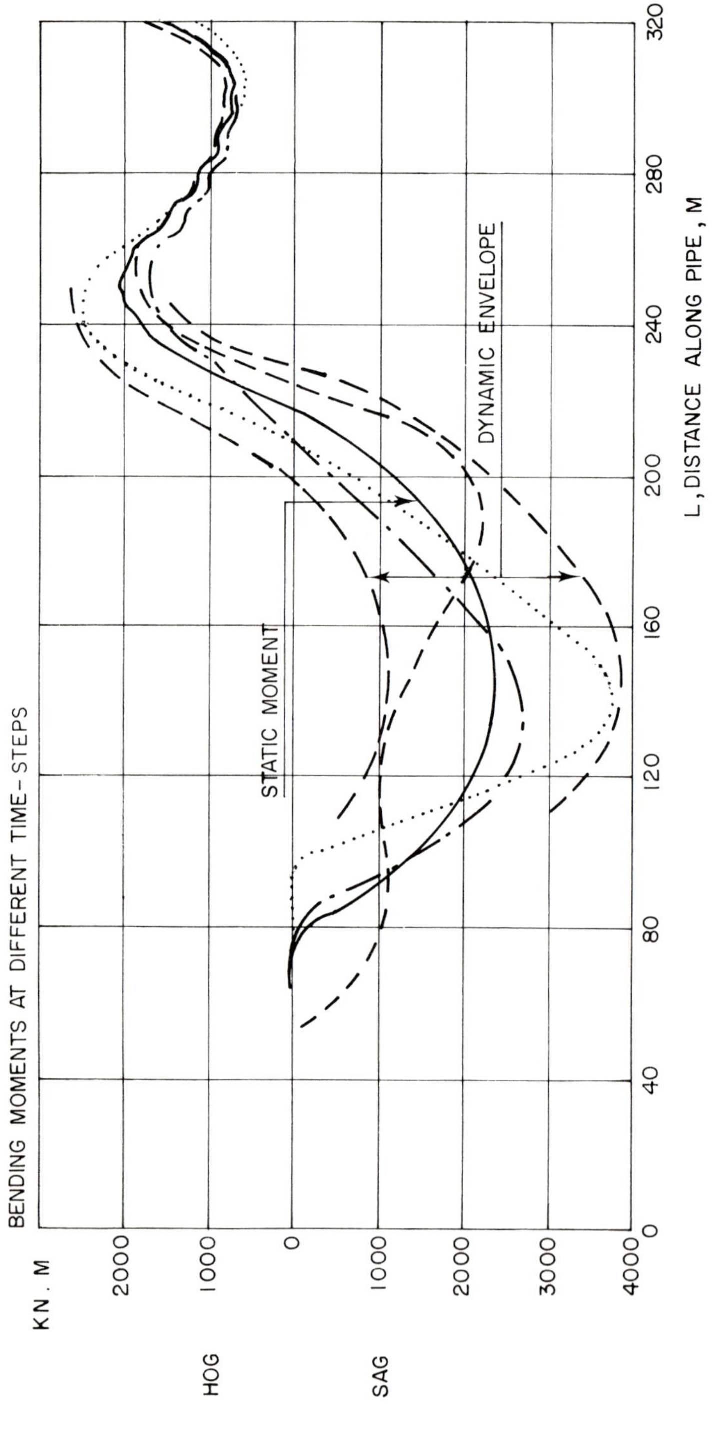

Figure 2.2.1 Shallow Water Laying of 30-inch Pipe

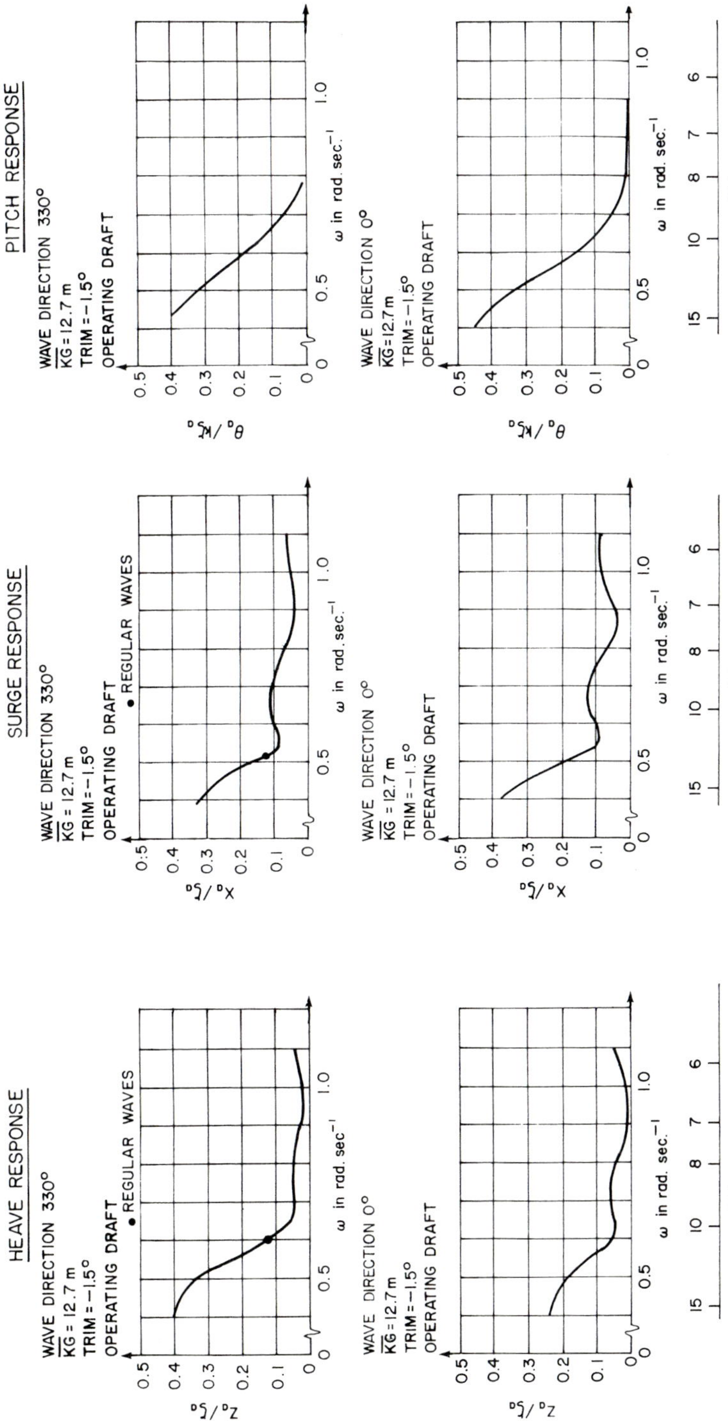

Figure 2.2.2 Typical RAO's from *Viking Piper*

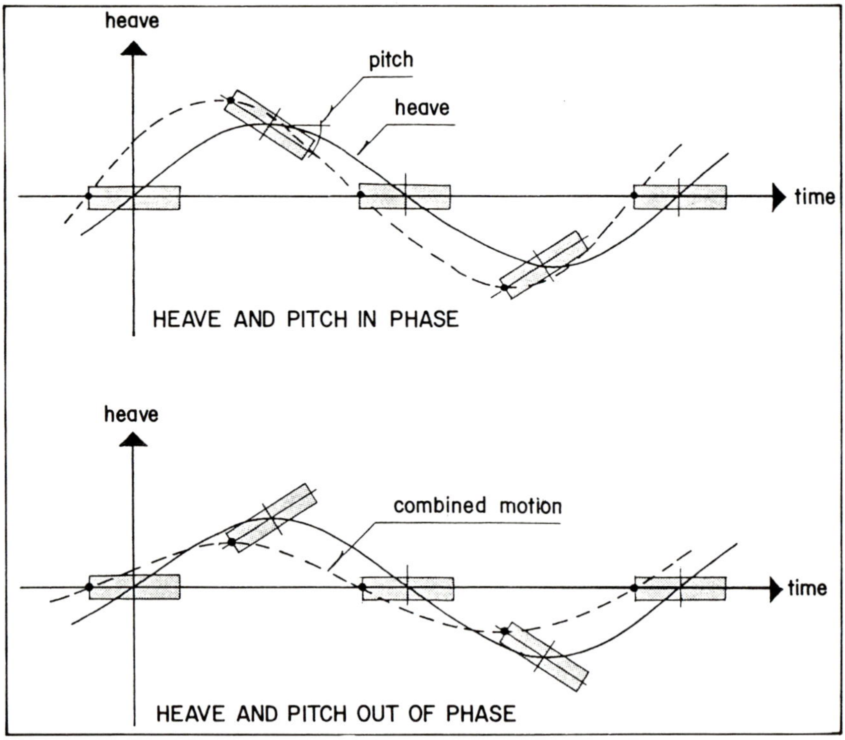

Figure 2.2.3 Effect of Heave and Pitch Phases

irregular sea state is computed by applying the RAO's to the individual components of a Jon-swap spectrum synthesised from the given sea state. A typical example of a vertical motion time series embracing the combined effects of heave and pitch is shown in Figure 2.2.4.

2.3 An Example of Deepwater Pipelaying: Flow-lines in the Mediterranean

A typical deepwater application of this computer program involved laying 4–6 inch diameter flow-lines at depths exceeding 700 metres in the Mediterranean using a J-ramp barge or drill-ship. This study demonstrated that deepwater pipelaying differed significantly from shallow water pipelaying in that the dynamic response in the sagbend region was less pronounced and the increase in stress relative to the static case was only 20–30 per cent, under the influence of simple harmonic heave and surge motion of the vessel of amplitude 1 metre and period 6 seconds respectively. This result was due to the considerable damping effect of the water on the long suspended pipestring. On the other hand, the near vertical pipestring is very vulnerable to current forces or surge motions of the

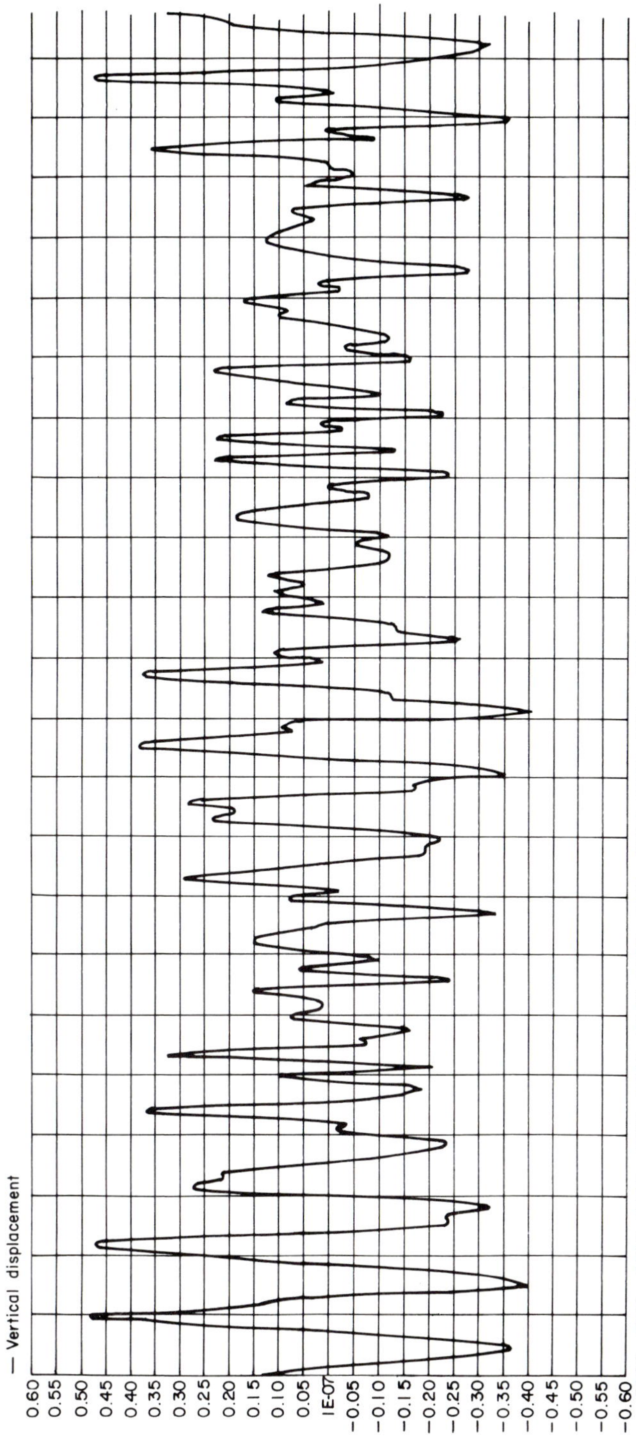

Figure 2.2.4 **Typical Time Series of Vertical Motion at Hitch Point on a Modern Semisubmersible.** ($H_S = 6.5$ metres)

vessel. In these situations, the critical location of the pipestring is at the end of the ramp or pipeguide where a bellmouth is preferred to mitigate these effects.

Typical computerised results for this situation are shown in Figure 2.3.1. These relate to a 6-inch diameter, half-inch thick pipeline, laid in 460 metres of water and subjected to a harmonic heave motion at the barge with amplitude 1 metre and period 6 seconds with no surge. Under these conditions the increase in bending moment due to dynamic effects is seen to be much more dramatic.

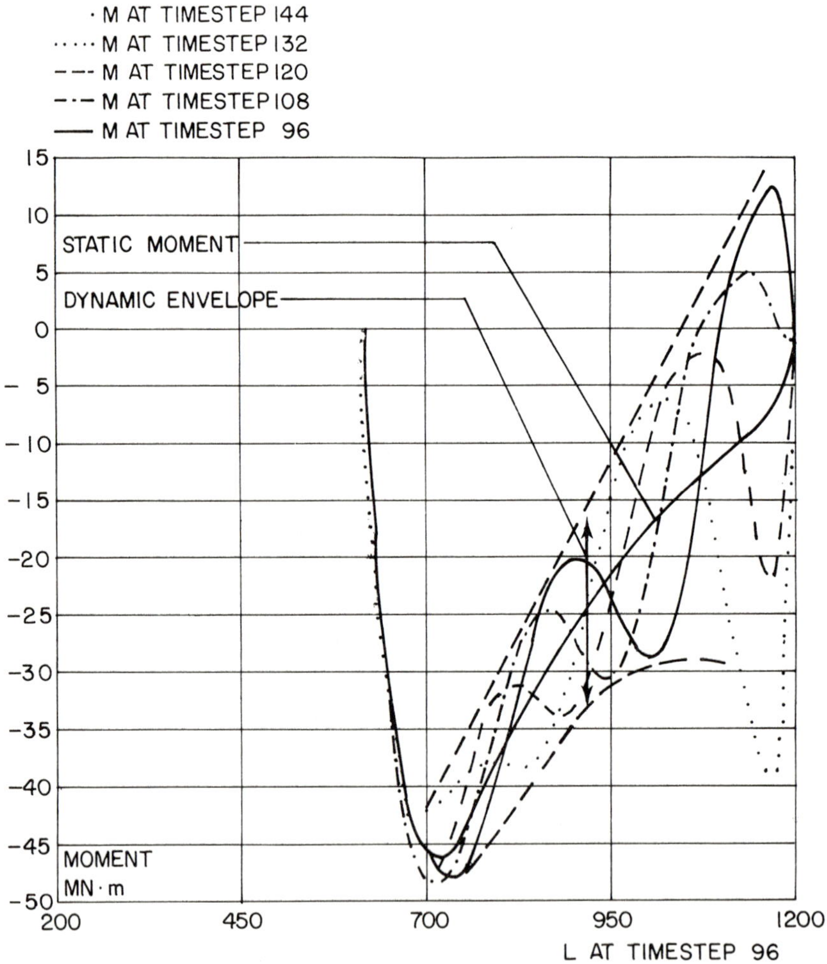

Figure 2.3.1 Dynamic Simulation with Regular Motions of Barge. (Period 6 seconds, pipe diameter 6 inches, wall-thickness ½ inch, WD 1,500 feet)

In addition to the effects brought about by dynamic stresses and fatigue, it is important to remember the increasing effect of tensile hoop stress as water depth increases, the significance of which is indicated pictorially in Figure 2.3.2 and described at length by Strader and Nielsen (1982).

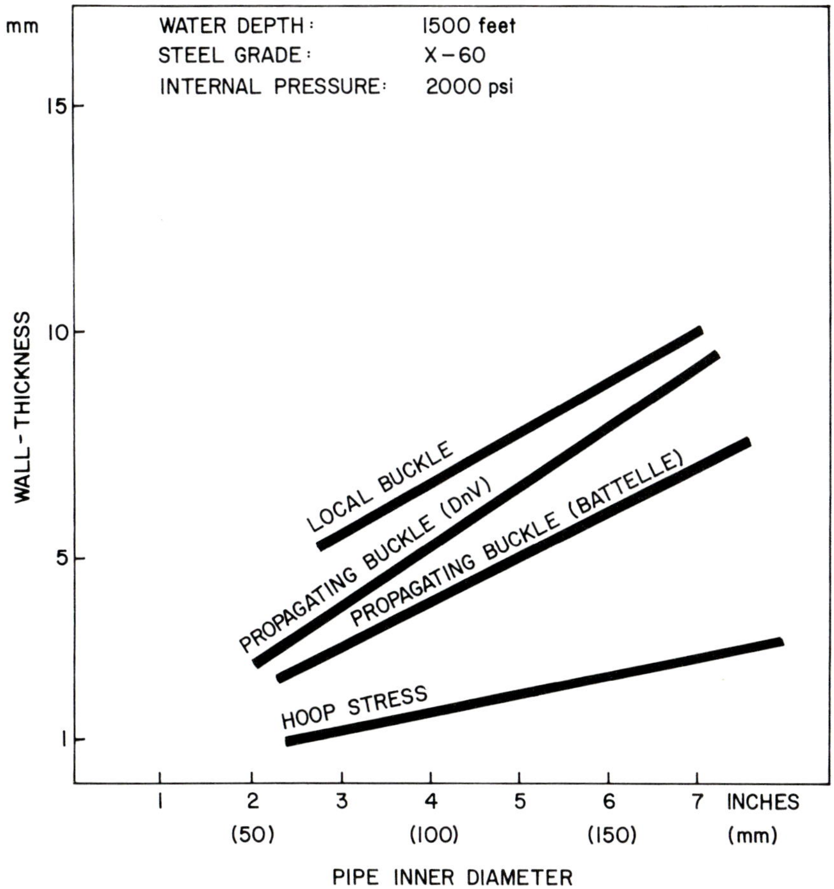

Figure 2.3.2 The Effect of Pressure on Pipeline Diameter and Wall-thickness

2.4 An Example Involving a Submerged Tow-out

Another application of this computer program involved the sub-surface towing of 3 kilometre lengths of sealed and pressurised 26-inch diameter flow-line bundles. To model this procedure satisfactorily it was necessary to extend the set of equations to cope with large deflections due to high axial elasticity resulting from the use of towing lines incorporating sections of nylon hawser to protect the pipes against shock loading.

Figure 2.4.1 shows the configuration used during the simulated tow, in which the leading tug maintained a constant tow speed on a prescribed course, whereas the trailing tug maintained a constant tension in the cable whilst maintaining the system in a straight line.

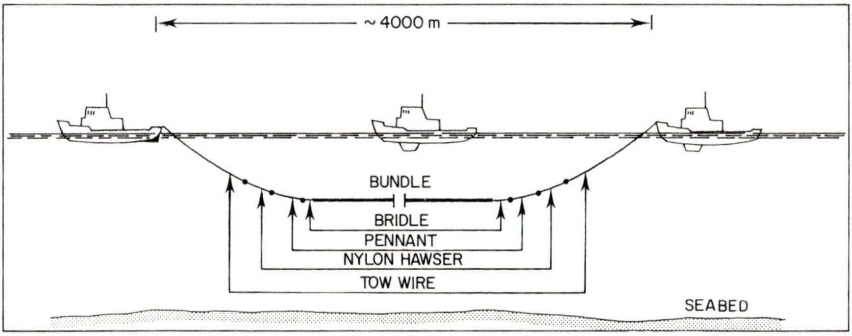

Figure 2.4.1 Configuration Used in the Pipeline Towing Simulation Analysis

Some of the more interesting results of this study are shown below. Figure 2.4.2 shows the moment distribution along the pipe at a typical time-step. Overleaf, Figure 2.4.3 shows the moment envelopes for a given period and for the first quarter of this period. The development of nodes and antinodes can be seen. Figure 2.4.4 shows the envelope of vertical displacement along the pipe and Figure 2.4.5 shows the envelope of axial tension along the pipe. From these results it was confirmed that for a sea state comprising a significant wave height of 5 metres with period 11.43 seconds the pipe would neither be overloaded, fatigued, nor tend to dive or surface.

2.5 Other Applications of the Seapipe Program

Other applications of this program include the dynamic stress analysis of submarine cable spans subjected to resonant vibration due to vortex shedding in high currents, the fatigue analysis of deepwater mooring cables and SPM hoses, the simulation of pipeline thermal expansion devices exposed to dynamic environmental loading, and the dynamic analysis of marine riser systems.

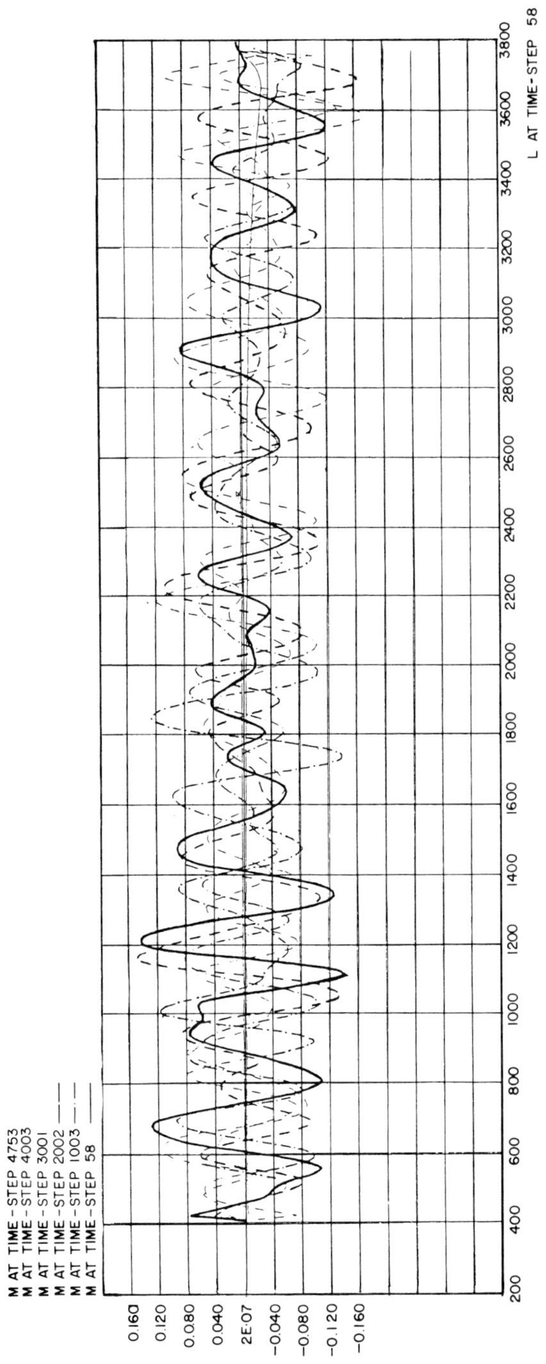

Figure 2.4.2 Moment Distribution along Pipe at Time-step 1003

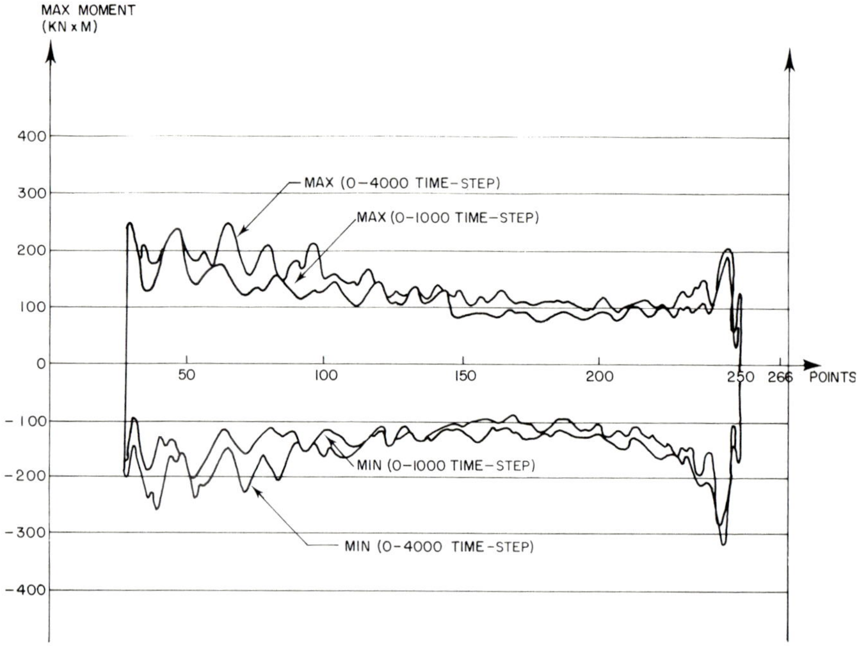

Figure 2.4.3 Envelope of Moments

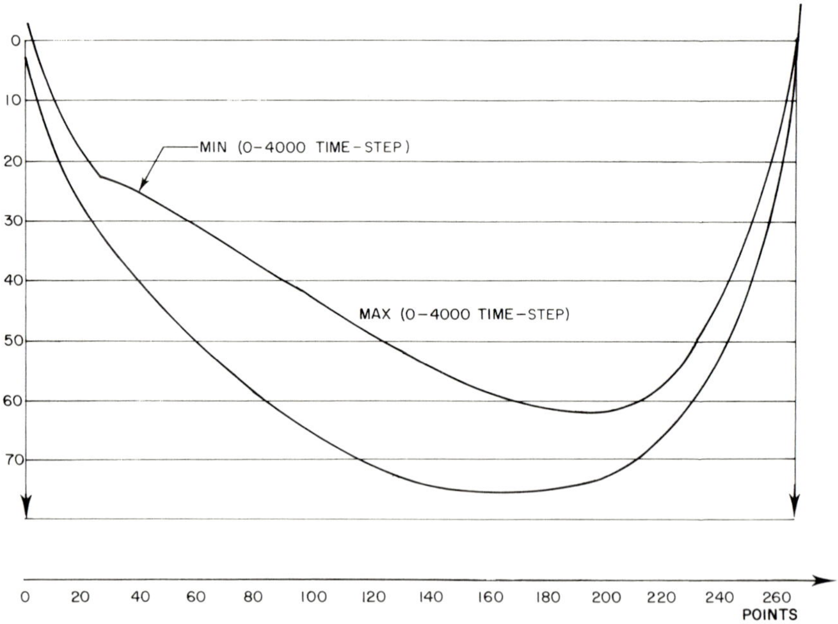

Figure 2.4.4 Envelope of Vertical Displacements

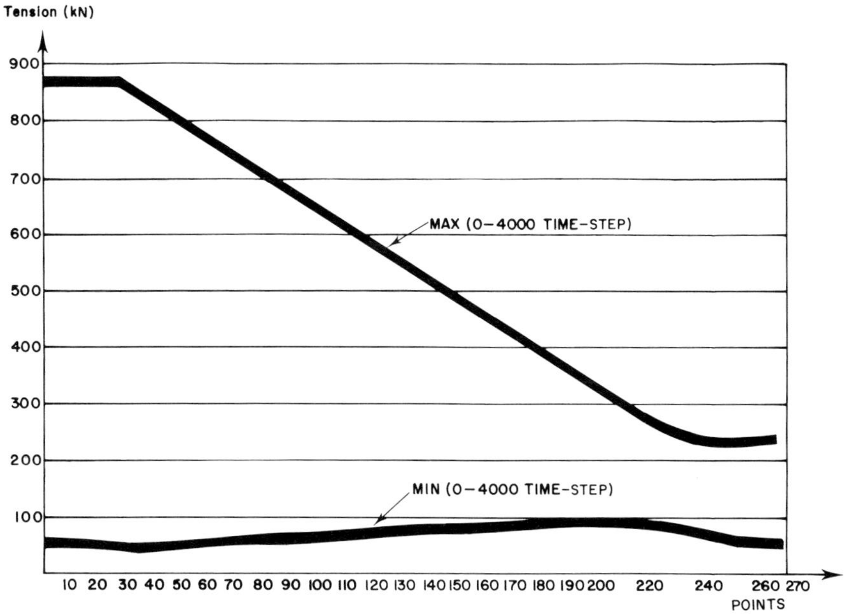

Figure 2.4.5 Envelope of Tension

2.6 Bibliography

Bryndum, M. B., Colquhoun, R. S. and Verley, A., 'Dynamic Laystresses for Pipelines', Paper OTC 4267, *Proceedings, Offshore Technology Conference*, (Houston, May 1982).

Michelsen, F. C. and Adler, W., 'Rough Water Operation of a Semi-submersible Pipelayer in the North Sea', *International Offshore*, (1976).

Rasmussen, H. J., Thygesen, J. E., Gravesen, H. and Damsgaard, A., 'Design Aspects of the Danish Offshore Transmission Lines', *Proceedings, 1982 European Seminar, Offshore Oil & Gas Pipeline Technology*, (Basel, 1982).

Strader, J. B. and Nielsen, R., 'Design Aspects of Subsea Flow-lines in over 1,500 Feet Water Depth', *Proceedings, 1982 European Seminar, Offshore Oil & Gas Pipeline Technology*, (Basel, 1982).

M B ABBOTT
M B BRYNDUM
R S COLQUHOUN

3

THE MECHANICS OF PIPELINE REELING

The oldest method of submarine pipeline construction is to weld long lengths of pipe together onshore, to wind them onto a reel, to transport the reel to the site of the pipeline, and to unwind the pipe onto the sea-bottom. In principle, the technique is elegantly simple, and has the particular advantage that it reduces the need for offshore operations close to the irreducible minimum. In practice, it has not achieved the degree of acceptance that might have been predicted. The reasons are partly technical, partly commercial and partly political. This chapter sets out to explore some of the technical aspects of reeling.

3.1 The Development of Reeling Methods

In the last year of the Second World War, there was a need for pipelines from England to France to supply petrol to the allied armies. PLUTO was a bold and imaginative project to meet this need, by a fast and economical method which could continue in the face of air attack. 3-inch pipe was laid in 40-kilometre lengths from floating reels ('conundrums') 15 metres in diameter, towed behind tugs and barges. Two types of pipe were used, one conventional steel pipe and the other made by the same process as cable sheathing. Although the pipe was bent some way beyond yield, it was not straightened as it left the reel. Since the lines were not expected to be needed for long, there was no corrosion protection system, and no attempt to protect them against damage from fishing gear. Interestingly, the PLUTO lines encountered some problems which were later rediscovered by submarine pipeline engineers. At least one line collapsed by buckle propagation: this was cured by filling subsequent lines with petrol before they were laid.

In 1961, Gurtler Hebert, a Louisiana contractor, converted a landing-ship hull into a reel barge U-303 (later RB-1). It laid pipe up to 6 inches in diameter from a vertical axis reel. The pipe was bent plastically in the horizontal plane, and had to be straightened before it was laid, since otherwise the suspended span would bow sideways and the pipe would form a series of kinks on the seabed. A second barge, RB-2 (later *Chickasaw*) was constructed in 1970, and again laid pipe from a vertical axis reel. The technology was acquired first by Fluor and later by Sante Fe, who went on to build the reel ship *Apache*, which went into service in 1979. The original concept for the reel ship was more ambitious than the vessel actually constructed, which has a horizontal axis reel with a hub diameter of 16.5 metres and an outer diameter of 25 metres. The maximum pipe weight on the reel is approximately 1,800 tonnes, and the capacity is 22 kilometres of 10-inch pipe, and proportionately more for smaller diameter pipe. The reel ship has been used successfully to construct a number of pipelines in the North Sea and the Gulf of Mexico, while reels and straightener systems intended for smaller pipe have been mounted on a number of vessels.

The above relates to steel pipe. Flexible pipe has followed a separate development path, and has invariably been laid from reels. Since flexible pipelines can accept tight bending radii, handling and laying methods have much in common with cable-laying. On the *Flexservice 1* layvessel, for instance, the pipe is stored in baskets with an outside diameter of 15 metres.

3.2 Reeling Mechanics

Consider first the form taken up by a pipe as it is wound onto a cylindrical reel. If the pipe had no flexural stiffness, and a negligible self-weight, it would take up the form shown in Figure 3.2.1a: on the reel, the pipe would be bent to the reel radius R, and when it was out of contact with the reel it would be straight. However, a pipeline with a finite flexural rigidity cannot take up the form of Figure 3.2.1a, because that would imply a step change of bending moment at the point of contact between the pipe and the reel. Figure 3.2.1b shows the configuration taken up by an initially straight, weightless, linear elastic pipe with flexural rigidty F, reeled onto a drum of radius R under a tension T. The bending moment in the pipe on the drum must be F/R, the product of the flexural rigidity and the curvature 1/R. Since the bending moment must be continuous at the contact point, the tension acting through the lever arm Δ (the offset between the contact point and the line of action of the tension) must generate this moment, and so the offset must be:

$$\Delta = F/RT$$

Figure 3.2.1c indicates the distribution of curvature in the pipe: here, and throughout the chapter, the sign convention for curvature is that a pipe conforming to the reel has positive curvature, and that negative curvature corresponds to reverse bending. The configuration of the pipe can be determined by integration of a simple equation relating curvature to offset from the line of action of T.

If the pipe is to take up the form shown in Figure 3.2.1b, the offset must

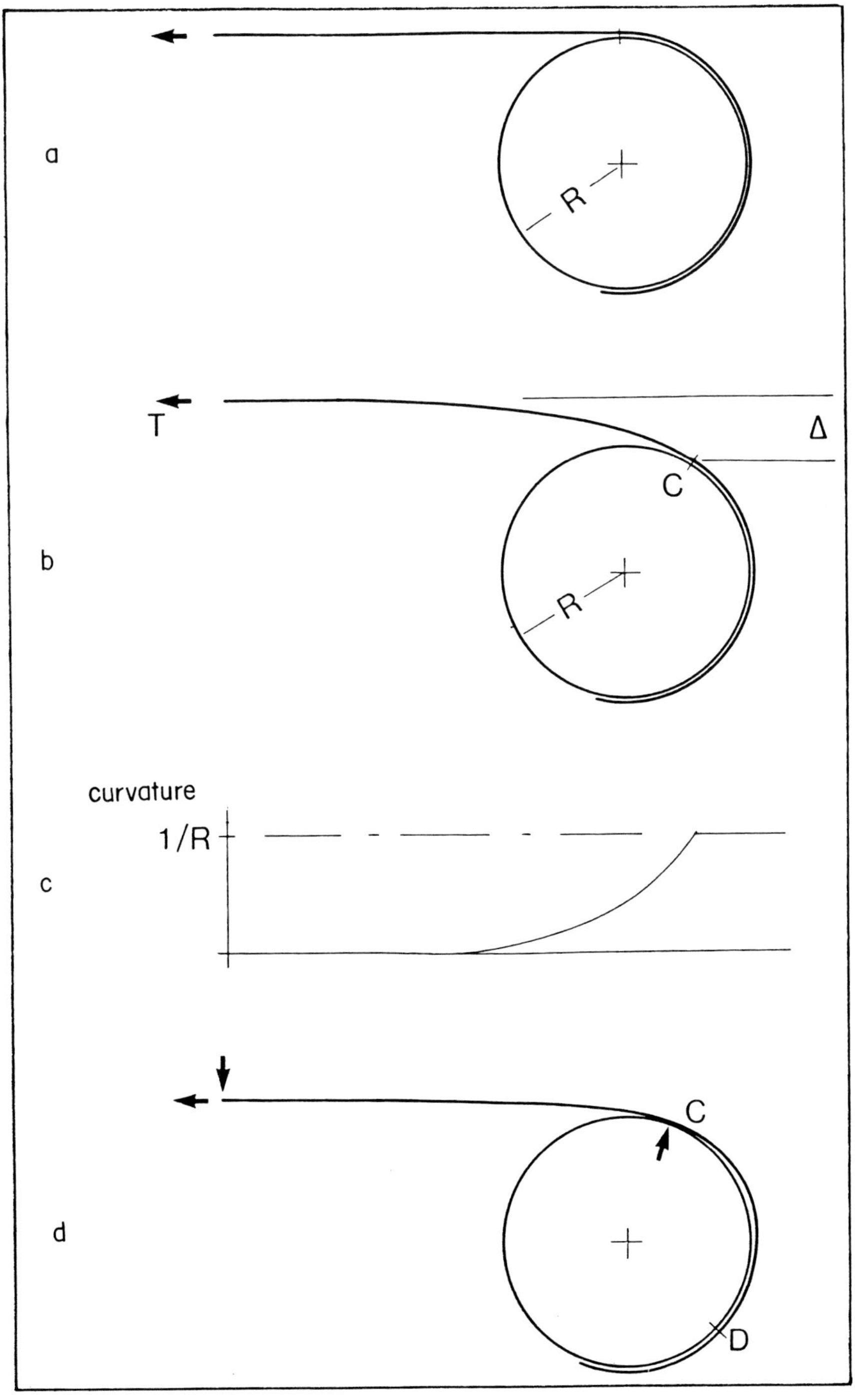

Figure 3.2.1 The Effect of Rigidity on Pipeline Reeling

be F/RT. If the offset is smaller than F/RT, the pipe takes up the form shown in Figure 3.2.1d: it then touches the reel at one point C, and bends away on the right-hand side of the drum before coming back into contact some way further round. Geometry then implies that the maximum curvature occurs between C and D, and that it is greater than the drum curvature 1/R. If the offset is greater than 1/R, on the other hand, the shear force in the free span is of opposite sign, and the contact point moves until the moment at the contact is again F/R.

Most pipelines bend beyond the elastic range as they are wound onto a reel. Plastic yield occurs if:

$$r/R > Y/E$$

where r is the pipe radius, R the reel radius (more precisely, the radius from the reel centre to the pipe axis), Y the yield stress and E the elastic modulus, so that a 10.75-inch diameter X-60 pipe will yield if it is wound onto a reel of less than 70-metre radius (which any practical reel must be). When a pipeline is bent plastically, its moment-curvature relation has the form shown in Figure 3.2.2, in which the inset diagrams represent the form of the distribution of longitudinal stress over the pipe cross-section. As plastic yield begins, the curve bends over, as a larger fraction of the cross-section yields plastically and as strain-hardening progresses in the yielded region. Ultimately, a buckle forms on the compression side, but this does not happen until a very large curvature has been reached, particularly with the relatively large wall-thicknesses usually used for reeled pipe. If the curvature is increased monotonically to a certain level (point J

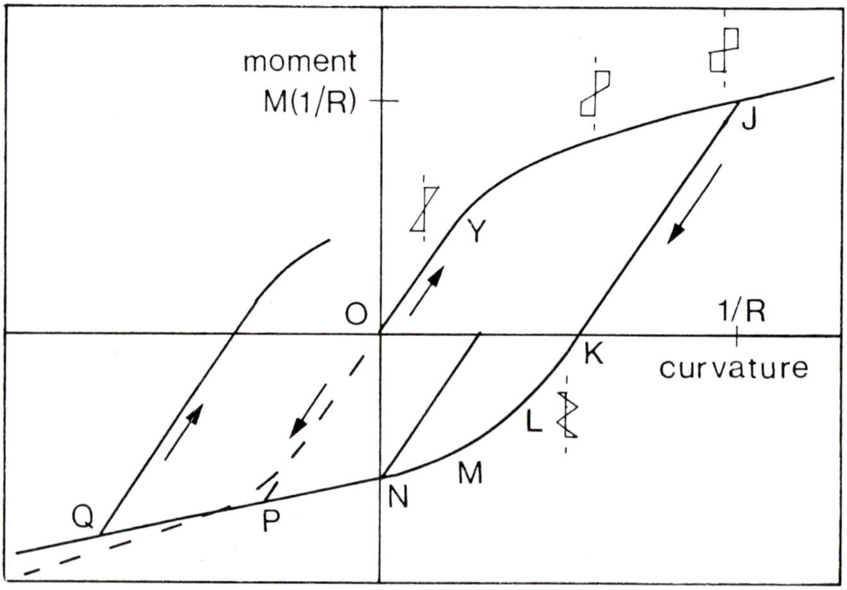

Figure 3.2.2 Pipeline Reeling in the Plastic Regime

in Figure 3.2.2), and then decreased, the initial unloading is linear, through point K to L, and then the unloading curve again bends round under the influence of reversed plastic strain and continues towards Q, with progressive yield in compression of regions which had previously yielded in tension. The moment-curvature relation for continued reverse bending does not exactly coincide with the relation for the bending of undeformed pipe in the same direction, because of the influence of the Bauschinger effect, which reduces the yield stress in compression of material that earlier yielded in tension.

If the direction of bending is again reversed, from state Q for instance, so that the curvature increases, the moment-curvature relation first returns along a line parallel to the initial loading line OP, and then again bends over as reversed deformation begins.

What has to happen if the pipe is to be restraightened after plastic bending? Figure 3.2.3 is part of Figure 3.2.2 redrawn. A straight pipe has zero curvature, and is represented by a point on the M-axis, not necessarily with zero moment. The diagram shows a number of unloading paths which all lead to zero curvature. Only path PO leads to both zero residual curvature and zero residual moment, though not to zero residual stress. Unloading from points on NP leaves a negative residual moment, while unloading from points on PQ leaves a positive residual moment.

It follows from this argument that a pipe plastically bent on a reel and then laid on a flat seabed always has a residual bending moment, unless its loading history is precisely OYJKLMNPO on Figure 3.2.2, and that

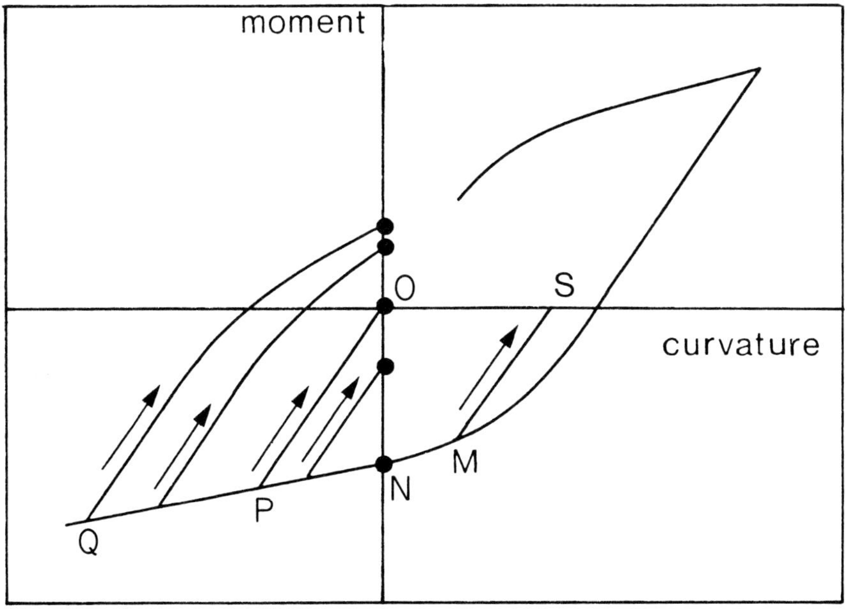

Figure 3.2.3 The Effect of Restraightening after Plastic Bending

there are always residual longitudinal stresses, which are of the same order as the yield stress Y. Concern is sometimes expressed about the influence of these stresses on subsequent behaviour in operation. However, if the pipeline is subsequently hydro-tested to a pressure which generates a hoop stress which is a large fraction of the yield stress, residual longitudinal stresses are at least partially erased. Appendix B outlines the theoretical basis of this effect, which is well known to the designers of pressure vessels, who often regard the hydrostatic test as final stage of manufacture rather than as the first stage of service. Hydrostatic testing has a major effect in reducing stress concentrations by localised plastic flow, and significantly lengthens fatigue life. Pipeline designers less often make conscious use of the effects of hydrostatic testing, but they are significant, for example, in reducing the very high residual stresses in cold bends and in generating favourable stress redistribution in spans. The more sophisticated codes implicitly recognise the relevance of redistribution of longitudinal stress in continuously supported pipelines, by replacing a longitudinal stress criterion by a strain criterion, based on the avoidance of buckling and of excessive plastic strains which might adversely affect ductility.

The moment-curvature relationship has important implications for reeling and unreeling. Figure 3.2.4 shows a pipe being reeled under tension T. The moment on the reel is M(1/R), and the offset must be (1/T)M(1/R). If the offset is less than this, the pipe will bend away from the reel on the far side, as in the elastic case, and the maximum curvature and moment will be larger.

Unreeling is more complicated. Figure 3.2.5 shows a profile of the pipe

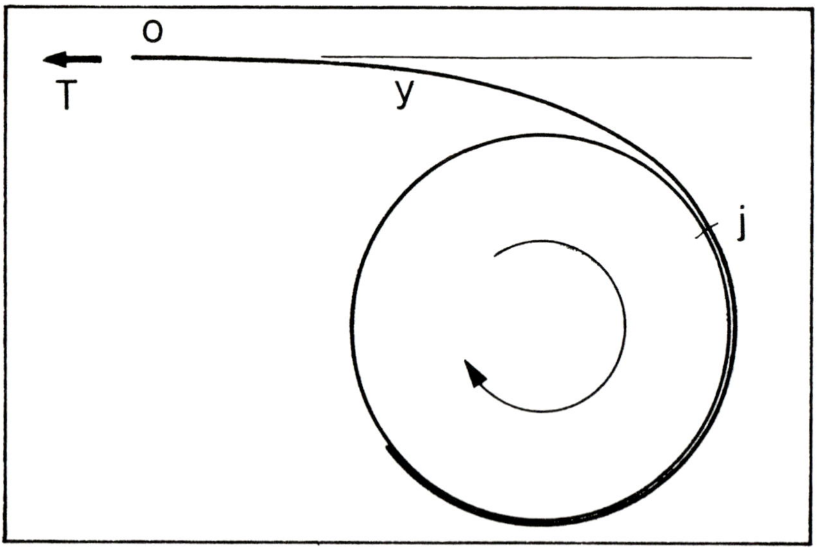

Figure 3.2.4 Pipeline Reeling under Tension in the Plastic Regime

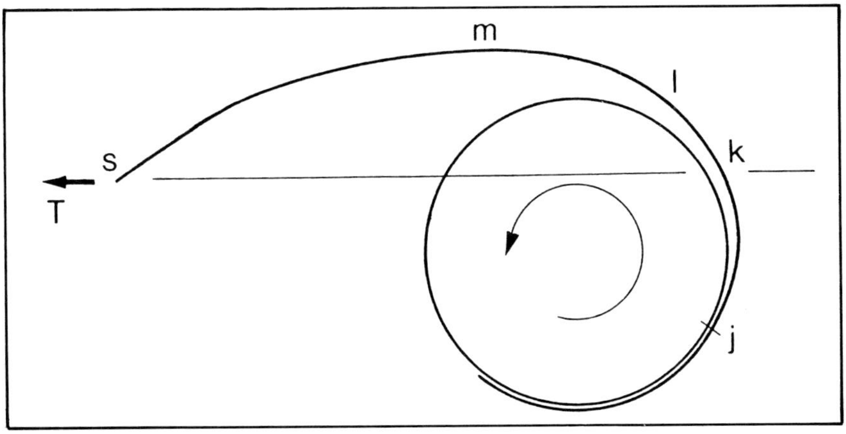

Figure 3.2.5 Pipeline Unreeling in the Plastic Regime

as it is unreeled; the lower-case letters on the profile correspond to the points on the moment-curvature diagram in Figure 3.2.2. At the point j at which the pipe loses contact with the reel, the moment is M(1/R) and the offset from the line of action of the tension is again (1/T)M(1/R). If any reverse plastic curvature is to occur, the bending moment must be negative, and so the profile must lie above the line of action of the tension force. At l reverse plastic bending begins, and at m the bending moment reaches its largest negative value. To the left of m, the moment is small, and the pipe responds elastically, with increasing positive curvature.

A profile of this type cannot correspond to complete straightening, which would return the state to point O in Figure 3.2.2. As we have seen, that would require a loading path OYJKLMNPO, and a largest negative moment corresponding to P, and simultaneously the corresponding negative curvature. At m in Figure 3.2.5, however, the moment is negative but the curvature must be positive. It follows that point m in Figure 3.2.5 is represented by a point on LN in Figure 3.2.2. A subsequent reduction in bending moment will return the state along an unloading path from m, and the residual curvature at zero moment will be positive. In that sense, the pipe cannot be completely straightened by an unreeling process of the type shown in Figure 3.2.5, in which the pipe is pulled from the reel by pure tension.

Conventional reeling practice recognises this difficulty by passing the pipe through a straightening device after it leaves the reel. Straighteners were introduced on the original vertical axis reel barges to avoid horizontal snaking that was found to occur if the pipe was laid with a residual horizontal plastic curvature. In its simplest form shown in Figure 3.2.6 (overleaf) a straightener consists of three rollers which together induce a negative moment and curvature corresponding to point P in Figure 3.2.2.

It is not essential that the straightening be perfect, though if it is not the pipe will have a residual moment after it is laid. That is what happens in floating-reel laying without a straightener. A negative moment and concurrent negative curvature in the sagbend are generated by the weight of the pipe, and the pipe is laid with a residual moment.

In some reeling applications, the process involves both flexure in the plane perpendicular to the reel axis and simultaneous lateral flexure in a transverse direction. This happens if the reeling system does not traverse across the drum, and straightening then requires reverse bending in both directions.

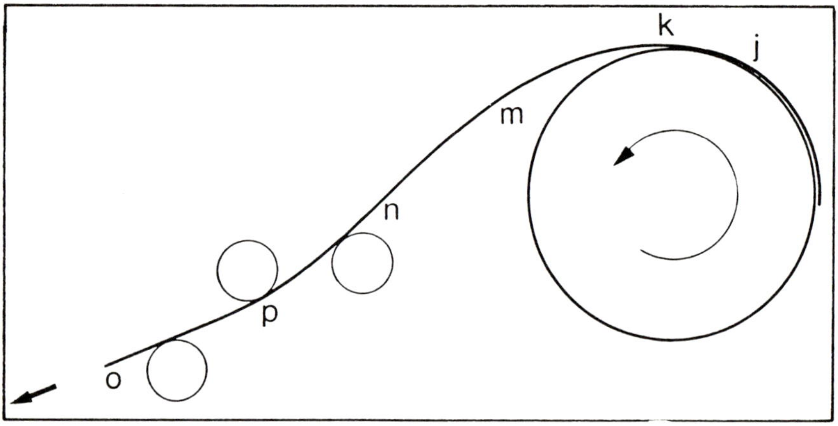

Figure 3.2.6 A Schematic Presentation of a Reel-pipe Straightener

3.3 Pipeline Ovalisation

Bending a pipe induces ovalisation, into a roughly elliptical cross-section, with the minor axis of the ellipse in the plane of bending. Ovalisation is a nonlinear effect, and markedly increases in the plastic range. Excessive ovalisation is not acceptable, because of incompatibility with pigs and TFL tools.

Analyses of ovalisation have been given by Brazier (1926), Reddy (1979) and Gellin (1980) and others. Figure 3.3.1a shows the longitudinal

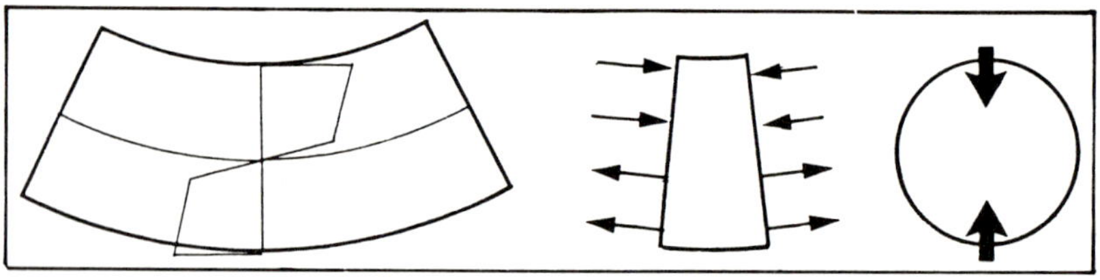

(a) (b) (c)

Figure 3.3.1 The Longitudinal Stress Distribution in a Tube Bent beyond Yield

stress distribution in a tube bent beyond yield, and Figure 3.3.1b the forces on a short element. Since the ends of the element are not in line, the longitudinal stresses induce resultant transverse forces on the cross-section, and these forces tend to induce ovalisation. The ovalisation results from the interaction of the curvature and the longitudinal stress distribution, and both are necessary for the ovalisation to occur.

Consider now a pipe, as shown by Figure 3.3.2, which has been bent plastically to a positive curvature, and then bent part of the way back, so that the curvature is reduced. The longitudinal stresses reverse, but the curvature is still positive. The interaction now generates forces which tend to flex the cross-section back towards a circular form, and indeed to ovalise it in the opposite direction. However, as the curvature is further reduced the effect weakens, and when the pipe is again straight there is no tendency to restore the circular cross-section. When the curvature becomes negative, interaction again tends to increase ovalisation in the original direction. The effect of these interactions is that most of the ovalisation that occurs during bending is recovered during straightening, but that some remains. Tests show that the residual curvature is generally acceptable. If it is not, further deovalisation can be achieved by the application of opposed external forces through rollers as part of a straightening operation, though this requires relatively large loads which may threaten to damage the pipeline coating.

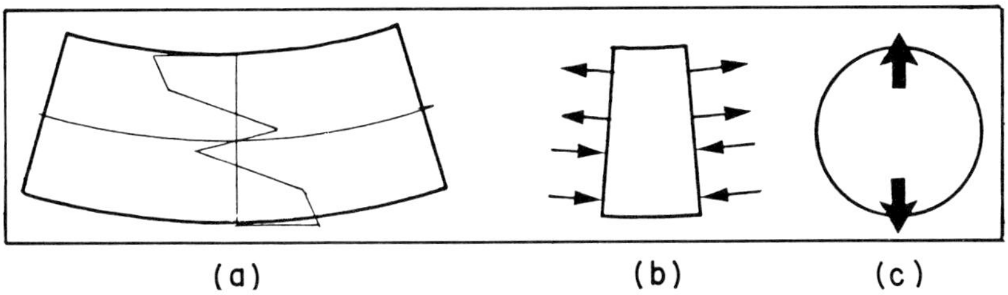

Figure 3.3.2 The Longitudinal Stress due to Bending Reversal

Experiments show that in some materials reversed plastic deformation can reduce resistance to fracture. The effect is related to the growth of micro-cavities and other defects, which can initiate ductile crack propagation. However, published results of tests on reeled pipe indicate that the effect is quite small, at the plastic strain levels reached during reeling, and that it can be taken care of by proper specification of the pipeline steel.

3.4 The Problem with Anti-corrosion Coating

Submarine pipelines generally carry an external coating of concrete, which protects the anti-corrosion coating against mechanical damage and gives the line enough weight for it to be stable on the seabed. Conventional concrete coatings cannot be reeled, and so reeled pipe has its anti-corrosion coating exposed to damage during reeling and straightening, as it reaches the seabed, and later during trenching and operation. Moreover, inventional concrete coatings cannot be reeled, and so reeled pipe has its anti-corrosion coating exposed to damage during reeling and straightening, as it reaches the seabed, and later during trenching and operation. Moreover, in order to give a reeled pipeline enough weight for stability, its wall-thickness has to be increased, often beyond the level required by other design factors, such as collapse and resistance to internal pressure.

This factor has important implications for reel construction, because of its impact on project schedules. A decision to adopt reel construction has to be taken before the pipe is ordered. If, as is much more usual, the schedule demands that pipe be ordered before construction tenders are received, it may be too late to select reel construction, whatever its other advantages.

The problem would be resolved by development of a reelable weight coating, not necessarily concrete. The last few years have seen significant progress in the development of tough and abrasion-resistant protective coatings for submarine pipelines, based on polymers and on polymer-modified and fibre-reinforced cement systems. These materials have so far been used as barrier coats to protect thin-film coatings, and in new field joint systems, but should have wider applications. A reelable weight coating ought to be within reach.

3.5 Conclusion

Reeling is an extremely attractive method of submarine pipeline construction. It transfers as many as possible of the operations involved from the high-cost and weather-sensitive environment offshore to a relatively low-cost protected factory environment onshore. The pipeline can be carefully inspected and tested onshore, in conditions which allow time for thoughtful decisions. Time offshore is reduced close to a practical minimum, and the reel vessel can take advantage of brief weather windows and opportunities to work in congested areas around platforms. The full application of the method has been delayed by concerns about the mechanical and metallurgical consequences of reversed plastic deformation, and about coating damage. The first concern seems to have been exaggerated, and the second is on the way to being solved.

3.6 Bibliography

Brazier, L. G., 'On the Flexure of Thin Cylindrical Shells and other "Thin" Sections', *Proceedings, Royal Society, Series A,* 116, 104–114, (1926).

Gellin, S., 'The Plastic Behaviour of Long Cylindrical Shells under Pure Bending', *Intl. J. of Solids and Structures,* 16, 397–407, (1980).

Reddy, B. D., 'Plastic Behaviour of a Cylindrical Shell in Pure Bending', *Intl. J. of Mechanical Sciences,* 21, 671679, (1979).

Reddy, B. D., 'An Experimental Study of the Plastic Buckling of Circular Tubes in Pure Bending', *Intl. J. of Solids and Structures,* 15, 669–683, (1979).

R J BROWN
A C PALMER

4

THE BUCKLING OF SUBMARINE PIPELINES

When a submarine pipline is bent severely, it buckles. This usually makes the pipeline unfit for service so it then has to be repaired. Construction must therefore be carried out in a way which avoids the risk of buckling. Many buckles are the result of excessive bending which occurs when control of tension is lost, laybarge anchors drag or the stinger breaks. Another frequent cause of buckling failure is hooking by construction vessel anchors accidentally pulled across a pipeline. In this chapter we examine the buckling of a pipeline under bending alone and then go on to consider buckling under more complex loading conditions.

4.1 Introduction

An offshore pipeline is usually subjected to the most severe loads during the laying operation when large bending loads are augmented by the external hydrostatic pressure. The highest bending stresses occur in the pipeline sections in contact with the stinger (the overbend region) and in the section situated a small distance above the seabed (the sagbend region), as shown in Figure 4.1.1 (overleaf). According to Pedersen (1978) and Kimure *et al* (1980) the maximum sagbend moment often approximates the maximum overbend moment so the combined influence of bending and hydrostatic pressure will be most critical for pipe sections in the sagbend region close to the ocean bottom.

If the yield stress of the pipe material has been exceeded in the overbend or sagbend region, residual stresses will exist in the straight pipe as laid on the ocean bottom. However, the results of Toma and Chen (1979) and Toma *et al* (1982) show that the residual stresses in most cases have a negligible influence on the ultimate load carrying capacity.

In the analysis of pipeline buckling and collapse, one can consider the length of the critical pipe section to be infinite with constant curvature and ovalisation in the longitudinal direction. For a purely elastic material, Axelrad *et al* (1983) and Saal *et al* (1979) have shown that the influence of boundary conditions on the buckling strength of a pipe subjected to bending can be neglected if the pipe length is larger than the critical length l* where:

$$l^* \simeq 3R\sqrt{R/t} \qquad (4.1.1)$$

where the radius and the wall-thickness of the pipe are denoted by R and t respectively. For pipelines the radius-thickness ratio R/t is between 8 and 40, so the length l* is so small that no significant variation in the loads occurs over this length of the pipe. Analysis of short plastic pipes support this conclusion.

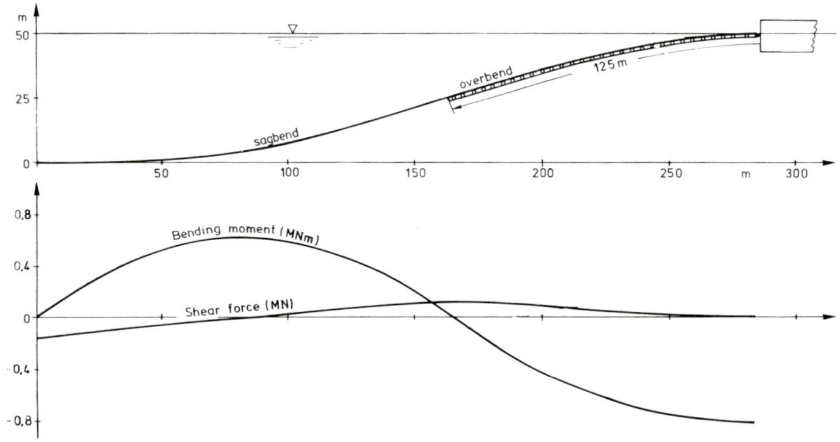

Figure 4.1.1 Numerical Analysis of Pipe-laying Procedure with a Flexible Stinger

In this chapter some theoretical results are compared with published experimental results. Then design curves are presented together with a proposed interation formula for collapse of long pipes subjected to combined bending and external pressure loads. The effect of a positive additional axial tensile force is not considered since it is normally of minor importance for the actual collapse loads. However, reference can be made to Miller (1982) for information on pipes subjected to axial tension and external pressure loads.

Furthermore, the possibility of collapse due to propagating buckling is not treated here, but reference can be made to a recent paper by Chater and Hutchinson (1983), which yields an elegant theoretical solution to this important problem.

4.2 Theoretical Predictions and Experimental Results

Because the radius-thickness ratios used in offshore pipelines are below

40 for steel pipes, the collapse behaviour of such pipes inevitably involves large plastic deformation. By contrast, the theoretical results of Fabian (1981) show that a purely elastic behaviour until collapse of pipes made of usual steel qualities requires radius-thickness ratios greater than about 500.

However, the possible use in the future of new carbon-fibre reinforced plastic pipes can make a purely elastic approach useful. To this end, a review of collapse formulae for elastic pipes follows before similar results are presented for elastic-plastic pipes.

4.2.1 Elastic Pipes

Buckling and collapse of long pipes in combined bending and pressure have been studied in detail by Fabian (1977), Emmerling (1982) and Kyriakides and Shaw (1982). Their work takes into account the ovalisation of the cross-section of the pipes, resulting in a geometrically non-linear load-deflection behaviour. The results obtained by the former researchers are almost identical, whereas differences are found when comparing their results with those of the latter.

The main result from the elastic analysis is the relation between the applied bending moment M and the corresponding curvature κ of the pipe for various values of the difference p between the internal and external pressures. The bending moment M is given by:

$$M = m \frac{\pi E R t^2}{\sqrt{12(1-v^2)}} \qquad (4.2.1.1)$$

the curvature as:

$$\kappa = c \frac{1}{\sqrt{12(1-v^2)}} \frac{t}{R^2} \qquad (4.2.1.2)$$

and the pressure as:

$$p = \gamma \frac{E}{12(1-v^2)} \left(\frac{t}{R}\right)^3 \qquad (4.2.1.3)$$

where the computed relations between the dimensionless quantities m, c and γ are shown in Figure 4.2.1.1 (overleaf). The material properties enter through the modulus of elasticity E and Poisson's ratio v. Negative values of γ correspond to nett external pressures.

In the subsequent analysis, the subscripts and superscripts have the following meanings:

(i) 'c' refers to the value at the point of collapse;

(ii) 'o' refers to the value in the absence of other loads;

(iii) 'e' refers to the elastic value; and

(iv) 'y' refers to the fully plastic value.

Based on the results shown in Figure 4.2.1.1, interaction diagrams between bending moment M^c and pressure p^c, and between curvature κ^c and pressure p^c can be constructed at the limit point, as shown in Figure 4.2.1.2.

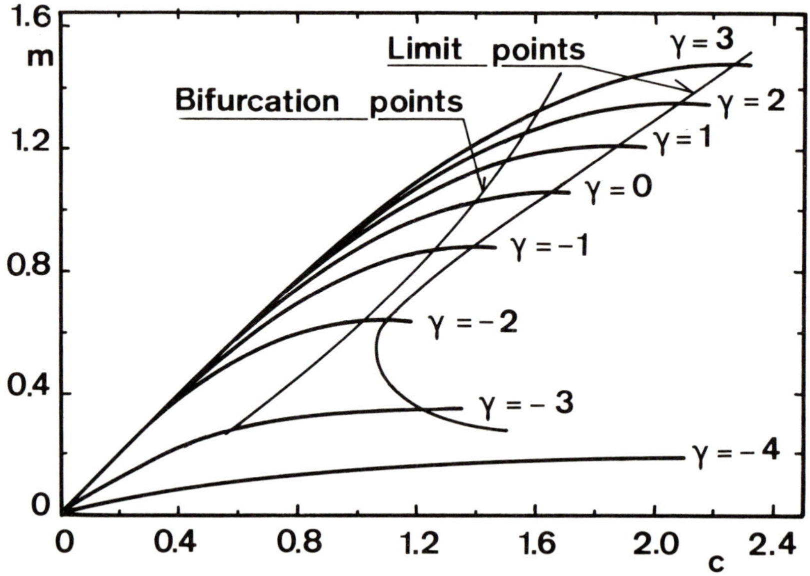

Figure 4.2.1.1 Dimensionless Moment-curvature Relation. *[after Fabian (1977)]*

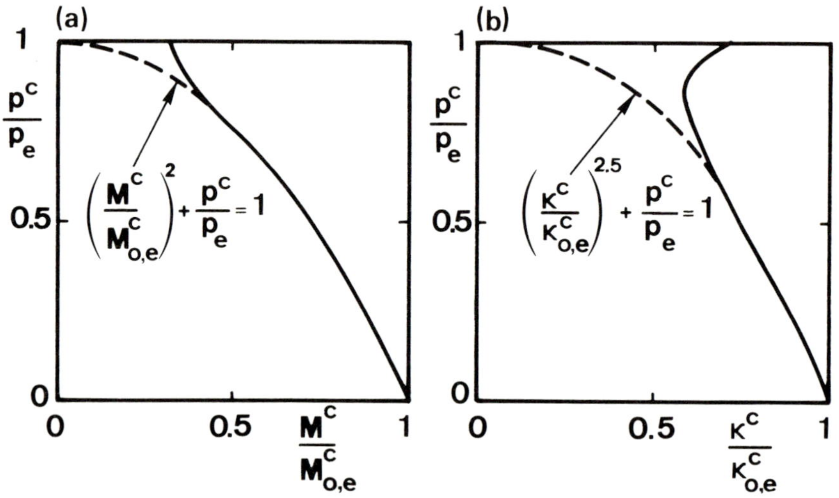

Figure 4.2.1.2(a) Limit Moment M^c *versus* Pressure p^c

(b) Limit Curvature κ^c *versus* Pressure p^c

The bending moment $M^c_{o,e}$ and the corresponding curvature $\kappa^c_{o,e}$ at the limit point without pressure are given by Equations 4.2.1.1. and 4.2.1.2 with $m=m^c_{o,e}$ and $c=c^c_{o,e}$. Quoted values of $m^c_{o,e}$ range from 1.03 to 1.09, whereas the values of $c^c_{o,e}$ range from 1.56 to 1.65. With Poisson's ratio $v = 0.3$ the bending moment $M^c_{o,e}$ and the curvature $\kappa^c_{o,e}$ thus become:

$$M^c_{o,e} = 1.02\,ERt^2 \tag{4.2.1.4}$$

and:

$$\kappa^c_{o,e} = 0.49\,\frac{t}{R^2} \tag{4.2.1.5}$$

using the values provided by Fabian (1977) and Emmerling (1982). The pressure p^c is conveniently normalised by the elastic buckling pressure:

$$p_e = -\frac{E}{4\,(1-v^2)}\left(\frac{t}{R}\right)^3 \tag{4.2.1.6}$$

for a straight tube subjected to external pressure only. If the external pressure is raised above the pressure p_e the ovalisation of the cross-section becomes large, as demonstrated by Kyriakides and Babcock (1981). Collapse occurs when two opposite points on the internal surface of the cross-section make contact. Numerical results for elastic rings show that this occurs when the external pressure approximates 1.75 p_e. This result explains the somewhat strange behaviour of the interaction diagrams in Figure 4.2.1.2 for large values of the external pressure p_c. From a design point of view, however, ovalisation must be limited to small values so it would be prudent to avoid pressures larger than the elastic buckling pressure p_e. With this modification the following empirical interaction formulæ:

$$\left(\frac{M^c}{M^c_{o,e}}\right)^2 + \frac{p^c}{p_e} = 1 \tag{4.2.1.7}$$

and:

$$\left(\frac{\kappa^c}{\kappa^c_{o,e}}\right)^{2.5} + \frac{p^c}{p_e} = 1 \tag{4.2.1.8}$$

represent a close approximation to the numerical results presented by the previous three researchers.

Equation 4.2.1.7 is well-known, whereas the results of De Winter (1981) lead to an equation similar to 4.2.1.8 except that an exponent equal to 2 is obtained instead of 2.5.

These theoretical results were derived under the assumption that the cross-sectional flattening takes place along the whole length of the tube. This is in disagreement with the localised deformation pattern, with a few axial wrinkles on the compression side found in experiments. This difference can be explained by the theoretical work of Tvergaard and Needleman (1983), who studied a general case where bifurcation buckling occurs just before the limit point. In this case it is found that a periodically varying bifurcation mode will change to a deformation pattern with only one or a few buckles when the load approaches the limit load. As shown by Figure 4.1.1, bifurcation actually occurs just before the limit point for pipes subjected to combined bending and pressure and thus localisation of the deformation pattern should be expected.

In the previous discussion the initial pipe geometry was assumed to be perfect, but real pipes possess initial out-of-roundness as well as axial wrinkles. According to Fabian (1977) however, pipes are only slightly sensitive to such imperfections so only modest reductions of the load carrying capacities should be expected due to axisymmetric, periodic imperfections. This conclusion is also supported by further work by Fabian (1981) whereby a reduction of the limit moment M^c, equal to 17 per cent and 13 per cent respectively, was found for pure bending and bending with a superimposed external pressure equal to 0.67 p_e. The imperfection amplitude was taken to be ten per cent of the pipe wall-thickness t. In effect therefore, evidence suggests that an initial ovalisation is found to have little influence on the load carrying capacity of the tube. It should, however, be stressed that this conclusion is only valid for a purely elastic behaviour.

4.2.2 *Elastic-Plastic Pipes*

Accurate theoretical models of the collapse of elastic-plastic pipes under combined bending and pressure have been developed by Kyriakides and Shaw (1982) and Fabian (1981). Both references make use of the principle of virtual work to derive a set of governing equations for the problem in question, which is nonlinear with respect to geometry and material. The main difference in the two approaches is found in the choice of plasticity theory. The former researchers use the Ramberg-Osgood approximation of the experimental stress (σ) – strain (ϵ) curves, where:

$$\varepsilon = \frac{\sigma}{E}\left[1 + \frac{3}{7}\left(\frac{\sigma}{\sigma_0}\right)^{\bar{n}-1}\right] \qquad (4.2.2.1)$$

The three parameters describing the pipe material are here the modulus of elasticity E, the nominal yield stress σ_0 and the strain hardening parameter n. The analysis presented by Fabian (1981) makes use of a small-strain J_2-flow theory with isotropic hardening together with a modified power law description of experimental stress-strain curves:

$$\varepsilon = \begin{cases} \dfrac{\sigma}{E} & ; \quad \sigma \leq \sigma_p \\ \dfrac{\sigma_p}{E}\left[\dfrac{1}{n}\left(\dfrac{\sigma}{\sigma_p}\right)^n - \dfrac{1}{n} + 1\right] & ; \quad \sigma > \sigma_p \end{cases} \quad (4.2.2.2)$$

where the material parameters are the modulus of elasticity E, the proportional stress σ_p and the strain hardening parameter n.

According to measurements by Reddy (1979), the ovalisation of the cross-section caused by the loading is slightly unsymmetrical about the neutral axis of bending, thus making the assumption of a doubly symmetric deformation pattern used by Kyriakides and Shaw (1982) somewhat doubtful.

Kyriakides and Shaw (1982) provide a large number of theoretical results for pipes made from aluminium with a ratio between the nominal yield stress σ_0 and the modulus of elasticity E of 0.0043. These results were found to be in agreement with their experimental results, but they cannot be compared with the numerical results of Fabian (1981), which relate to pipes made of the steel qualities X52, X65 and X77 as defined by API. However, the former researchers have made a comparison between some of their theoretical results and some experimental results for steel pipes, reported by Johns (1975). The steel used in these experiments was very similar to the steel quality X77, except for a lower strain-hardening

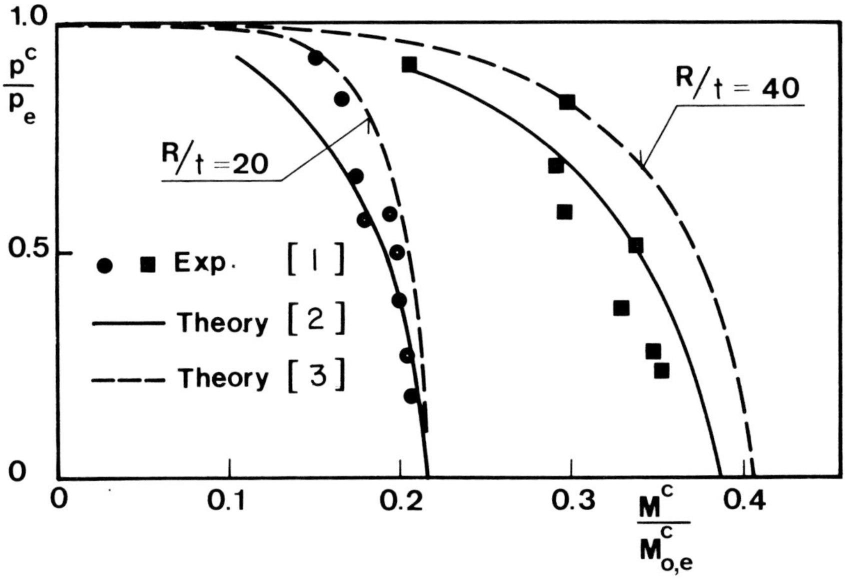

Figure 4.2.2.1 Steel Tubes in Combined Bending and External Pressure. Comparison between Experimental and Theoretical Results. *[[1] Johns (1975); [2] Fabian (1981); [3] Kyriakides and Shaw (1982)]*

behaviour. These results are shown in Figure 4.2.2.1 where it is seen that very good agreement is found for the thick-walled tubes (R/t = 20), whereas the results of Fabian (1981) yield better agreement than Kyriakides and Shaw (1982) with the experiments for the thin-walled tubes (R/t = 40). However, it should also be noted that if a higher strain hardening exponent n had been used by Fabian then an even closer agreement with the experiments would have been achieved.

Unfortunately, no further experimental data for the collapse of pipes subjected to combined bending and pressure have been found in the open literature, but some experimental results are available for pipes exposed to either pure bending or to external pressure.

The results from a large number of pipes tested in pure bending have been reported by Khalig and Schilling (1979) and Sherman (1976). The steel materials used in these experiments have various uni-axial stress-strain curves with yield stresses between 290 and 420 MN/m^2 and also different strain hardening behaviour. These experimental results are shown in Figure 4.2.2.2 in the same way as they were presented by Sherman. The ultimate bending moment M_0^c is normalised by the plastic moment My defined as the stress $\sigma_{.005}$ at a total strain equal to 0.005, multiplied by the plastic section modulus Z_p for the circular (undeformed) pipe where:

$$M_y = Z_p \sigma_{.005} \simeq 4R^2 t \sigma_{.005} \qquad (4.2.2.3)$$

The abscissa is the radius-thickness ratio multiplied by the factor $\sqrt{\sigma_{.005}/E}$. The theoretical results due to Fabian (1981) provide a single curve independent of steel qualities X52, X65 and X77, thereby supporting the findings by Sherman that the relevant material parameter here is $\sqrt{\sigma_{.005}/E}$ rather than $\sigma_{.005}/E$.

It should be noted that the effect of the strain hardening behaviour on the limit moment cannot be estimated from Fabian's work as the three steel qualities used have nearly the same strain hardening exponent n equal to 9.05, 9.65; and 10 for X52, X65 and X77 respectively. However, theoretical results due to Kyriakides and Shaw (1982) and Gellin (1980) indicate that the degree of strain hardening only has a minor influence on the ultimate moment.

Figure 4.2.2.2 shows that the theoretical results are in fair agreement with the experimental results. However, these exhibit a rather large scatter which can to some extent be explained by unavoidable geometrical imperfections. For tubes, loaded in pure bending, a small initial ovalisation will not reduce the ultimate bending moment M_0^c, but an axisymmetric imperfection with a periodic variation in the longitudinal direction will reduce M_0^c by approximately ten per cent if the imperfection amplitude is equal to ten per cent of the thickness t and the wavelength is between $\sqrt{Rt}$ and $2\sqrt{Rt}$. These results are included in Figure 4.2.2.3 (overleaf).

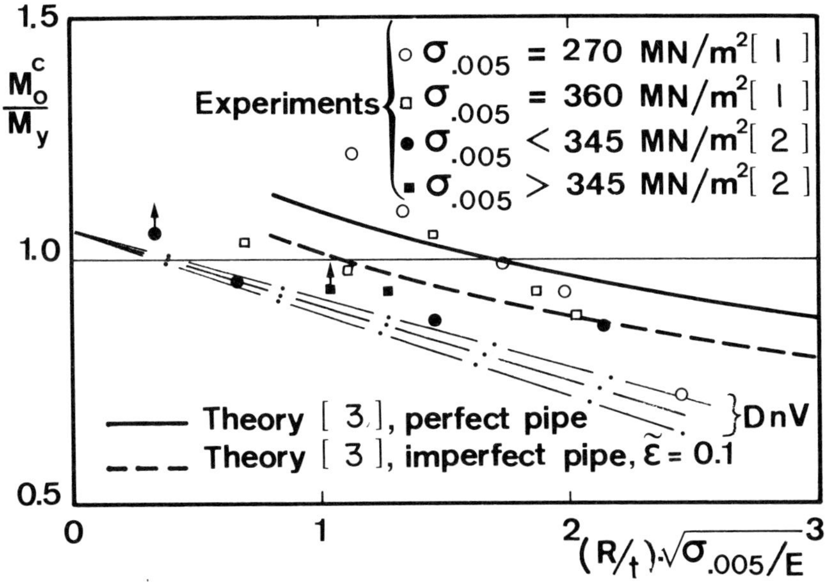

Figure 4.2.2.2 Comparison between Experimental and Theoretical Results for the Limit Moment in Pure Bending.
[[1] Khalig and Schilling (1979); [2] Sherman (1976); [3] Fabian (1981)]

In the design of pipelines, the curvature κ^c at collapse is a more important parameter than the ultimate moment M^c, as κ^c governs the choice of stinger geometry and applied tension in the pipelaying procedure.

The results obtained by Fabian (1981) and Gellin (1980) for pure bending show that the curvature κ_o^c can be written as:

$$\kappa_o^c \simeq \frac{1.2}{\sqrt{12\,(1\text{-}\nu^2)}}\ \frac{t}{R^2} = 0.36\,t/R^2 \qquad (4.2.2.4)$$

for medium strain hardening steel materials (X52, X65, X77). However, a small axisymmetric periodic imperfection along the length of the pipe will reduce the critical curvature by about 20 per cent. These results compare reasonably with the experimental results obtained by Reddy (1979), Sherman (1976) and Korol (1979) as shown in Figure 4.2.2.3 (overleaf).

For tubes loaded by an external pressure only, the maximum pressure p_o^c is often estimated by the Timoshenko formula:

$$\left(p_o^c\right)^2 - \left[p_y + \left(1 + 6\Delta_o\frac{R}{t}\right)p_e\right]p_o^c + p_y p_e = 0 \qquad (4.2.2.5)$$

where the elastic buckling pressure p_e is given by Equation 4.2.1.6 and the pressure p_y by the expression:

$$p_y = -\sigma_{.005} t/R \qquad (4.2.2.6)$$

and where the initial ovalisation Δ_o is defined as:

$$\Delta_o = \frac{D_{max} - D_{min}}{4R} \qquad (4.2.2.7)$$

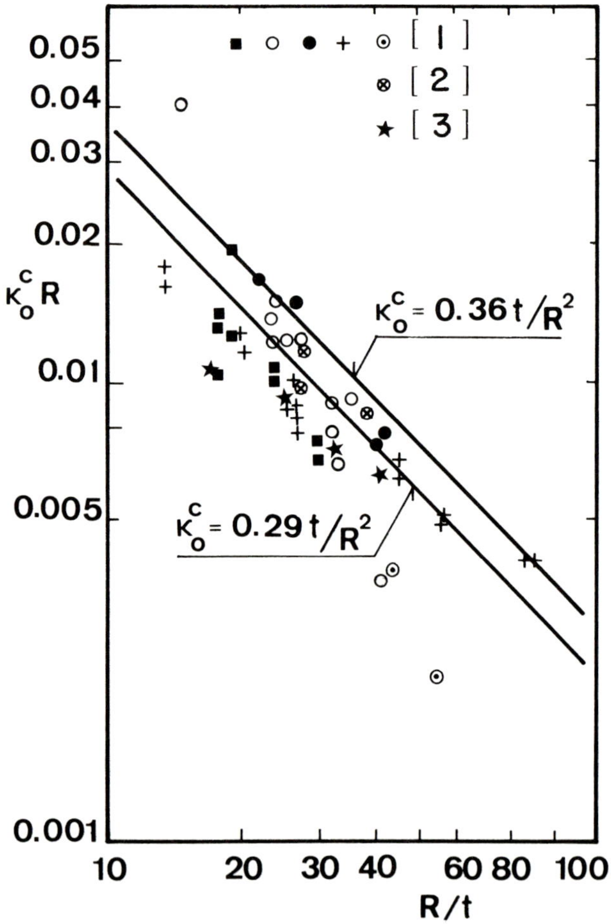

Figure 4.2.2.3 Comparison between Experimental and Theoretical Results for the Limit Curvature in Pure Bending.
[[1] Reddy (1979); [2] Sherman (1976); [3] Korol (1979)]

in which D_{max} and D_{min} is the maximum and the minimum diameter, respectively, in the initial deformed cross-section. This equation is derived assuming small deflection elastic behaviour and defines p_o^c as the pressure which causes first yield. Equation 4.2.2.5 has been shown by Kyriakides and Babcock (1981) to be a good approximation if $|p_e| \lesssim 0.25|p_y|$ or, from Equation 4.2.1.6 and 4.2.2.6:

$$\frac{R}{t}\sqrt{\frac{\sigma_{.005}}{E}} > 1 \qquad (4.2.2.8)$$

Another formula for the critical external pressure without bending has been proposed by Haagsma (1981), based on the formation of four plastic hinges in the cross-section. The experimental results quoted by Haagsma actually yield a perfect agreement with Timoshenko's formula.

Equation 4.2.2.5, as well as the more accurate theoretical results given by Kyriakides and Babcock (1981), shows that an elastic-plastic pipe subjected to external pressure is very sensitive to geometrical imperfections in the form of an initial ovalisation of the cross-section. Fabian (1981) has examined the combined effect of external pressure and bending on pipelines with imperfections. It was found that the imperfection sensitivity is highly dependent of the radius – thickness ratio R/t and the level of external pressure p/p_e.

4.3 Practical Buckling Design of Steel Pipes

4.3.1 Design Charts

The principal loading parameters in pipeline design are the curvature and the hydrostatic pressure which the pipe sections have to sustain simultaneously. As a result, design charts showing the relations between material properties, pipe sectional geometry and the load parameters at the limit point are valuable. In Figure 4.3.1.1 a design chart is proposed for pipes made of the steel material API–X52 based mainly on the results of Fabian (1981).

The external hydrostatic pressure p is given through the associate water depth $H = -p/\rho g$, where the mass density of water $\rho = 1{,}025\text{kg/m}^3$ and the acceleration of gravity $g = 9.81\text{m/s}^2$. The imposed curvature κ is normalised by 1/R. Two sets of curves with constant values of κR are drawn. One set (the full lines) are valid for perfect pipes and the other set (the broken lines) are valid for pipes with an initial ovalisation $\Delta_0 = 0.005$. The effect of an additional axisymmetric, periodic imperfection is not shown in these charts, but according to Fabian (1981) a ten per cent to 20 per cent additional reduction in the curvature κ^c at the limit point would provide a reasonable estimate of the influence of such imperfections, the highest value being used in pure bending. Figure 4.3.1.1.(b) (overleaf) shows examples of the effects of steel quality on the collapse strength of the pipes.

It should be stressed that these design charts do not include safety factors.

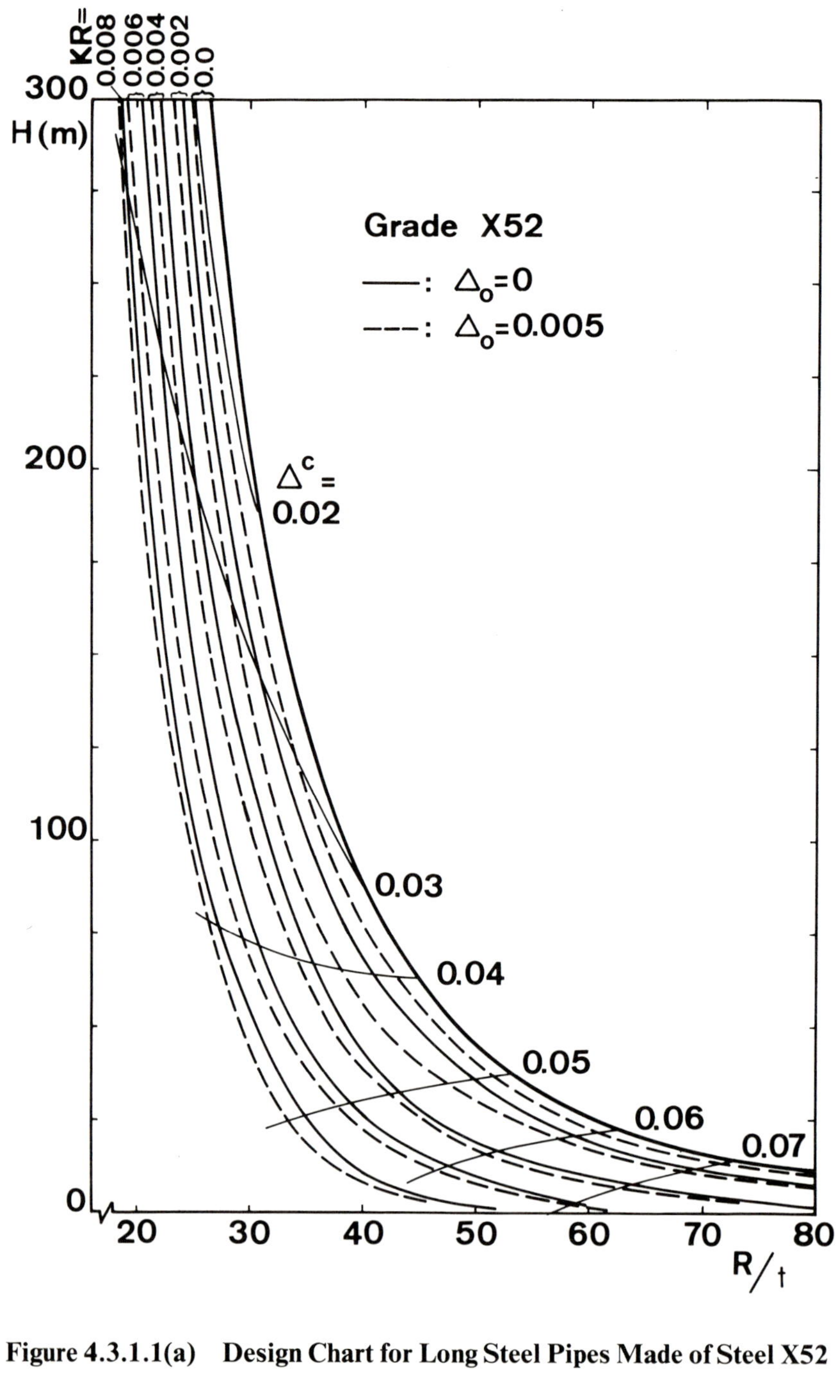

Figure 4.3.1.1(a) Design Chart for Long Steel Pipes Made of Steel X52

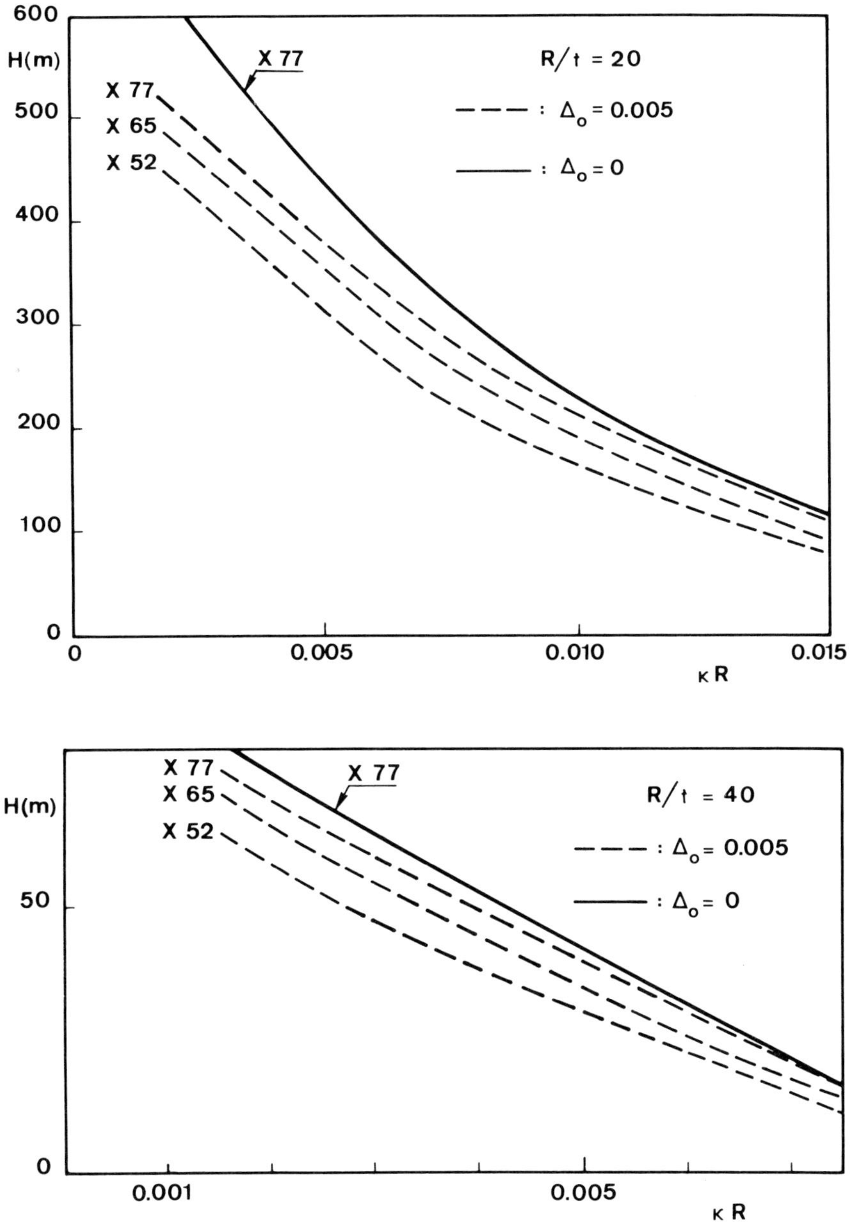

Figure 4.3.1.1(b) Design Chart for Long Steel Pipes Showing the Influence of Steel Quality and Out-of-roundness

In Figure 4.3.1.1(a) the amount of ovalisation Δ^c at the limit point is also indicated for an initial perfect round pipe.

4.3.2 Design Formula

For elastic pipes, simple but accurate interaction formulae were presented by Equations 4.2.1.7 and 4.2.1.8. In the plastic range, similar empirical formulae will now be sought in the following form:

$$\left(\frac{\kappa^c}{\kappa_o^c}\right)^{\alpha} + \frac{p^c}{p_o^c} = 1 \qquad (4.3.2.1)$$

The collapse value of the external pressure p_o^c without bending is taken from Timoshenko's formula for perfect and imperfect pipes.

The curvature κ_o^c at the limit point in pure bending could be taken from Equation 4.2.2.4 but in order to include the effect of strain hardening, an alternative expression based on the following simplified arguments will be used.

For a purely elastic pipe, Brazier (1926) showed that the instantaneous value of the ovalisation Δ and the curvature is related through:

$$\Delta = (1\text{-}v^2)\left(\frac{R^4}{t^2}\right)\kappa^2 \qquad (4.3.2.2)$$

A similar analytical formula does not exist for elastic-plastic pipes. However, the experimental results by Reddy (1979) indicate that Equation 4.3.2.2 can be used as a good approximation for elastic-plastic pipes too, if the factor $1-\nu^2$ is replaced by a number between 0.5 and 0.6:

$$\Delta = \psi \frac{R^4}{t^2}\kappa^2 \qquad (4.3.2.3)$$

If the ovalisation Δ represents an elliptic deformation of the cross-section, then the nonlinear moment-curvature relation for the pipe can be written as:

$$M = \sigma Z_O\left(1\text{-}\beta\psi\left(\frac{R}{t}\right)^2\varepsilon^2\right) \qquad (4.3.2.4)$$

to first order in Δ. Here σ denotes the stress at the outer fibre and the section modulus of the ovalised section is given as the section modulus Z_O of the undeformed pipe multiplied by a reduction $(1 - \beta\Delta)$ due to the ovalisation. For a pipe of elastic-plastic material with some degree of strain hardening, these section moduli will take values between the corresponding elastic and fully plastic section moduli. For a purely elastic material $\beta = \frac{1}{2}$ and for a perfectly plastic material (no strain-hardening) $\beta = \frac{2}{3}$.

In Equation 4.3.2.4, the nominal strain ϵ, defined as:

$$\varepsilon = \kappa R \qquad (4.3.2.5)$$

has been introduced. In the elastic case, the relation between the stress σ and the strain ϵ becomes:

$$\sigma = (1\text{-}\Delta)\,E\varepsilon \qquad (4.3.2.6)$$

whereas the stress-strain curve 4.2.2.2 will be used for elastic-plastic pipes.

The curvature κ_0^c at limit point now follows by eliminating the stress σ from Equation 4.3.2.4 and imposing the condition $dM/d\epsilon = 0$. This yields:

$$\kappa_0^c = \frac{\sqrt{1+\delta^2}+\delta}{\sqrt{\eta\,(2n+1)}} \cdot \frac{t}{R^2} \qquad (4.3.2.7)$$

where:

$$\delta = \sqrt{\eta}\left(\frac{n-1}{\sqrt{2n+1}}\right)\frac{\sigma_p}{E}\frac{R}{t} \qquad (4.3.2.8)$$

The material parameters were defined previously and the constant η is given by:

$$\eta = (\beta+1)\,\psi = \frac{3}{2}(1\text{-}v^2) \quad \text{for elastic pipes}$$

$$\eta = \beta\psi \sim \frac{1}{3} \quad \text{for elastic-plastic pipes}$$

For the purely elastic case where the value of n, equals unity Equations 4.3.2.7 and 4.3.2.8 conform to Brazier's Equation 4.2.1.5, whereas for a perfectly plastic material, where the value of n is infinite, then:

$$\kappa_0^c\,(n \rightarrow \infty) = \frac{\sigma_p}{E}\frac{1}{R} \qquad (4.3.2.9)$$

corresponding to collapse when the yield stress is reached.

For pipe material possessing a medium strain hardening behaviour and $\eta = 0.5$, Equation 4.3.2.7 can be reduced to:

$$\kappa_0^c \simeq \frac{1}{\sqrt{n}}\frac{t}{R^2} + \frac{1}{2}\frac{\sigma_p}{E}\frac{1}{R} \qquad (4.3.2.10)$$

A comparison of Equation 4.3.2.10 with the theoretical results given by Kyriakides and Shaw (1982), Fabian (1981) and Gellin (1980) has shown that the equation is slightly conservative for high strain hardening exponents ($n \gtrsim 50$).

Now, turning to the interaction Equation 4.3.2.1, the remaining unknown parameter is the exponent . An inspection of the results by Kyriakides and Shaw (1982) and Fabian (1981) for pipes subjected to combined bending and external pressure indicates that this exponent is virtually independent of the grade of pipe material and the size of any initial ovalisation and thus depends only on the radius-thickness ratio R/t.

Subject to the condition that $10 \leq R/t \leq 200$, the following interaction formula between the curvature κ^c and the pressure p^c is found to yield slightly conservative predictions for the collapse strength of long pipes:

$$\left(\frac{\kappa^c}{\kappa_o^c}\right)^{0.5+0.01R/t} + \frac{p^c}{p_o^c} = 1 \qquad (4.3.2.11)$$

As mentioned previously, the interaction formula is derived on the basis of experimental results for long steel pipes. But since Equation 4.3.2.11 takes into account variations of ratios between Young's modulus and yield stress then the formula should also give reasonable results for pipes made of different materials.

In Figure 4.3.2.1 the interaction Formula 4.3.2.11 is compared with experimental results due to Kyriakides and Shaw (1982). The results covers pipes with a radius to thickness ratio $R/t = 17.35$ and made of an aluminum alloy having a stress-strain curve which can be modelled by the Ramberg-Osgood Equation 4.2.2.1 with $\sigma_0/E = 4.3.10^{-3}$ and $n = 35$. It is seen that the interaction Formula 4.3.2.11 yields a reasonable accurate prediction of the ultimate strength of the pipe under combined bending and external pressure.

Finally, interaction formulas relating the bending moment M^c and

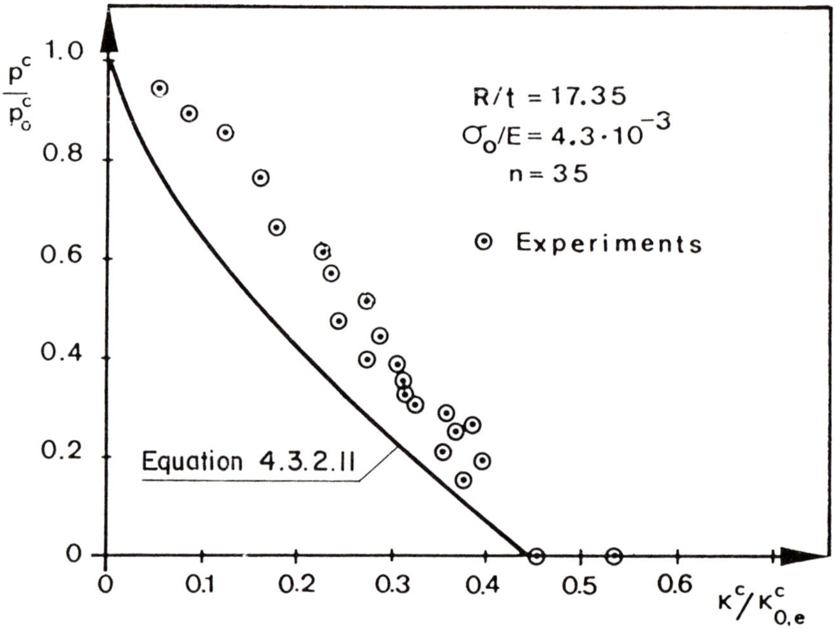

Figure 4.3.2.1 Comparison between the Interaction Equation 4.3.2.11 and Experimental Results by Kyriakides and Shaw (1982) for Collapse of Pipes

the external pressure p^c at collapse are considered. For elastic pipes such a formula is given as Equation 4.2.1.7 previously. Using the numerical results by Fabian (1981) as a basis it is found that the formula:

$$\left(\frac{M^c}{M_o^c}\right)^4 + \frac{p^c}{p_o^c} = 1 \qquad (4.3.2.12)$$

provides a very good approximation to all the results given by Fabian (1981), see Figure 4.3.2.2. It is noted that no significant variation of the exponent on the moment term in Equation 4.3.2.12 with the choice of steel material can be deduced from Fabian's results. Also, for radius to thickness ratios in the range 20 to 80, the exponent seems to be nearly independent on R/t too. A further validation of Equation 4.3.2.12 is given in Figure 4.3.2.3 (overleaf) where the experimental and numerical results, given by Kyriakides and Shaw (1982) for pipes made of an aluminium alloy are compared with Equation 4.3.2.12.

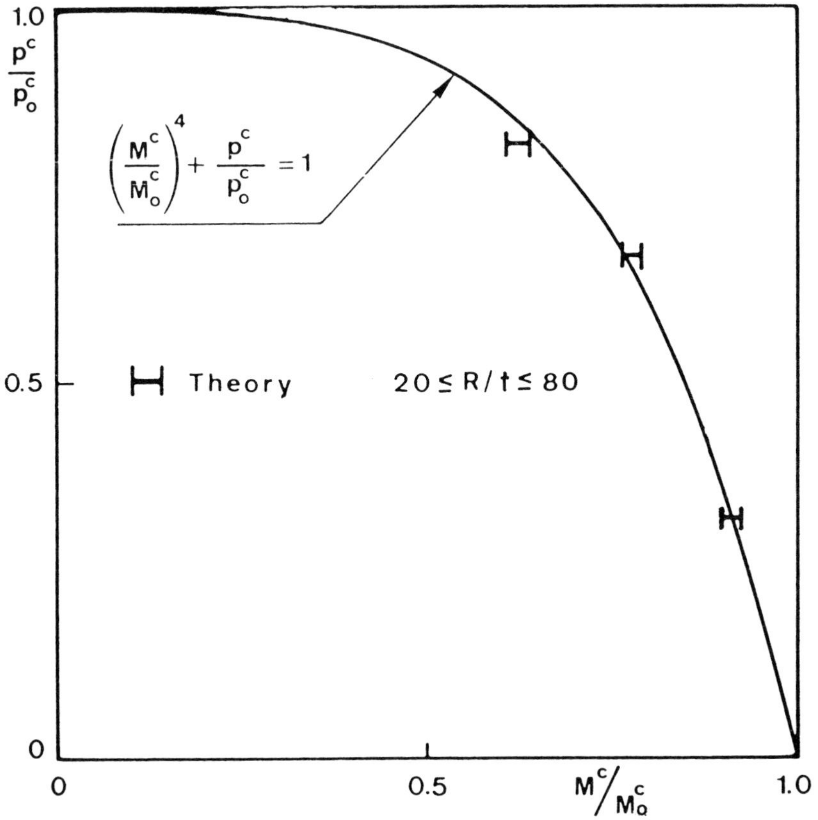

Figure 4.3.2.2 Comparison between the Interaction Equation 4.3.2.12 and Numerical Results for Pipes with 20 ≤ R/t ≤ 80 and Made of the Steel Qualities X52, X65 and X77

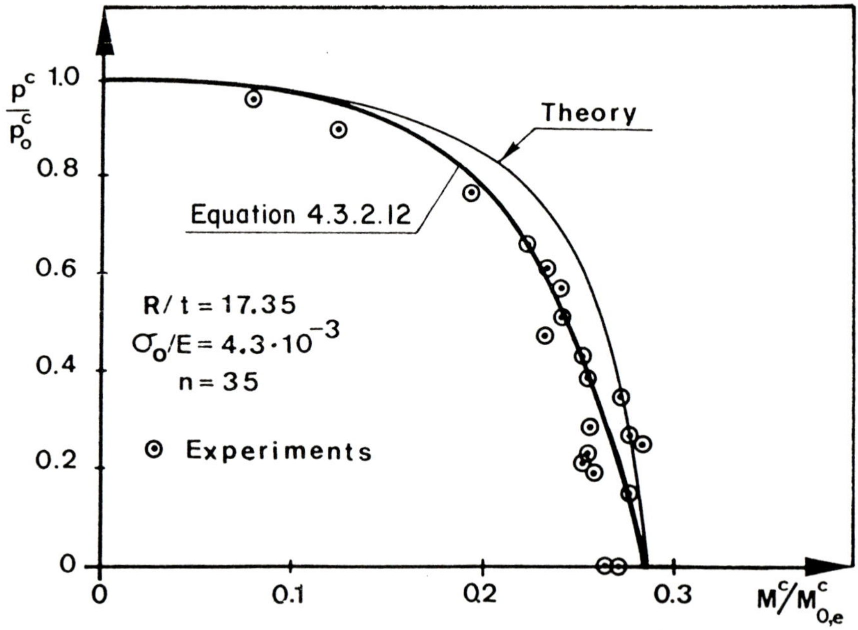

Figure 4.3.2.3 Comparison between the Interaction Equation 4.3.2.12 and Numerical Results by Kyriakides and Shaw (1982) for Collapse of Pipes

4.4 Bibliography

Axelrad, E. L. and Emmerling, F. A., 'Grosse Verformungen und Traglasten Elastischer Rohre unter Biegung und Aussendruck', *Ingenieurarchiv*, 53, 41–52, (1983).

Brazier, L. G., 'On the Flexure of Thin Cylindrical Shells and other Sections', *Proceedings, Royal Society, Series A*, 116, (1926).

Calladine, C. R., 'Plastic Buckling of Tubes in Pure Bending', *Proceedings, IUTAM Symposium on Collapse*, Cambridge University Press, (Cambridge, 1983).

Chater, E. and Hutchinson, J. W., *On the Propagation of Bulges and Buckles*, Report MECH–44, Division of Applied Sciences, Harvard University, (Cambridge, Mass., June 1983).

De Winter, P. E., 'Deformation Capacity of Steel Tubes in Deep Water', Paper OTC 4035, *Proceedings, Offshore Technology Conference* (Houston, 1981).

Emmerling, F. A., 'Nichlineare Biegung und Beulen von Zylindern und krummen Rohren bei Normaldruck', *Ingenieurarchiv*, 52, 1–16, (1982).

Fabian, O., 'Collapse of Cylindrical, Elastic Tubes under Combined Bending, Pressure and Axial Loads', *Int. J. of Solids and Structures*, 13, 1257–1270, (1977).

Fabian, O., 'Elastic-Plastic Collapse of Long Tubes under Combined Bending and Pressure Load', *Ocean Engineering*, 8, 295–330, (1981).

Gellin, S., 'The Plastic Buckling of Long Cylindrical Shells under Pure Bending', *Int. J., of Solids and Structures*, 16, 397– 407, (1980).

Haagsma, S. C. and Schaap, D., 'Collapse Resistance of Submarine Lines Studied', *Oil & Gas Journal*, 86–95, (1981).

Johns, T. G. *et al.*, 'Inelastic Buckling of Pipelines under Combined Loads', Paper OTC 2209, *Proceedings, Offshore Technology Conference* (Houston, 1975).

Khalig, A. A. and Schilling, C.G., *Plastic Buckling Strength of Seamless Steel Tubes under Bending Loads*, Research Bulletin, United States Steel Corporation, (1979).

Kimure, T. *et al,*. 'Experimental and Analytical Studies of Elasto-plastic Behaviour of Offshore Pipelines during Laying', Paper OTC 3737, *Proceedings, Offshore Technology Conference* (Houston, 1980).

Korol, R.M., 'Critical Buckling Strains of Round Tubes in Flexure', *Int. J. of Mechanical Sciences*, 21, 719–730, (1979).

Kyriakides, S. and Babcock, C. D., 'Large Deflection Collapse Analysis of an Inelastic Inexternal Ring under External Pressure', *Int. J. of Solids and Structures*, 17, (1981).

Kyriakides, S. and Shaw, P. K., 'Response and Stability of Elasto-plastic Circular Pipes under Combined Bending and External Pressure', *Int. J. of Solids and Structures*, 18, (1982).

Miller, C. D., 'Tension and Collapse Tests of Fabricated Steel Cylinders', Paper OTC 4218, *Proceedings, Offshore Technology Conference* (Houston, 1982).

Pedersen, P. T., 'Equilibrium of Marine Pipelines during Laying', *ZAMM,* 58, T228–T230, (1978).
Reddy, B.D., 'An Experimental Study of the Plastic Buckling of Circular Cylinders in Pure Bending', *Int. J. of Solids and Structures,* 15, 669–682, (1979).
Saal, H. *et al.*, 'Experimentelle und Theorische Untersuchungen an Beulgefährdeten Langen Kriesrohen', *Der Stahlbau,* 12, 353–359, (1979).
Sherman, D. R., 'Tests of Circular Steel Tubes in Bending', *J. of Structural Division, ASCE,* 102, No. ST11, (1976).
Toma, S. and Chen, W. F., 'Analysis of Fabricated Tubular Columns', *J. of Structural Division, ASCE,* 105, No. ST11, (1979).
Toma, S. *et al.*, 'External Pressure and Sectional Behaviour of Fabricated Tubes', *J. of Structural Division, ASCE,* 108, No ST1, (1982).
Tugcu, P. and Schroeder, J., 'Plastic Deformation and Stability of Pipes Exposed to External Couples', *Int. J. of Solids and Structures,* 13, 643–658, (1979).
Tvergaard, V. and Needleman, A., 'On the Development of Localised Buckling Patterns', *Proceedings, IUTAM Symposium on Collapse*, Cambridge University Press, (Cambridge, 1983).

J JUNCHER JENSEN
P TERNDRUP PEDERSEN

5

ADVANCES IN SUBMARINE PIPELINE TRENCHING BY PLOUGHING

The principal techniques for the trenching of submarine pipelines comprise jetting, fluidisation, milling and ploughing. Each method has its own particular features and constraints. The report by E and P Forum (1979), on the problems associated with pipeline burial, provides an excellent means for assessing initially the relative merits and costs of these four different techniques. To some extent, that report was prompted by the damage inflicted on North Sea pipelines by jet trenching machines, evidence of which is contained in detail in the report by MATS (1979). As a result, there currently exists new interest in ploughing as a less risky trenching technique.

5.1 Introduction

As described by Lawton (1939), the first offshore cable plough buried 65 kilometres of telegraph cable to the west coast of Ireland in 1938. Since 1955, the development of the cable plough has been continuous, prompted by the high incidence of damage to telephone cables by trawlers, so nowadays these ploughs are fairly sophisticated and they use advanced instrumentation and control systems. Examples of the routine nature of cable ploughing are given by Duncan (1969), Baxter (1971) and Pretat (1977). Despite this advancement however, the development in the shape of cable ploughs, the plough form, seems to have stopped at an early stage suggesting possibly that their geometry is optimal.

By contrast, pipeline trenching ploughs have been in use for a much shorter period, primarily to excavate trenches across tidal flats and in shallow water. Early designs were simple and a lack of appreciation of the problems involved with ploughing sometimes led to difficulties with high pull-forces, failure to trench to the correct depth and plough seizure in the seabed. As a consequence, ploughs fell out of favour and attention was concentrated instead on trenching systems based on jetting. In recent years however, there has been a growing awareness of the high cost of jetting and a realisation that a jetted trench does not always protect a pipeline efficiently. A resurgence of interest in ploughs has followed, therefore, and several have been built to new designs which are reviewed in this chapter.

5.2 The Justification for Pipeline Trenching

Trenching is a means of pipeline protection when there exists the possibility of damage to a pipeline if it is installed simply on the natural seabed. Damage may be caused by waves, tides or turbidity currents and hazards resulting from human activity. Some of these influences are reasonably well understood, and are discussed in greater detail elsewhere in this book, whereas others are much more difficult to evaluate precisely.

Trenching may involve lowering a pipeline below the natural seabed into an open trench or, alternatively, it can entail its complete burial. The merits of either method depend on their practicality, the nature and likelihood of the potential hazard and their costs.

Environmental forces are usually greatest in shallow water where complete burial is preferred. However, complete backfilling of an open trench often occurs naturally in shallow water, induced by bottom currents. As discussed in Chapter 1, the effects of waves and currents are usually least in deep water and can be withstood by suitable design of the pipeline and its coatings.

Man-induced hazards to offshore pipelines include vessel anchors, fishing gear and dumped solid waste. Protection against damage due to large anchors can seldom be justified except possibly in harbour approaches where its likelihood is highest and distances are sufficiently short to make the costs acceptable. In the open sea however, anchoring by merchant ships is usually minimal but construction vessel anchors do pose a serious hazard, especially to pipelines under construction. Evidence provided by E and P Forum (1979) shows that anchors can penetrate several metres into the seabed and exert forces approximating 100 tonnes. No economically acceptable trenching system can protect a pipeline against such a risk, which is best minimised by rigorous control of anchoring procedures. The same form of control applies to the dumping of solid wastes.

Fishing gear can present a potential hazard to offshore pipelines. Trenching may eliminate this hazard if contact between the heavy components of fishing gear, such as trawl boards and beams, and the pipeline is prevented. This requires the pipeline to either be buried with a covering of material greater than the scouring depth of the gear or, alternatively, for the pipeline to be lowered into an open trench of sufficient depth and configuration to prevent contact occurring. A significant contribution to understanding the damage to pipelines due to trawl collisions is reported by SINTEF (1974).

In effect therefore, burial is desirable on beaches, tidal flats and in shallow water since this removes pipelines from the direct action of waves and tidal currents and reduces the risk of scour. It also protects against fishing and the possibility of stranded vessels, retains the beauty of the environment and reduces the temptation to malicious damage. Increased depth of burial, in these situations, gives additional security against storm induced changes to the profile of the beach and the phenomena of liquefaction, which is covered in greater detail in Chapter 1.

In deep water however, if trenching is justified, then a completely buried pipeline is preferred to one which is partially buried or exposed. As advocated by Broussard (1979) and Erling (1979), if lowering within an open trench is adopted then a steep-sided trench is preferred to one with gentle sloping sides. Deep burial or a deep trench is not necessarily preferred to shallow burial or a shallow trench, unless significant changes to the seabed are likely to occur.

5.3 The Rudiments of Plough Design

Ploughs are very simple devices. In comparison with other seabed excavating systems, such as cutter-section dredgers, jetting machines, crawler excavators with cutting wheels and blasting equipment, ploughs are rugged and fast. Furthermore, they are relatively simple so that little can go wrong mechanically. They also use power efficiently and do minimal environmental damage. Despite these advantages, however, they suffer several problems. The first problem is concerned with the control of the depth and direction of the trench. A steeply inclined plough-share tends to lift out of the ground, because the soil force on it has an upward component. By contrast, a shallowly inclined share tends to dig further into the ground.

The mechanics of these phenomena are discussed in detail by Palmer *et al* (1979a). The net effect is that without some positive control a ploughshare either fails to penetrate the seabed or, alternatively, it drives itself into the ground, digging more and more deeply until the required pulling force becomes too much for the traction vehicle or vessel. Closely related to this problem is the design of anchors, which must penetrate the seabed and yet not drive so far that they become impossible to break out. Paradoxically, therefore, a good plough has attributes not unlike those of a good anchor.

The force required to pull a plough is a major factor in its physical design and costs. Pull forces can be relatively large: a 1 metre deep trench with sloping sides at 45 degrees requires an effective force approximating 100 tonnes, whether in sand or in overconsolidated clay, though somewhat less in soft clay. Some research has been done on the effects of different share shapes and methods for reducing the motive force, such as supplementary jetting, vibration and multiple share heads have also been investigated. Another possibility, which has been investigated using models, is to cut the trench progressively in a number of passes of the plough.

From the preceding material, it follows that a plough must be steered and it must have directional stability. However, if it is overstable, then a misalignment caused by striking a boulder or encountering a side-slope, will cause it to remain off-course, and it may prove extremely difficult to bring back to its intended path. On the other hand, if it possesses poor directional stability, it will tend to wander and generate a kinked and uneven trench. In these respects, therefore, the design of the plough-share is somewhat critical, because a prerequisite in trench design is for it to follow a straight line or reasonably smooth curve. Furthermore, if a plough is pulled along an existing pipeline, it must follow the pipeline precisely to avoid damaging it. Cable ploughing avoids this risk by laying and ploughing-in the cable simultaneously, so the cable is fed through the plough as it advances. Studies by Brown and Palmer (1978) have been made of the possibility of doing the same thing with rigid pipelines, by towing a plough behind the laybarge and laying the pipe through the plough into a trench, and this method is being developed further.

One major factor in plough design is the distinction between ploughs making a stable trench which remains open after the plough has passed, and those which cut a steep-sided trench, which is unstable in cohesionless soils and caves in eventually. The second concept is used to bury cables and flexible pipelines, which can be fed to the bottom of the trench within the length of the plough, so the collapsing sides of the trench leave the cable or flexible pipeline buried. Unfortunately, this technique cannot be applied generally to rigid pipelines because they cannot bend to the bottom of the trench within an acceptable plough length.

5.4 The Basic Configuration of Ploughs

Plough designers have responded in a variety of ways to the problems outlined above. It is, however, possible to distinguish four basic configurations, which are described below. An inventory of recent full-scale ploughs is given in Table 5.4.1 and can be used in conjunction with the following descriptions. This table does not include the many model ploughs which have been used in development work.

5.4.1 Configuration 1: Sleds with Projecting Shares

This design concept is illustrated by Figure 5.4.1.1 (overleaf). The share is carried on a sled, which runs on skids at either side. The share projects downwards into the seabed, and is raked so that the soil force on it is inclined downwards. The vertical component of the soil force on the share is balanced by the upward reaction on the skids, and the horizontal component is the pulling force. The Coflexip plough, which is illustrated in Figure 5.4.1.2 (overleaf) and described in detail by De Panafieu

Table 5.4.1 Plough Data

YEAR OF CONSTRUCTION	OWNERS	CONFIGURATION; CABLE (C), OR PIPELINE (P); BEFORE (B), DURING (D) OR AFTER (A) LAYING	MAXIMUM TRENCH DEPTH (M)	TRENCH CROSS-SECTION	SOIL	MAXIMUM PULL FORCE (TONNES)	REFERENCE
1959	Land & Marine	1 p b	na	na	silty clay	na	Reynolds (1974)
1969	ATT (Seaplow III)	1 c d	0.7	□	all	45	Duncan (1969) Baxter (1969) Pretat (1977)
1972	Nippon Telegraph	1 c d	1.5	na	all	na	
1973	KDD ('KS-1')	4 c d	0.7	□	all	18	
1976	Mobil	2 p b	1.2	na	hard clay	na	Brown (1978) Abel (1978)
1977	Panarctic Oils	2 p b	1.5	▽	soft clay	50	Palmer (1979a)
1977	Coflexip	1 p b	1.0	□	all	40	De Panafieu (1979)
1977	Esso Australia	1 p b	na	▽	sand	na	Brown (1978)
1978	Smit International	3 p a	1.2	▽	all	100	

na = not available, proprietary information.

(1979), is an advanced version of this concept. It is designed specifically for use with flexible pipe. The share cuts a trench with almost vertical sides, and a short shield behind the share holds back the sides of the trench. The pipe is guided through the plough into the trench by an adjustable shoe so that it reaches the bottom of the trench at the rear end of the shield. In cohesionless bottom material, the sides of the trench cave in after the share has passed, leaving the pipe buried.

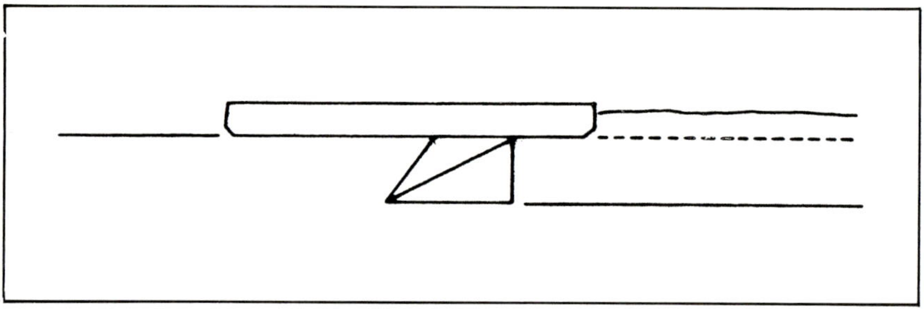

Figure 5.4.1.1 Sled with Projecting Share

The Coflexip plough uses jets on the face of the share, which loosen the bottom soil and reduce the pulling forces. For example, in consolidated clay, cutting a trench 1 metre deep at the rate of 6 metres per minute requires a towing force ranging from 16 to 25 tonnes with jetting and from 20 to 40 tonnes without jetting. Although jetting increases the complexity of this plough, the jetting power is usually quite small by comparison with that used on traditional jet barges, typical examples being 240 hp for the former and 20,000 hp for the latter.

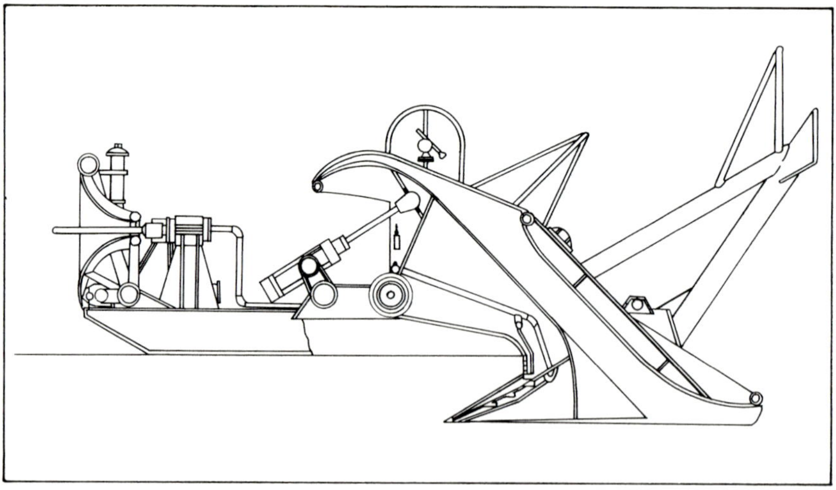

Figure 5.4.1.2 Coflexip Plough

5.4.2 Configuration 2: Long-beam Ploughs

This design is illustrated by Figure 5.4.2.1 and its mechanics are described by Palmer *et al* (1979b). Such ploughs are often used in large forestry and land drainage projects. The frame has the form of an inverted L. A share is fixed to the lower end of the vertical arm and beneath it is a rigidly attached heel. The leading end of the horizontal arm of the L – the beam – is held at a fixed distance above the ground, by wheels or skids. The advantage of this configuration is that the presence of the heel gives it good depth control. If the share digs in too far, the heel presses into the bottom of the trench and generates a large reaction which lifts the share and restores its cutting depth. If the share is cutting too shallow, the heel loses contact with the trench bottom, and the downward soil force on the front of the share is not balanced by the heel reaction, so the share digs in deeper until the heel is again in contact with the trench bottom. The same principle can be used to give the plough directional stability: if the share has rigid sides, they press against the sides of the trench, and give directional stability in the same way as vertical depth control.

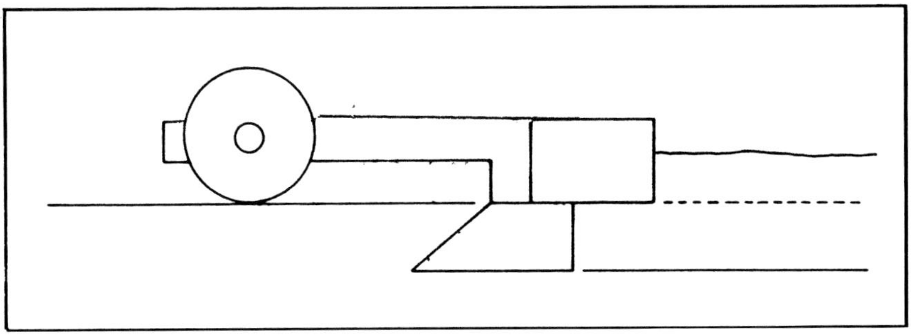

Figure 5.4.2.1 Long-beam Plough

This concept has been adopted recently by several marine pipeline ploughs designed by R J Brown and Associates. The first of this series, which is illustrated in Figure 5.4.2.2 (overleaf) was used in 1977 to excavate a 1 metre deep trench to a length of 2 kilometres in hard clay at the Statfjord oilfield in the North Sea, as part of the bottom tow installation of a 36-inch loading line. This operation, which is described by Brown (1978), Abel (1978) and Palmer (1979b) had a major effect in encouraging renewed interest in ploughs. A second plough of the same general form, but with a different share and skids instead of wheels, as shown in Figure 5.4.2.3 (see over), was used in 1978 to excavate a trench in soft clay at the Drake gasfield, in the Canadian Arctic offshore Melville Island. This operation is described by Palmer (1979a) and Palmer (1979c).

5.4.3 Configuration 3: 'Post-trenching' Ploughs

Trenching under an existing pipeline raises additional problems because

Figure 5.4.2.2 Mobil Statfjord A Plough

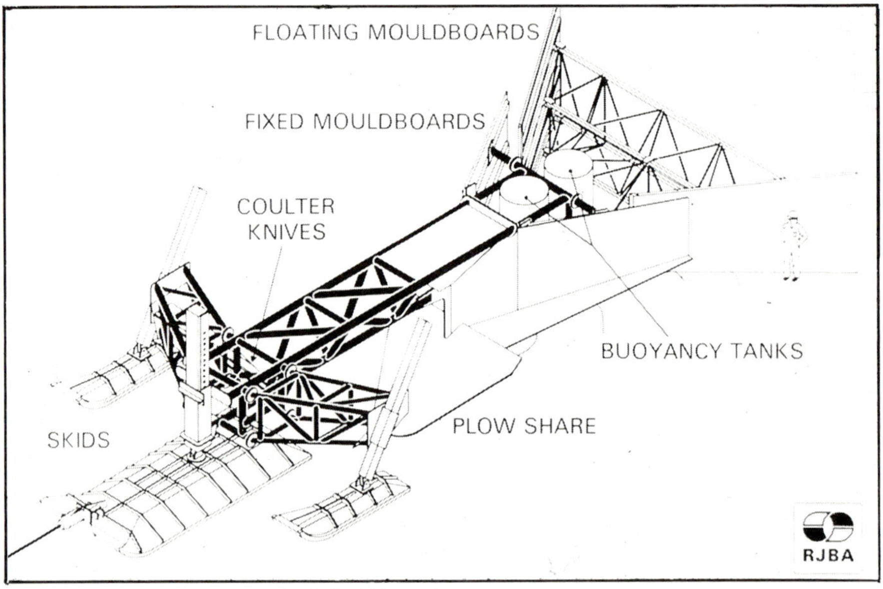

Figure 5.4.2.3 Panarctic Oils Plough

the plough must follow the pipe safely, be deployed in position at the start and recovered at the end of the operation. Many alternatives have been examined. One concept is an asymmetric plough which runs along one side of the pipeline excavating a vertical-sided trench which partially collapses behind the plough, thus lowering the pipe into the trench. A second concept is to apply one of the configurations described previously, but to reduce the plough height to a minimum, and pull it along under the pipeline. Guide rollers would then carry the pipeline over the plough, lifting it off the seabed as the plough advances, so that it settles back into the trench after the plough has passed. This operation is quite simple but it poses problems for placement and removal of the plough at the beginning and end of the trenching operation.

A third concept is a symmetrical plough with articulated sections which close under and around the pipeline. The Smit International post-trenching plough is one realisation of this concept. Figure 5.4.3.1 is a schematic representation, and Figure 5.4.3.2 a photograph of such a plough. It has two beams, hinged together at the front end which is carried on skids. The plough spans the pipe and is guided along it by rollers. Each

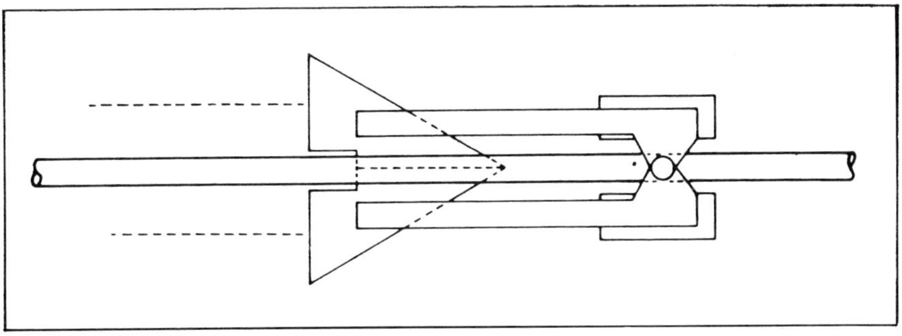

Figure 5.4.3.1 Post-trenching Plough (Plan)

Figure 5.4.3.2 Smit International Post-trenching Plough

beam carries a half-share, which cuts one half of the trench, and is extended to form a mouldboard. The plough is lowered in an 'open' position, with the front end over the pipe and one half-share on each side. When it is pulled forwards, the soil reaction on each half-share pushes them together, and they close under the pipe. The share cuts a triangular trench with side-slopes which remain stable for a wide range of soil conditions. Tests show the closing operation to be smooth and reliable. At the end of the trench, the two halves of the plough are jacked apart by a ram until the plough can be lifted clear.

This plough has recently been tested in shallow water in the North Sea and offshore tests were conducted in early 1980.

5.4.4 Configuration 4: Multi-bladed Ploughs

This design concept is shown in Figure 5.4.4.1 and it is similar to Configuration 1, except that it employs several shares in tandem. The advantage of such a system is that it can reduce the towing force. Tests show that the pull force to move a single share is proportional to the cutting depth to a power m greater than unity. It follows that the total pull force can be reduced by having several shares each cut part of the trench. However, there is a penalty in structural complexity, overall plough length and interference between the flows of soil over consecutive shares.

This configuration is used in some Japanese cable ploughs, which have as many as eight shares.

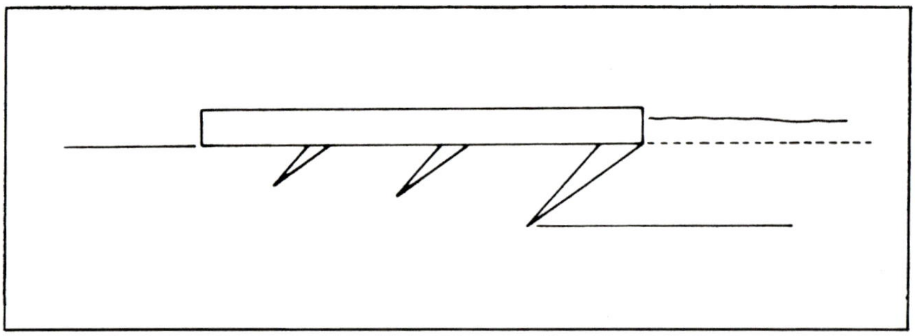

Figure 5.4.4.1 Multi-blade Plough

5.5 The Instrumentation and Control of Ploughs

Many pipeline ploughs have little or no on-plough instrumentation and their behaviour can only be monitored from the tow vessel or winch.

In shallow water, this is satisfactory because the track of the plough follows the pull cable. If the trench needs to be curved, the tow vessel is steered along a pre-calculated path, using suitable navigational aids, and the plough follows.

In deep water however, it is desirable to monitor directly the behaviour of the plough during the trenching operation, so as to control the intended path and cutting depth. This is particularly important in post-trenching or simultaneous laying and trenching operations. To this end, some recent pipeline ploughs are provided with relatively advanced instrumentation systems. In this respect, pipeline ploughing is catching up with cable ploughing practice, where it has long been recognised that sophisticated monitoring systems are essential in deep water. Duncan (1969) and Baxter (1971) provide excellent examples of such systems. Also, Table 5.5.1 lists the instrumentation functions available on the ATT Seaplow III cable plough and on the Coflexip plough and a comparison shows that the two systems are quite similar. The only significant difference of emphasis is that the Seaplow III system assigns great importance to direct measurement of plough speed. An analog computer on the Seaplow support ship calculates the peak cable tension which would result if the plough were to stop suddenly, and presents this information to the helmsman on a single meter. This illustrates an important principle with instrumentation systems of this type, namely that so much information is generated, it has to be presented

Table 5.5.1 Instrumentation Comparison

	ATT SEAPLOW III		COFLEXIP PLOUGH	
FUNCTION	Measured on plough	Measured on ship	Measured on plough	Measured on ship
Plough speed	yes (3)*	—	—	—
Tow-wire tension	—	yes	yes	yes
Tow-wire payout	—	yes	—	yes
Umbilical tension	—	yes	—	yes
Umbilical payout	—	yes	—	yes
Cable tension	—	yes	—	yes
Television camera	yes (3)	—	yes (3)	—
Share position	yes	—	yes	—
Trench depth	yes	—	—	—
Lowering cable interlock	yes	—	na	—
Water depth	yes	yes	yes	yes
Plough pitch	yes	—	yes	—
Plough roll	yes	—	yes	—
Plough heading	—	—	yes	—
Jet pressure	na	—	yes	—
Relative position of ship & plough	—	—	—	yes

na = not applicable
* numerals in parentheses indicate the number of independent redundant systems, if there is more than one.

to the operator in a form that he can easily assimilate and understand, since otherwise he may overlook significant changes. Likewise, the Coflexip system incorporates a minicomputer, on the support ship, which provides a real-time readout on a display unit.

It would clearly be feasible to extend the use of mini and microcomputers to provide some degree of automatic control. For example, the system could be arranged so that the tow ship's engines would be throttled immediately if the tow cable tension reached a preset level, and the cable could be released completely if a higher cable tension were reached. Further developments in this direction are expected, but substantial development time and resources are required, because such systems will have to operate reliably in a hostile environment.

5.6 Operational Problems with Marine Ploughs

Lowering and recovering a plough to and from the seabed poses major difficulties. A pipeline plough may be 25 metres long, weigh up to 75 tonnes in air and be awkwardly shaped. It must be lowered precisely in position so that it is pointing in the desired direction, or alternatively lowered in one place and then manoeuvred across the seabed until it is correctly aligned. If the plough is to trench an existing pipeline, then its positioning is particularly critical. Ideally, it should be possible to lower and recover a plough without awaiting calm seas. In deep water, accurate positioning of a plough will require a depth positioning system on the launch vessel, possibly supplemented by thrusters on the plough. During the development of the Kvaerner-Myren cutter-suction pipeline plough, close attention was given to positioning, and a sophisticated thruster system was developed, as described by Biberg (1979) and Morgan (1979). In trials, this system successfully located the plough on a 36-inch pipe in 320 metres of water, without diver assistance: with a typical deployment time of approximately 45 minutes.

Launching and recovery systems for particular ploughs have been studied extensively but it is too soon to arrive at general conclusions. Ideally, one vessel should transport the plough to site, launch it, tow it, and carry out the control and monitoring functions and the associated survey. In practice however, this has not always been possible.

In post-trenching operations, it is essential to minimise interaction between the plough and the pipe so that the engineer can be sure that the operation does not damage the pipeline. This is a further reason for reducing the pull forces. If the normal ploughing force is 100 tonnes, one cannot risk the possibility that such a force will suddenly be applied to the pipe, so elaborate control systems are essential to obviate this risk. On the other hand, if the ploughing force can be reduced to only 10 tonnes, it can be applied to many pipelines without harm and the control problem is correspondingly less critical.

5.7 Future Developments in Plough Design

It is clear that pipeline trenching ploughs are still at an early stage of development and that much remains to be done. Among the areas which need further investigation are the following:

(i) *Share shape*
The forces required to pull ploughs become unacceptably large as trench depths exceed 1.5 metres in most kinds of soil. These forces could be reduced by more advanced share designs and by the use of cutting knives and auxiliary shares, although a large reduction seems unlikely. A better understanding of basic soil mechanics and soil deformation in front of the share could also lead to economies.

(ii) *Combinations of ploughing and jetting*
There is a range of intermediate options, between the pure ploughs without jets and the jetting sleds without ploughs. The Coflexip plough is but one example of these intermediate systems. The Harmstorf jet plough is another which has been used successfully in shallow water and in Alaska, as described by Mellor (1978).

(iii) *Share vibration*
It is known from work on land ploughs that vibrating the share can reduce significantly the ploughing force in suitable soils, but this has not yet been investigated for offshore application.

(iv) *Ploughing soft rock*
All the remarks in this chapter refer to ploughs working in uncemented soil. However, some areas of seabed consist of soft rock such as chalk, conglomerate, sandstone and shale, and it seems conceivable that a plough-like device might work in such material, but again this has not yet been demonstrated.

(v) *Steering*
The directional stability and steering of ploughs is another area which requires further study, although a start has already been made on the development of better steering systems.

(vi) *Improvements in the control*
Launching, positioning and recovery of ploughs could conceivably be made cost effective.

5.8 Conclusion

In appropriate conditions, ploughs offer a straightforward and economical means of burying and trenching pipelines. Much remains to be done to develop their full potential but useful systems are already available. The most urgent need is for wider offshore experience of the operational aspects of ploughing in a variety of sea states and bottom conditions.

5.9 Bibliography

Abel, W. and Brown, R. J., 'Offloading Line Made up Onshore, Towed to Site', *Oil & Gas Journal*, 76, 67–72, (1978).

Baxter, H. A. and Mueser, R. E., 'The Development of Ocean-cable Plows', *IEEE Transactions on Communications Technology*, COM-19, 1233–1241, (1971).

Biberg, O. and Åril, B., 'A New Trenching System for Deepwater Pipelines', *Proceedings, Eleventh Annual Offshore Technology Conference*, 3, 1939–1952, (Houston, 1979).

Broussard, D. E., 'FLAGS Gasline – Seabed Stability and Safety', *Proceedings, 1979 European Seminar on Offshore Oil & Gas Pipeline Technology*, (London, 1979).

Brown, R. J. and Palmer, A. C., 'Plow for Pipeline Burial', *Noroil*, 6, (1978).

Cutler, R. C. and Beale, P. A. H., 'Twin Submarine Pipelines to Khor-al-Amaya', Paper 7445 S, *Proceedings, Institution of Civil Engineers*, (1972).

Duncan, C. C., 'Plowing Cables under the Sea', *IEEE Transactions on Communications Technology*, COM-17, 74–82, (1969).

Erling, S. and Andersen, T., 'Pipeline Trenching in the Future', Paper SPE 8174. 1, *Offshore Europe 1979*, Society of Petroleum Engineers, (1979).

E. & P. Forum, *Pipeline Burial*, (London, 1979).

Lawton, C. S., 'The Submarine Cable Plow', *AIEE Transactions (Power Applications Systems)*, 58, 675–686, (1939).

Marien Technologisch Speurwerk (MATS), *A Survey of Pipelines in the North Sea. Incidents during Installation, Testing and Operation*, MATS Report, Project P1–6, (1979).

Mellor, M., *Undersea Pipelines and Cables in Polar Waters*, Report 78–22, Cold Regions Research and Engineering Laboratory, US Army Corps of Engineers, (Hanover, NH, 1978).

Morgan, M. J. and Swanson, T. L., 'A Control System for Deepwater Trencher', *Proceedings, Eleventh Annual Offshore Technology Conference*, 1279–1290, (Houston, 1979).

Palmer, A. C., 'Bottom Tow of Submarine Pipelines', *Proceedings, 1979 European Seminar on Offshore Oil & Gas Pipeline Technology*, (London, 1979b).

Palmer, A. C., Baudais, D. J. and Masterson, D. M., 'Design and Installation of an Offshore Flowline for the Canadian Arctic Islands', *Proceedings, Eleventh Annual Offshore Technology Conference*, 765–772, (Houston, 1979c).

Palmer, A. C., Kenny, J. P., Perera, M. R. and Reece, A. R., 'Design and Operation of an Underwater Pipeline Trenching Plough', *Geotechnique*, 29, 305–322, (1979a).

De Panafieu, D. and De Bailliencourt, B., 'The Simultaneous Installation and Burial of Flowlines – An Innovative and Cost-Effective Solution to Line Protection', *Proceedings, Eleventh Annual Offshore Technology Conference*, 1401–1415, (Houston, 1979).

Pretat, G. H., 'TAT-6: 4000 Circuits at One-third Cost', *Civil Engineering (ASCE)*, 57, (1977).

Reynolds, J. M., 'Submarine Pipeline Burial and Protection', Paper T-I/iii–3, *Proceedings, Offshore North Sea 1974 Conference*, (Stavanger, 1974).

Vassdrags og Havnelaboratories (VHL) SINTEF, *Influences of Bottom Trawl Gear on Submarine Pipelines, Phase 1,* SINTEF Report number STF60 A74027, (1974).

A C PALMER
P DAVIES

6
THE AUTOMATED CONNECTION OF OFFSHORE PIPELINES

In recent years there has been a steady upsurge in development activity to reduce the costs of installing, connecting and stabilising pipelines. This activity is now in the process of being expanded to include the deepwater connection of pipelines to other pipelines, to risers, to subsea manifolds and to wellheads. New methods, tools and equipment are being developed for this purpose and, with the benefit of increasing experience, these are gaining acceptance by pipeline proprietors.

The resulting trend is moving away from the use of large and expensive construction spreads for accomplishing these various operations, and is tending towards the application of a more detailed and planned engineering approach aimed at combining new techniques with existing hardware. The objective here is to minimise the need for intervention, either from the surface, or by man during pipeline installation.

6.1 Introduction

One costly element in the installation, maintenance and repair of offshore pipelines is their connection to platform risers, manifolds and other pipelines. Experience has shown that conventional techniques for connecting pipelines involving large surface spreads can be vulnerable to weather and the adverse effect of sea state on the efficiency of operation. For offloading lines connecting platforms to towers, the cost of the connections alone can approximate 60 to 70 per cent of the cost of the entire installation, regardless of the method by which the lines are installed. Using the innovative engineering methods, which are described herein, these connections can be installed for considerably less money and in a shorter time than by conventional methods, and the rectification work that is frequently required after a pipeline goes into operation can be virtually eliminated.

The new technique involves accurately positioning the lead-end of the pipe within a prescribed target area on the seabed and its remote mating with a wellhead or other subsea installation. Important among the novel features of this method has been the detailed use of model simulation techniques for a step-by-step development of the connection procedures and related contingency plans. These methods were employed on the full-scale installation for Panarctic Oils, which was successfully completed after closely following model procedures; the design of the pipeline and connection control system was also based on simulation studies.

6.2 A Case History Involving High Arctic Experience

The connecting of subsea completions for either flow control or water injection lines can be enhanced by grouping several types of service pipelines and associated control/power cables in a single bundle. Figure 6.2.1 shows the pipe bundle that was made up for Panarctic Oils on Melville Island in the Canadian High Arctic. The bundle was subsequently pulled 4,000 feet under the ice in 180 feet of water, using a mating sled on the lead-end. The bundle consisted of two 6-inch pipelines with heat tracing systems, four 1-inch lines for control purposes and glycol transmission, one 2-inch annulus access line, an instrumentation line and a power cable. The entire group of lines was contained within an 18-inch carrier pipe.

On the offshore end of the bundle, which is shown in Figure 6.2.2, a sled was attached for use in navigating the pipeline end into its final posi-

Figure 6.2.1 The Pipeline Bundle Used for the Panarctic Project

tion by pulling into a receiver on a subsea wellhead. The receiver, shown in Figure 6.2.3, acted as the guide into which the bundle lines and sled connectors were inserted. The receiver could in fact be adapted to suit

Figure 6.2.2 Navigating Sled Used on Panarctic Project

Figure 6.2.3 Sled Mating Receiver

numerous different combinations of pipeline to wellhead, to riser or manifold, or to another pipeline. The function of the sled and receiver was to place the components to be connected within close proximity of each other.

Based on the results of a model simulation, which is described later, sheaves and anchors were placed on the seabed and light lines were run from a central control cabin, through sheaves, to the deflecting sled at the lead-end of the pipe. Approximately 1,000 feet of the front end of the pipe bundle were buoyed off the seabed using pontoons counterbalanced by drag chains, as shown by Figure 6.2.4. This arrangement minimises the loads during the deflection operation, with the chains providing sufficient drag to maintain the lateral stability of the pipe in the maximum design current.

For the majority of pipeline deflect-to-connect projects, either three or four winches are adequate. Winch 1 deflects the pipe, Winch 4 acts as holdback, and Winches 2 and 3 are used for pulling the sled to the receiver.

For the full-scale control room used on the Panarctic project, all the necessary equipment was built into a single air-transportable cabin, as shown in Figure 6.2.5. Controls were mounted within the control room, and the winches used for deflecting the pipe were mounted externally on the control cabin frame. The interior of the control room is shown in Figure 6.2.6. On the right hand side, this shows the track plotting sheet

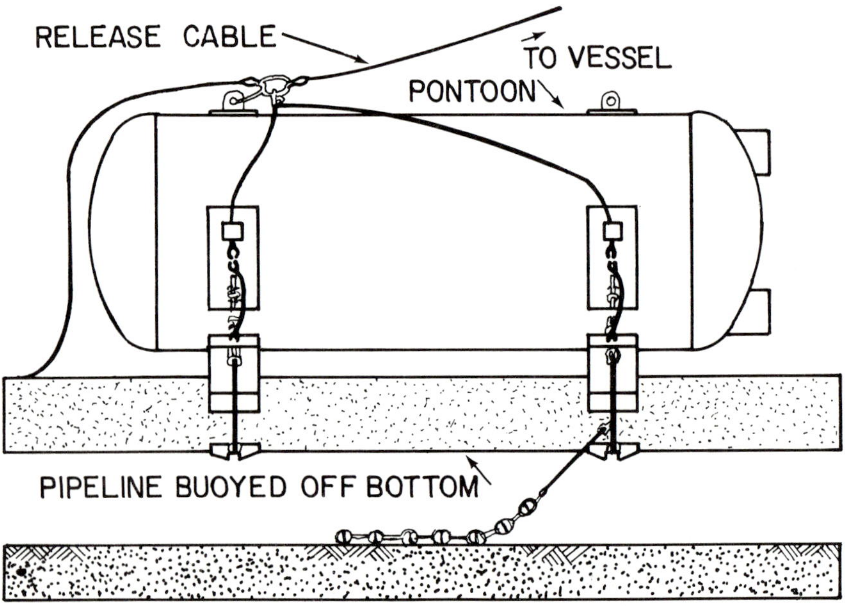

Figure 6.2.4 Schematic Presentation of a Pontoon Used for Manoeuvring the Pipeline into its Tie-in Position

Figure 6.2.5 Control Room for Remote Deflect-to-Connect Projects

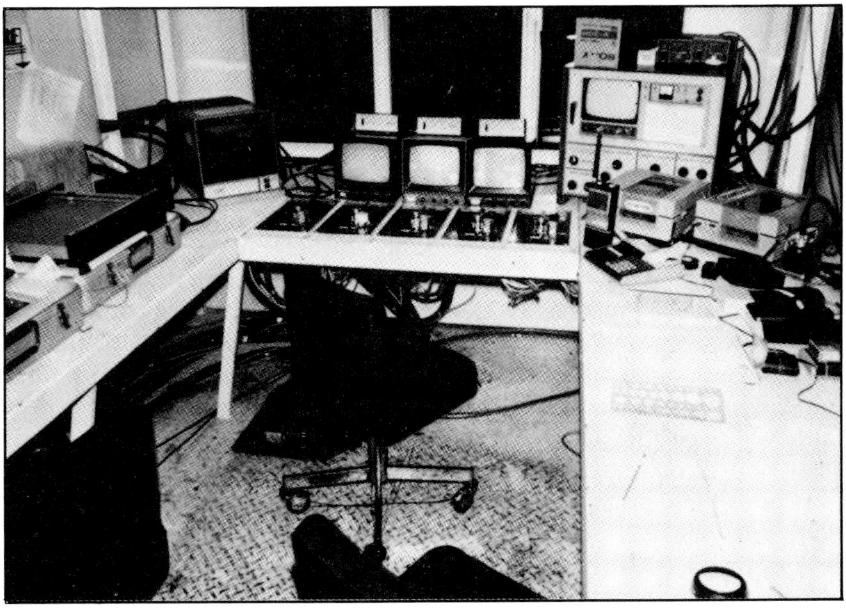

Figure 6.2.6 Interior of Control Room Cabin

which was developed from the simulation experiments and, on the left, the sled navigation system.

After the winches had been connected, the navigation system positioned, the TV cameras installed on the wellhead, and all systems had been confirmed as operational, the deflecting process began. When the initial deflection of 33 feet had been completed, the process was temporarily halted. This enabled the navigation system to interrogate the various transponders on the pontoons to confirm that the deflecting process was being executed as planned. This procedure was repeated a number of times as deflection of the line proceeded.

Eventually, the sled reached a position which was sufficiently close to the wellhead for a visual sighting on TV to be made. The process was then continued until the sled and pipeline were directly aligned over the final approach line.

At this juncture, a completely diverless operation was initiated in which the pull-in cables were released from the sled and, *via* a threading operation at the ice surface, were positioned through the receiver on the wellhead. The pull-in cables were then slowly spooled in as the deflecting and holdback cables were simultaneously slackened off. The mating of the sled with the receiver was completed when locking wedges were set. Finally, connection of the bundled lines was executed *via* mechanical connectors.

The equipment requirements for an actual installation by the deflection method are very modest, costing about 1/10th to 1/15th of a normal spread. The crew executing the tie-in can be located on the platform, with the advantage that there is much reduced sensitivity to weather conditions. This consideration, coupled with the lower risk inherent in the deflection method, greatly enhances the chances for success as well as reducing the potential for cost overruns.

The method for completing the final pipe-to-pipe connection is optional, so that any one of a variety of hardware may be used. The key feature is for the selected method to be capable of use with a mating sled and receiver. The system described here was a diverless connection involving seven pipelines in a bundle, *via* a sled and receiver, and achieving the final mating by means of a mechanical connector. Although this method was originally engineered on behalf of Panarctic Oils Ltd for under-ice use in the Canadian High Arctic, it has direct application to other offshore areas of the world and similar procedures can be used for any of the other combinations of pipelines to terminating structures, as previously mentioned. The system has been designed for use at depths down to 1,200 feet, and plans exist to extend this to 4,000 feet.

For bundled pipelines where thermal expansion presents a problem, a technique has been devised whereby the inner lines are pre-tensioned relative to the outer casing. This arrangement relieves the stresses which are imposed in the pipes during the start-up period, and considerably reduces the amount of expansion occurring in the individual lines. As a general rule-of-thumb, and depending on the relative steel cross-sections of the inner lines and outer casing, the extent of the expansion can be reduced by a factor of between 10 and 15.

6.3 Remote Installation Using Collet Connectors

Once a pipeline or bundle has been brought to within close proximity of its mating structure, use can be made of any one of the variety of types of connection which are currently at various stages of development. When properly used, these can considerably simplify the installation and connection of pipelines, and can assist in reducing operation and maintenance costs. The connections can be by flange or mechanical connector, or by explosive, hyperbaric or atmospheric welding. Mechanical connectors offer the most simple solution, especially collet connectors which are available in a range of sizes and ratings.

The specific advantage of collet connectors is that they can be coupled to platform risers by cold springing of the riser until mating is achieved. For collet connections to subsea manifolds or other pipelines, use can be made of hydraulic rams or jacking screws to perform the necessary connector movement and remote diverless connection. Sufficient background has been gained with this system to confirm that it offers considerable promise and, with adequate and detailed attention paid to the procedures for connecting the pipeline, a low-cost and maintenance-free connection can be achieved.

The method for installing the pipeline on the seabed can be selected according to the nature of the project. It can include the use of a reel barge for small pipelines, or a laybarge for larger diameter lines, for bundled and/or insulated lines. Alternatively, on-bottom or off-bottom tow procedures can be employed. For pipes transporting either high pour point oil, or gas with hydrate problems, the lines can be insulated, heat-traced and placed in a casing to enable the entire system to be pulled directly into a receiver.

6.4 Remote Connection of Large Diameter Pipelines in Deep Water

Pipelines of 36-inch diameter have been installed by laybarge to a maximum depth of 530 feet. Pipeline strings of this diameter have also been bottom-towed to a depth of 1,400 feet across the Norwegian Trench on behalf of Mobil. By internally pressurising lines to resist collapse, large diameter pipes can be installed at much greater depths. Using this technique with plugs fitted on the pipeline ends, it is now technically feasible to perform a mechanical or collet tie-in of two or more pipe strings or a welded connection using an atmospheric chamber.

6.5 The Benefits of Model Simulation Techniques

Using the simulation model which is illustrated in Figure 6.5.1 (overleaf), pipeline stiffness, weight, loads, stresses and deflection can be assessed realistically. This enables an entire operation to be engineered and planned correctly. For the deflect-to-connect technique for example, modelling can determine the stresses and loads occurring during pipeline deflection and connection. This facilitates the correct sizing of the equipment and, for the control module, winches of only 1 to 3 hp have been shown to be adequate for deflection of the pipelines and plug-in of the connector sleds.

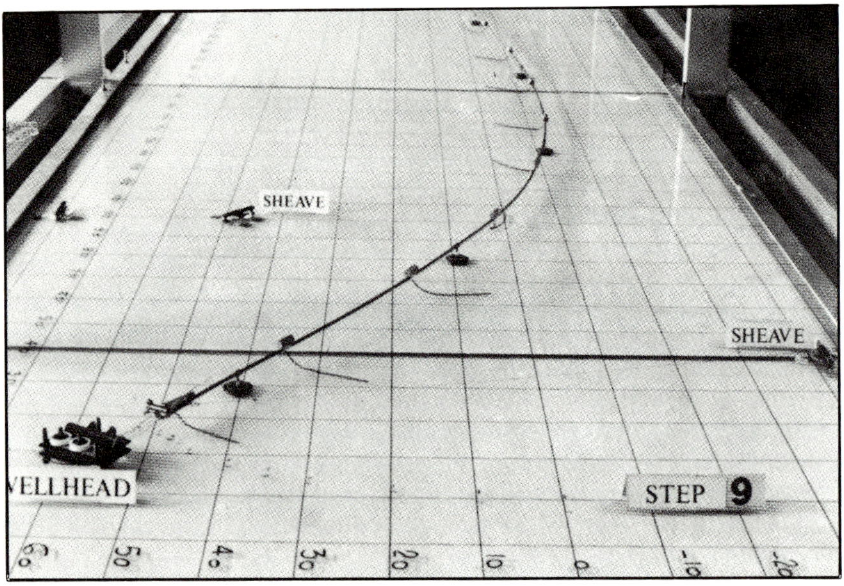

Figure 6.5.1 Pipeline Tie-in Simulation Model

As a consequence of this, the control module containing the winches can be compact in size, and weigh only around 10 tons, including the winch, navigation and computation equipment needed for controlling the deflecting and mating processes. This facilitates installation of the module on platforms or SPMs, and means that operation of the deflect-to-connect process can be rendered insensitive to sea state and perform with a minimum of surface support.

In addition, the ability to operate the deflect-to-connect process in the higher sea states during the off-season period offers further benefits. Not only are vessel hiring costs lower but pipeline connection can be performed when it does not interfere with the normal surface activities of production facility connection and hook-up on the platform.

These innovations were untried prior to the Panarctic project. The breakthrough from the engineering standpoint was the ability to develop the systems conceptually and adapt many aerospace industry simulation techniques to complete the design, prove the system and provide this simulation technique as a basis for training the contractor for minimising the risk during installation.

Since then, these techniques have advanced, and models and simulators can take account of the conditions experienced by the pipeline in its actual environment. These models are three-dimensional, and can simulate a long pipeline for end-connection to manifolds or risers, the installation of a pipe-to-pipe end mating and the performance of trenching equipment, such as ploughs.

6.6 The Cost Savings of Novel Tie-in Methods

A relatively large fleet, comprising a lay or derrick barge and various supply vessels, is normally required for a conventional pipeline tie-in connection. This costs between US $150,000–$200,000 per day plus mobilisation and remobilisation expenses. Furthermore, such tie-ins can be accomplished in as little as five days or take as long as five months depending on weather conditions.

By contrast, the deflection method, which was described previously, only requires an investment of approximately $100,000 for the central control cabin complete with its equipment. The only additional costs involved are $40,000 per day for a supply vessel to transport the cabin to its offshore location and a five man crew to supervise its installation and control.

Placement of the necessary anchors and sheaves on the seafloor can be accomplished by the supply vessel after detailed rigging and placement procedures have been developed. When the rigging position has been confirmed, execution of the deflection process can proceed.

One of the key features of the deflection process is the method of connection employed. If it is a diverless method using a mating mechanical connector, then the primary cost will relate to the connector sled and receiver. If flanges are employed, the sensitivity of control offered by the deflecting process provides for very precise alignment of the connecting faces. A saturation spread suspended from the platform can then be used to install the gasket and bolts.

The low cost of the flanges and approximately $4,000 per day for the saturation spread make this a relatively inexpensive method of connection. The cost savings attainable through use of the deflection process and simple flanges reach factors of between 10 and 50 when compared with conventional methods of connection. Where multiple connections, rather than a single connection, are involved the savings can be even greater.

6.7 Bibliography

Brown, R. J., 'Engineering Innovation for Subsea Pipeline Connections', *Pipeline Industry*, (August 1978).

Brown, R. J., 'Innovations Used in Pipeline Installation under Arctic Ice', *Oil & Gas Journal*, (November 20, 1978).

Brown, R. J., 'Pipeline Risers: Plan now for Cost Savings', *Noroil*, (April 1979).

Brown, R. J., 'Pipeline Consultant's Goal: Long-life Safe System at Lowest Cost', *Offshore Engineer*, (March 1979).

R J BROWN

7

CONCRETE COATINGS FOR SUBMARINE PIPELINES

In this chapter we review the technology of concrete coating for submarine pipelines. The first section is devoted to an examination of design requirements and the second to concrete design and the statement-of-the-art of concrete application. The third section discusses the need for further research into rational design methods and identifies the needs for future development.

7.1 Design Requirements in Concrete Coating

The primary functions of the concrete coating on a submarine pipeline are to give the pipeline additional weight to stabilise it on the seabed and protect the external corrosion coating against mechanical damage. The importance of these two factors varies, but both are significant.

Mechanical damage can occur at several stages during construction. If the pipeline is constructed by the laybarge method, for instance, the pipe joints are loaded onto supply vessels and shipped to the barge, where they are unloaded by cranes onto racks. Later, they are loaded onto conveyors and moved into position for alignment on the firing line. The pipe moves along the barge ramp, past welding stations and through a tensioner (which applies a localised circumferential force in the axial direction of the order of 100 tonnes), and over the stern across a stinger. If the sea is rough, wave forces acting on the pipe alternately lift it away from the stinger and let it fall back, so that heavy impacts can occur as the pipe comes back into contact with the stinger rollers. Some stingers are not rigidly attached to the barge, and there is relative pitch movement between the two, but this tends to flex the pipe in the region above the hitch between the stinger and the barge. As a result, this overbend region is subject to repeated dynamic bending stresses, and fatigue damage can occur if the concrete is insufficiently reinforced. The pipeline leaves the lower end of the stinger and continues to the seabed in a long suspended span until it reaches the touch-down point. If the barge is pitching in the waves, and the seabed is rocky, alternate lifting and lowering of the pipeline can induce further damage in this area. On the seabed the pipeline is subject to impact from trawl boards and less frequently from anchors and ships cables. Furthermore, trenching equipment can do further damage to a pipeline if it post-trenched.

Little has been done to quantify the forces that cause mechanical damage, except in the case of trawl board impacts as reported by Moshagen (1980), Bergan (1981) and De Groot (1979). Many pipeline operators and contractors carry out simple impact tests, designed to simulate the impact of a trawl board. A heavy striker hits the pipe repeatedly at the same point and the development of damage is observed. Coatings are compared in terms of the number of impacts required to reach the reinforcement or to reach the corrosion coating. Typical results are shown in Table 7.1.1 which demonstrates the level of resistance of modern designs.

Table 7.1.1 The Results of Typical Experiments to Test the Impact Strength of Pipeline Concrete Coatings

Based on experiments using a striker mass of 10,000 kilogrammes, with a velocity of 2.1 m/s, energy of 2.1 kJ and a striker with a semi-cylindrical nose 160 millimetres in diameter).

	PIPE DIAMETER (mm)		
	406.4	406.4	324
Coating thickness (mm)	27	51	37
Concrete density (kg/m^3)	2,250	2,250	2,250
Type of reinforcement	hexagonal wire mesh	1 layer welded wire mesh	2 layers hexagonal wire mesh
Number of blows to first exposure of reinforcement	3	16	15
Number of blows to reach corrosion coating	13	97	86

Although pipeline impact on laybarge rollers is very different from impact with a trawl board, the relative velocity immediately before impact is of the same order of magnitude, and this determines the dynamic stresses close to the point of impact. It seems reasonable to conjecture that if one concrete design resists simulated trawl board impact better than another, it will also have a better resistance to other kinds of impact.

Mishaps in the North Sea have emphasised the importance of the mechanical strength of pipeline concrete. Extensive concrete damage was observed in surveys of the West Sole pipeline, laid in 1967, and sections had to be restabilised by mattresses. In 1975, a 1.5 kilometre section of the Brent-Cormorant line lost more than half its weight coating and floated to the surface during construction in the Sound of Yell in the Shetlands. In 1976, half a kilometre of the Noordgastransport pipeline in the Netherlands sector of the North Sea floated 18 metres off the seabed, as a result of substantial loss of concrete, only two years after it had been laid. In this last incident, it appears that initial damage during laying led to the formation of a short span, which had oscillated significantly during a severe storm in January 1976, causing fatigue damage to the concrete and a progressive growth of the span. Observations from a submersible suggested that the hexagonal wire mesh reinforcement had done little to hold the concrete together. All these mishaps necessitated costly repairs and the third example led to a six-week shutdown of the pipeline.

In addition to impact resistance, the concrete must be strong enough to transmit to the pipeline the force applied by the laybarge tensioner, which has tracks or wheels that grip the concrete surface. As far as the concrete is concerned, this is not a major design factor, but the interface between the concrete and the corrosion coating is important, and this point is returned to later. Furthermore, if the pipeline has to be pulled along the seabed, as in a shore crossing, the concrete must have adequate resistance to abrasion, particularly if the seabed is rocky.

The other primary function of the concrete is to give the pipeline additional weight. The selection of pipeline wall-thickness is based on stress analysis and is mainly determined by the operating pressure and stresses induced during construction. The weight of steel and corrosion coating alone is usually insufficient to provide the pipeline with sufficient submerged weight to maintain its stability on the seabed. It is uneconomic to increase the steel pipe wall-thickness to provide this additional weight so this is one of the functions of the concrete. The density of weight-coating concrete is in the range 2,250 to 3,050 kilogrammes per cubic metre (kg/m^3). The advantage of using very dense or 'heavy' concrete is that the concrete coating can be made thinner, making it easier to reinforce. A further advantage is realised by using heavy concrete in that hydrodynamic forces increase with increasing pipe diameter, so that a pipeline with a larger outside diameter needs even more submerged weight to keep it stable. It follows, therefore, that if heavy concrete is used, the outside diameter is less, so the submerged weight is reduced making the pipeline easier to construct and handle.

Accurate control of submerged weight is crucially important. A pipe which is too heavy is difficult to lay, because high stresses will be induced in the sagbend and on the stinger, and a large tension will have to be applied by the laybarge. However, if the pipe is too light, it may not be adequately stable and large lateral deflections of a suspended span may occur in cross currents. These factors are particularly important in deep water, so pipeline design specifications give close attention to precise weight control. Careful monitoring is essential, because the submerged weight is the small difference between two large quantities, the total 'air' weight and the buoyancy. It follows that a small difference in total weight can produce a much larger proportional difference in submerged weight. For example, a one per cent change in total weight could lead to a ten per cent change in submerged weight. Modern concreting machines can control the submerged weight to within 5 kilogrammes per metre (5 kg/m) or 3 pounds per foot (3 lbs/ft). Specifications are set at this level of accuracy to minimise difficulties during pipelaying.

An additional consideration is the fact that when a pipeline is immersed in seawater, the concrete gains weight by water absorption. It is desirable to minimise this effect because it can significantly increase the submerged weight. For example, a five per cent water absorption could increase the submerged weight by approximately 6 kilogrammes per metre.

Concrete coating also gives a pipeline some degree of thermal insulation and increases its resistance to buckling. However, these are not usually important considerations. Studies of the effects of concrete coatings on buckle propagation suggest that the increase in propagation pressure, due to the presence of the concrete, approximates ten per cent. In practice however, design for buckling strength is invariably based on the steel strength alone, so the extra strength provided by the concrete can be considered an additional safety factor.

In summary, the most important design requirements are mechanical strength, particularly against impact, and weight. The levels of strength and weight that can be achieved by present-day practice are described in the next section.

7.2 Concrete Coating Practice

The methods of applying concrete to pipelines are generally quite different from those used in conventional structural concrete practice and have a strong influence on the properties of the concrete. Because of this, the design of the concrete cannot be considered in isolation from the method used to apply it. The alternative application methods are therefore described first, in Section 7.2.1, whereas the design of the concrete and the reinforcement are reviewed in Sections 7.2.2 and 7.2.3. Interaction between the concrete and the corrosion coating is an extremely important factor, so this is discussed in Section 7.2.4.

7.2.1 Methods of Concrete Application

There are four processes for applying concrete to submarine pipelines: casting, guniting, impingement and wrapping or 'compression'. Each is described in turn.

7.2.1.1 Casting

The casting method is analogous to the conventional method used to form structural concrete. The reinforcement is assembled around the pipe, to which the corrosion coating has been applied previously, and the pipe is placed in a rigid sectional mould, as shown by Figure 7.2.1.1, which is open at the top. Concrete is poured into the annular gap between the mould and the pipe and vibrated to compact it. When the concrete has hardened sufficiently, the mould is removed, and the concrete is allowed to cure. It is kept moist using water sprays, or by a curing membrane, until the concrete has reached the specified compressive strength.

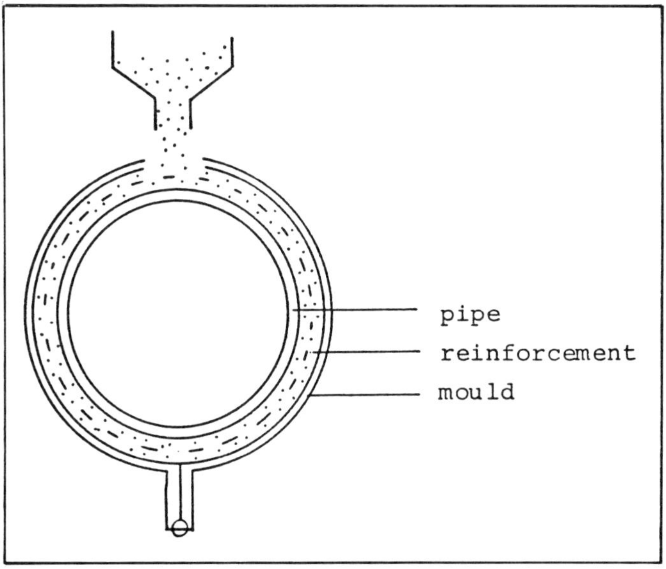

Figure 7.2.1.1 Concrete Application by Casting

This method is widely used for pipelines for river crossings. It was also used for a number of submarine pipelines constructed in shallow water in the Gulf of Mexico and the Arabian Gulf in the 1950s and 60s, and for some early lines in the North Sea. However, this method is labour-intensive, and most suitable for relatively short lengths of pipeline in conditions where labour costs are low and the design requirements are undemanding.

A difficulty with casting is that the narrow gap between the mould and the pipe means that the concrete has to flow easily if it is to surround the reinforcement properly, and avoid the formation of gaps and cavities. This implies the need for a high water/cement ratio, the use of plasticisers and a large proportion of sand aggregate. Unfortunately, these requirements tend to result in low strength and impact resistance, and lead to a porous and absorbent concrete which is subject to shrinkage. As a result of these deficiencies, casting is nowadays seldom used for offshore pipelines, either in the North Sea or elsewhere.

7.2.1.2 Guniting

This is analogous to conventional uses of gunite. It is slow, and subject to some of the same objections as casting, and for these reasons is rarely used nowadays for offshore pipelines.

7.2.1.3 Impingement

The impingement method of coating throws the concrete onto the rotating pipe by a system of rotating drums, paddles or brushes. The mix is relatively dry, so that the concrete remains on the pipe without slumping, and the pipe can be immediately transported to a curing yard.

This method has two variants. The high velocity vertical impingement system used by M K Shand is shown schematically in Figure 7.2.1.3(a). The concrete is mixed in a turbo-drum mixer and runs over a conveyor belt into the nip between a brush and a drum with paddles, both rotating with a peripheral speed approximating 40 metres per second (m/s). The pipe is held by its ends on a travelling carriage, which runs on rails parallel to the pipe axis, and carries a motor drive so that the pipe can be rotated. As the pipe rotates, it is transported axially, so that as it moves under the paddle/brush system, a layer of concrete is thrown onto the pipe, coating it continuously. A light scraper finishes the surface, which is left relatively rough. At the ends of the pipe joint, clamp-on segmented rings are used to form the 'cut-back' for welding. The pipe is then transferred to a steam curing bay, and remains there for eight to 12 hours, until the concrete reaches a compressive strength approximating 21 MN/m^2 (3,000 lb/in^2). It is then transferred to a curing yard.

The horizontal impingement process used by H C Price and its associated companies is shown schematically in Figure 7.2.1.3(b). The concrete is carried along a horizontal conveyor belt, and a rapidly rotating brush above the belt throws the concrete against the rotating pipe. The impingement velocity is lower than that used by M K Shand in vertical impingement, but otherwise the process is similar. It has been argued that the horizontal impingement process produces poorer compaction than vertical impingement, and that horizontally-impinged concrete has a higher variability of compressive strength and a higher water absorption, because of segregation of coarse and fine aggregate particles. However, the data required to test this conclusively are not available, and, if such a difference exists, it may be a consequence of different adjustments of the machines, rather than a basic difference between the two processes.

The properties of the concrete applied by the impingement process are discussed in Section 7.2.2.

Nowadays, impingement is by far the most widely used process. The vertical impingement system, as used at the M K Shand plants in Invergordon (Scotland), Brazil and elsewhere is described in more detail by Kiernan (1975 and 1979). The horizontal impingement system is used at Price plants in Delfzijl (Netherlands), Leith (Scotland), Sharjah (UAE),

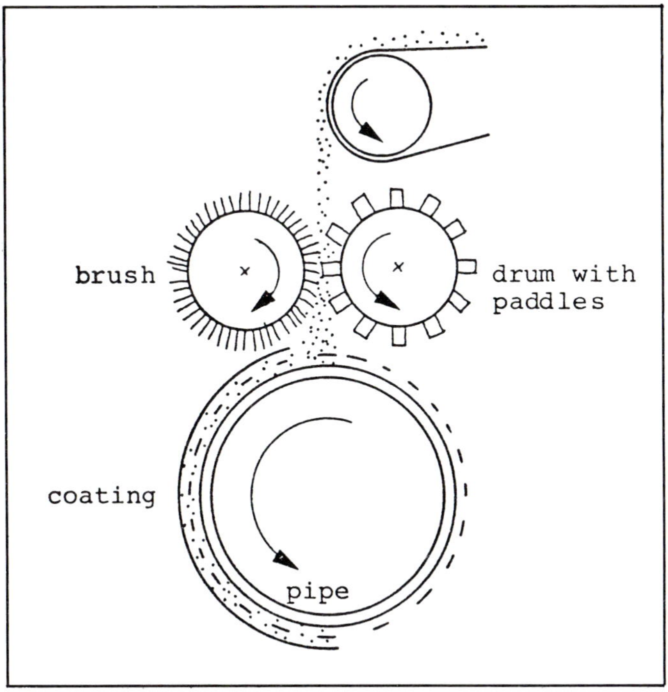

Figure 7.2.1.3(a) Concrete Application by Vertical Impingement

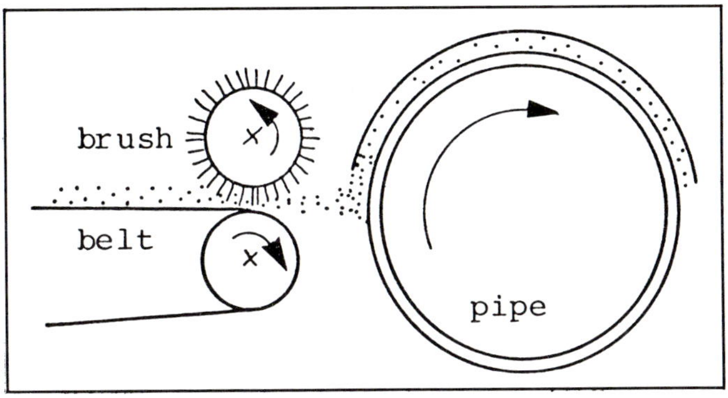

Figure 7.2.1.3(b) Concrete Application by Horizontal Impingement

Singapore, Texas and elsewhere, and by Anchor Wate Inc in the USA. Together, the plants at Invergordon, Delfzijl and Leith have coated the vast majority of deepwater pipelines in the North Sea. The equipment for the process is available in modular forms and can be set up in any location.

7.2.1.4 Wrapping

The wrapping or 'compression' process was developed by Compression Coat Inc and is usually referred to by the company name. It is represented schematically in Figure 7.2.1.4. Concrete flows from a mixer through a hopper and onto a polyethylene outerwrap supported by a belt. The level of concrete on the belt is controlled by a gate. The pipe rotates and moves axially. The concrete is compressed between the pipe and the outerwrap on the carrier belt, so that the wrap and the concrete together are wound helically onto the pipe, and the concrete is compacted in the nip between the pipe and the belt. The outerwrap, which is under tension, remains on the pipe, holding the concrete in position and forming a smooth and impervious outer curing membrane. In most applications of this method, wire mesh reinforcement is wound in with the concrete.

This method has been used since 1973 for river crossings, and more recently offshore in the Gulf of Mexico and in Chile. It has also been used by Tennessee Gas for the 26-inch diameter South Pass pipeline in water depths up to 280 feet, and for a number of other lines, the largest being 48 inches in diameter with a concrete coating of 7 inches in thickness.

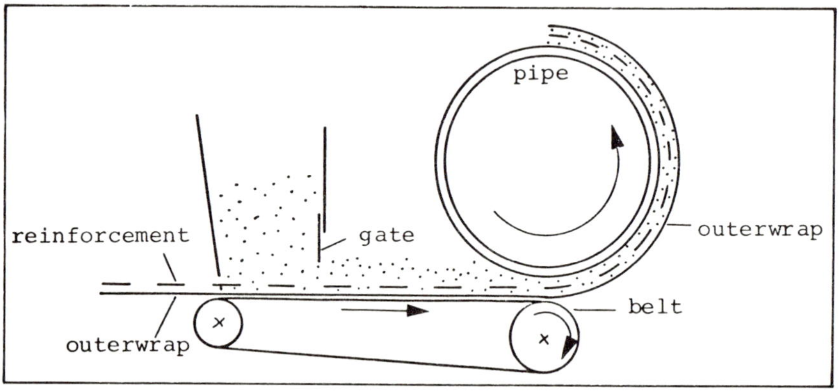

Figure 7.2.1.4 Concrete Application by Wrapping ('Compression')

7.2.2 Concrete Design

As mentioned previously, coating concrete needs to have high strength and impact resistance and to these ends it is often desirable to increase its density by incorporating heavy aggregate. A low and predictably consistent water absorption is also needed so that the submerged weight is accurately controlled and remains within the design tolerance as the pipeline moves through the suspended span. Furthermore, the concrete design must naturally be compatible with the application system.

Concrete made by the casting method generally has a relatively low strength because the high water content, needed to transport the concrete into the mould, leads to large pores between aggregate particles. Using a much drier mixture than the casting technique, the impingement method achieves good compaction from the high-speed impact of the concrete against the pipe and high compressive strengths are also achieved. Design specifications have increased steadily in recent years as shown by Table 7.2.2.1 (overleaf) which applies to a number of North Sea pipelines.

A typical recipe for the production of normal weight concrete by the impingement method is given below.

Component	Percentage by weight
Crushed granite	38.1
Sand	31.6
Cement	23.3
Water	7.0

This gives a water/cement ratio of 0.30 and an aggregate/cement ratio of 3. Each component of the mix is, of course, subject to a detailed specification. Typically, the maximum particle size of the crushed granite is 12.7 millimetres ($\frac{1}{2}$ inch); and 50 per cent of the aggregate passes through a 3.5 millimetre (0.14 inch) mesh. Concrete to this design, applied by the impingement method, has a measured mean compressive strength approximating 10,700 lb/in^2 after seven days and 12,200 lb/in^2 after 28 days. It should be noted that these figures are much higher than specification. The saturated mean density for this concrete is typically 2,460 kg/m^3 (158 lb/ft^3), and the water absorption approximates two per cent.

Heavy concrete includes iron ore or barytes, which have particle specific gravities approximating five times that of water. In the case of one specific pipeline, for example, the concrete, applied by impingement included 27 per cent of iron ore, had a water/cement ratio of 0.28 and an aggregate/cement ratio of 3.6. The mean compressive strength was 11,000 lb/in^2 at seven days and 12,600 lb/in^2 at 28 days. These compressive strengths were measured on cubes. Since it is impossible to compact the concrete into a cube mould in the same way as it impinges on a pipe, it is better practice to obtain compressive strengths from cores cut from the coated pipe by a diamond drill, and to specify a combination of core and cube tests for control purposes. The relationship between the compressive strength and the age of impinged concrete is shown graphically in Figure 7.2.2.1 (see over).

The remarks above apply specifically to vertical impingement. Concrete applied by horizontal impingement may have a slightly lower compressive strength, although the question is confused by the difference between core and cube strengths. For one recent contract for horizontally-impinged pipe concrete, measurements on cores gave a mean core compressive strength of 53.3 MN/m^2 (7,725 lb/in^2) at seven days, and over 55 MN/m^2 (8,000 lb/^{2}) at 28 days, in comparison with a specification of 41 MN/m^2 at 28 days. The three lowest seven-day strengths, from 23 cores in total, were 35.5, 40.7 and 43.1 MN/m^2 (5,150, 5,900 and 6,250

Table 7.2.2.1 A Comparison of Concrete Coating Specifications for Submarine Pipelines

YEAR	OPERATOR	LOCATION	DIAMETER		CONCRETE STRENGTH		REINFORCEMENT	
			mm	inch	MN/m²	lb/in²	type	percentage of cross-section of concrete (circumferential)
1973	Shell	Anglesey	1,016	40	24	3,500 (7 days)	hexagonal mesh (2 layers)	0.33
1973	BP	Forties-Cruden Bay	813	32	32	4,650 (28 days)	cage	0.57
1974	Occidental	Piper-Flotta	406	16	24	3,500 (28 days)	hexagonal mesh (2 layers)	0.33
1974	Shell	Cormorant	914	36	21–31	3,000–4,500 (28 days)	initially hexagonal mesh later welded mesh	0.17 0.50
1976	Shell	FLAGS (first 50 km)	914	36	34.5	5,000 (28 days)	cage	0.85
1977	Shell	FLAGS (remaining 400 km)	914	36	41	6,000 (28 days)	cage	0.90

lb/in²). The variability in the compressive strength of this concrete should be noted: on one pipe the first core gave a strength of 36.5 MN/m² (5,300 lb/in²) and the second core 68.9 MN/m² (10,000 lb/in²).

Unfortunately, less information is available on the wrapping method. Average compressive strengths of 50 MN/m² (7,250 lb/in²) were reported in tests by Tennessee Gas, for both normal density (2,230 kg/m³ : 139 lb/ft³) and heavy (3,240 kg/m³ : 202 lb/ft³) concrete. Specified compressive strengths over 34.5 MN/m² (5,000 lb/in²) are achieved routinely. It appears that the maximum strength reached by wrapping may be somewhat lower than that achieved by impingement, but this is conjectural and has not been documented.

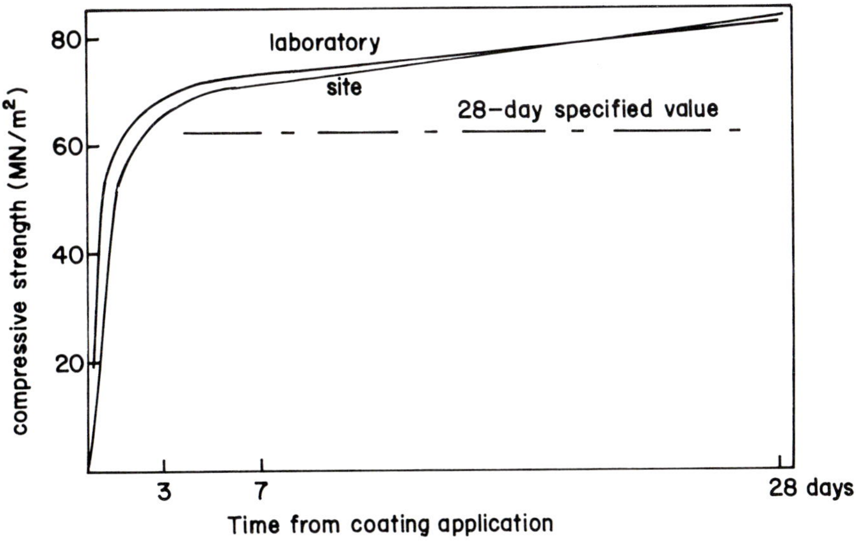

Figure 7.2.2.1 Impinged Concrete: Development of Compressive Strength with Age

7.2.3 Reinforcement Design

Pipeline coating concrete needs reinforcement to give it impact resistance and bending strength. Three types of reinforcement are in common use. They are hexagonal wire mesh or 'chicken wire', welded wire mesh and cage reinforcement. Each is based on the use of steel wire and bars. Systems based on other materials, such as the König system using polymer rope, have been used for river crossings but are considered unsuitable for offshore applications.

Hexagonal wire mesh consists of steel wire 1.35 millimetres (.053 inch) to 2.64 millimetres (.104 inch) in diameter, twisted to form meshes 38 to 51 millimetres (1.5 to 2 inches) across as shown by Figure 7.2.3.1(a), (overleaf). The variant, shown in Figure 7.2.3.1(b) has an additional wire across each mesh. The wire is supplied in rolls 200 to 500 millimetres (8 to 20 inches) wide; smaller widths are used for smaller pipes. When the concrete is applied by an impingement process, the wire is wound in spirally, immediately after the application of the first layer of concrete, so that there is an overlap of the order of ten per cent of the roll width. Figure 7.2.3.2 illustrates a section of the coating, showing consecutive layers of mesh. It can be seen that the right-hand edge of each width of mesh is on a larger diameter than the left-hand edge, and must therefore be stretched proportionately: this stretching is taken up by deformation of the hexagonal meshes.

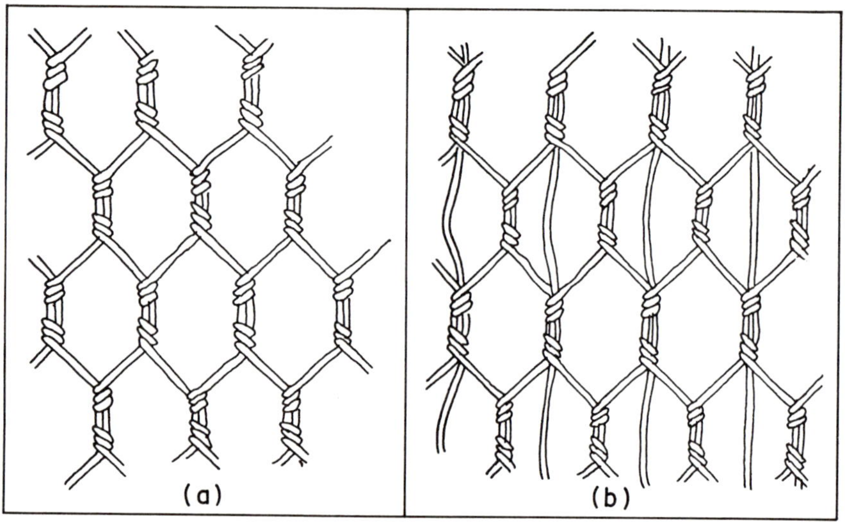

Figure 7.2.3.1 Hexagonal Wire Mesh Reinforcement
(a) Normal Configuration
(b) Configuration with Additional Wires

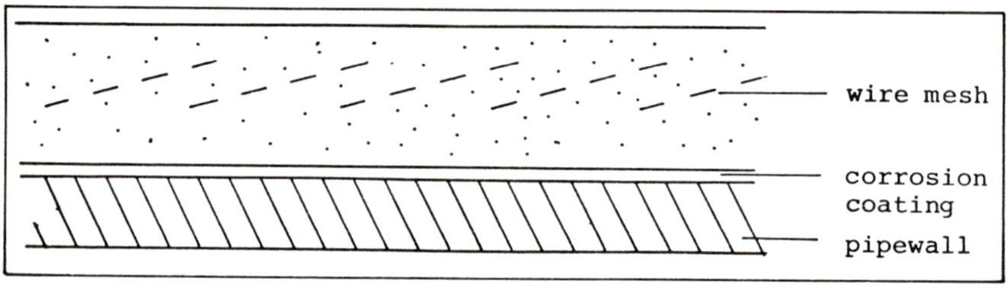

Figure 7.2.3.2 Wire Mesh Reinforcement: Longitudinal Section of Concrete Coating

Until the early 1970s, hexagonal wire mesh was used almost universally as the reinforcement for coating concrete. However, the observed partial failures of the concrete on the West Sole, Cormorant and Noordgastransport pipelines, and significant concrete damage on other lines as they were laid, led to a reaction against the use of hexagonal wire mesh for North Sea pipelines, although it is still widely used in the Gulf of Mexico, the Arabian Gulf and in south-east Asia.

An illustration of welded wire mesh reinforcement is given in Figure 7.2.3.3. It is spot-welded netting made from galvanised low-carbon steel wire. The cross wires, in the transverse direction across the net, are plain: the line wires, in the longitudinal direction, are heavily crimped. A typical mesh has line and cross wires 1.6 millimetres (.063 inch) in diameter, with 25.4 millimetres (1 inch) between consecutive line wires and 67 millimetres (2.6 inches) between consecutive cross wires. The line wires have an ultimate tensile strength of 400 to 550 MN/m^2 (58,000 to 80,000 lb/in^2), and the cross wires an ultimate tensile strength of 780 to 980 MN/m^2 (113,000 to 142,000 lb/in^2). Wire diameters up to 3 millimetres (1.2 inches) are available.

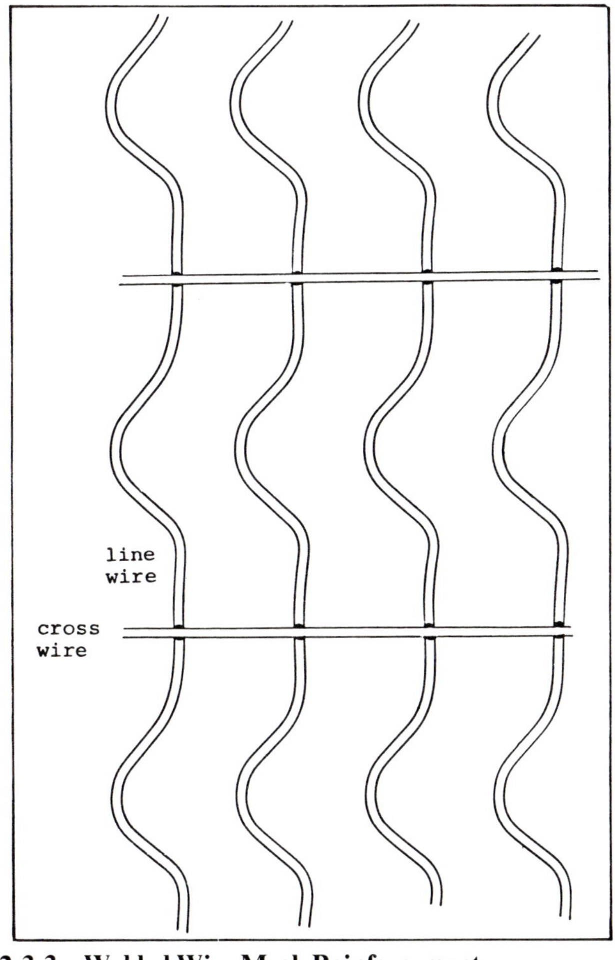

Figure 7.2.3.3 Welded Wire Mesh Reinforcement

One of the advantages of welded wire mesh is that it can be applied during the impingement process by the same machines as hexagonal wire mesh. The stretching needed to match the helical form of the layers of mesh is taken up by deformation of the crimps in the line wires.

Welded wire mesh has been available for many years and was used initially to reinforce concrete coatings applied by casting. It has been widely accepted and is generally thought to provide better reinforcement than hexagonal wire mesh. Among the more demanding projects, where it has been used, are the Gas del Estado crossing of the Strait of Magellan, pipelines in the Eldfisk and Albuskjell North Sea fields, 1,219-millimetre (48-inch) loading lines in Abu Dhabi, part of the Cormorant system (where it was adopted after the failure of concrete reinforced with hexagonal mesh), and a 914-millimetre (36-inch) pipeline at the Statfjord oilfield. The last of these was installed by bottom-tow from Norway, a particularly severe test of concrete, but a subsequent survey showed that no damage to the concrete had occurred.

Cage reinforcement is shown schematically in Figure 7.2.3.4 and in detail in Figure 7.2.3.5. It is similar to conventional reinforcement of structural concrete. It was developed by M K Shand to meet British Petroleum's requirements for a high-strength concrete coating for the Forties pipeline. The British Petroleum design for the 813-millimetre (32-inch) pipeline has 8-millimetre (0.31-inch) circumferential bars and 5-millimetre (0.20-inch) longitudinal bars. The meshes are rectangular having circumferential bars at 100-millimetre (4-inch) spacing and longitudinal bars at 400-millimetre (16-inch) spacing. In the fabrication of each cage, the wires are first spot-welded at each intersection, to form a flat sheet of mesh, then rolled to form a cage (with a substantial overlap on each circumferential bar), and finally welded to complete the cage. Before each pipe joint is coated with concrete, the cages are slipped over the ends and moved longitudinally into place. PVC chairs, resting on the

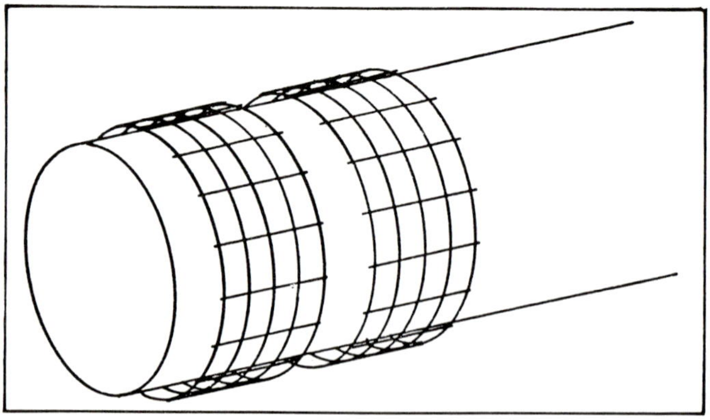

Figure 7.2.3.4 Cage Reinforcement: Schematic

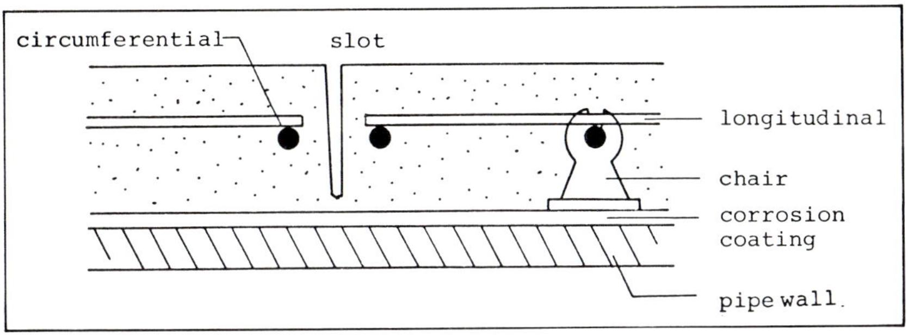

Figure 7.2.3.5 Cage Reinforcement: Section

corrosion coating, hold each cage concentric with the pipe, so that the reinforcement is midway between the concrete surface and the corrosion coating.

Cage reinforcement is used with the impingement method of application. The concrete is applied over the reinforcement, in one pass. The same sort of reinforcement can be used in combination with application by casting, but it does not seem to be compatible with application by wrapping, at least in its present form.

In the concrete coating designed by BP for the Forties line, each cage is 890 millimetres (35 inches) long. There is a gap between each cage and the next, and a circumferential slot 4 millimetres (0.16 inch) wide is cut in the concrete at each gap. The slots extend through most of the concrete thickness. The purpose of these slots was to divide the concrete into distinct segments to reduce the longitudinal stress induced in the concrete when the pipe was bent. The intention was to restrict the formation of circumferential cracks, so that any cracks that did form were at the slots. The slots were cut by a diamond-tipped saw wheel and precautions were taken to ensure that the wheel did not cut the corrosion coating or nick the pipe. In later applications of the same system, the slots were cut either by a water-jet or by a saw, or alternatively preformed by setting flexible plastic circumferential spacers on the pipe before concrete coating.

Not everyone agrees that circumferential slots are necessary or even desirable. By contrast, the 914-millimetre (36-inch) FLAGS gasline for Shell uses continuous cages, extending along the whole length of each joint, and a continuous 8-millimetre (0.31-inch) diameter circumferential bar in the form of a helix with 100-millimetre (4-inch) pitch. It has 12 equally spaced 5-millimetre (0.20-inch) longitudinal bars, welded to the circumferential bar at each intersection. At the end, and where the reinforcement reaches buckle arresters or anodes, the pitch of the circumferential bar helix is reduced.

Publicised tests intended to resolve the question of whether or not slots are necessary have so far been inconclusive, partly because of deficiencies in the design of the experiments. British Petroleum carried out further tests in 1980, in advance of the construction of the Magnus pipeline, but their results have not been published.

7.2.4 Interaction between the Concrete Coating and the Corrosion Coating

Most submarine pipelines have corrosion coatings of bitumen asphalt or coal tar enamel, usually reinforced with glass fibre and asbestos felt. Thin-film epoxy and polyethylene coatings have not been adopted to the same extent as for onshore pipelines, though this is now changing. One of the reasons for the slow introduction of thin-film epoxy coatings is the effect of the concrete application. The impingement process bombards the coating with aggregate particles, and trials have demonstrated that this causes severe perforation of the epoxy coating. The controlling factor is the impingement velocity: if the velocity is 25 metres per second, a 625-micron coating (much thicker than used on land) is not perforated, but if the velocity is as high as 40 metres per second, a coating thickness exceeding 1,000 microns is necessary if perforation is to be eliminated. Polyethylene coatings are much thicker than epoxy coatings but may nonetheless be damaged by impact. The wrapping method of concrete application is less agressive than impingement, but may still cause coating damage. A second area of concern with epoxy and polyethylene coatings is that the interface between the concrete and the smooth coating may not have enough shear strength to transmit tension applied by a laybarge, so the coating may slip along the pipe.

British Gas specified impinged concrete directly over epoxy in the design of their 42-inch pipelines for the Firth of Forth crossing. The epoxy coating was 635 microns thick so it was accepted that some damage might occur. However, it was argued that any damage would not initiate cathodic disbonding, and the pipeline would be adequately protected by the cathodic protection system. During construction, slip between the concrete and the epoxy coating caused some difficulties.

One solution to the problem of damaging the corrosion coating during the application of the concrete coating is to add a second layer of corrosion coating on top of the initial corrosion coating. As detailed by White (1980), the specifications for such a material are that it should have enough impact resistance to eliminate damage to the coating underneath, it should bond to both the initial coating and the concrete, and it should conduct electricity, so as not to shield the cathodic protection. The last requirement rules out polymeric materials such as butyl rubber and polyurethane. However, a relatively new material named FRC-80 has been developed to meet these requirements, and has been tested extensively by Tennessee Gas and others. Its composition is proprietary, but it contains a polymer and a cement, and is reinforced with chopped glass fibre. It is applied by a rotating-brush applicator in a layer 2 to 3 millimetres thick.

Since the application velocity is relatively low, and the material does not contain large particles, this does not damage the epoxy coating. It hardens in a few minutes, and cures in one or two hours, to a strong and resilient coating, which bonds to the epoxy and resists damage during concrete impingement, even at impingement velocities exceeding 40 metres per second. This material has been used by Tennessee Gas for offshore pipelines in the Gulf and by Marathon for the Brae pipeline.

Trials on concrete coating over polyethylene have been carried out by Mannesmann, Nippon Steel and H C Price. The Nippon Steel tests were on 609.6-millimetre (24-inch) outside diameter 15.9-millimetre (0.625-inch) wall steel pipe with a 6-millimetre (0.236-inch) polyethylene coating and 80 millimetres (3.15 inches) of 2,270 kg/m^3 (142 lb/ft^3) concrete reinforced by two layers of galvanised hexagonal mesh. Each sample was 500 millimetres long with 400 millimetres (16 inches) of concrete. The loading rig pushed the steel pipe through the concrete and the relative displacement was measured as a function of the total shear force transmitted through the interface. The results of this experiment are listed in Table 7.2.4.1. They are reported in terms of mean shear stress and a distinction is made between the stress at which relative motion begins ('adhesion') and the maximum stress ('friction'). Four types of extruded polyethylene were compared, one longitudinally-extruded and three side-extruded. Another test was conducted using a Western Gear model LPT 80-S tensioner. This test showed that an ultimate force of 320 kN (32 tonnes) could be transmitted to the 24-inch pipe but unfortunately further details of this test are not available.

Table 7.2.4.1 Results of Nippon Steel Tests on the Shear Stress between the Concrete and Polyethylene Coatings

COATING	ADHESION SHEAR STRESS*		FRICTION SHEAR STRESS†	
	mean (kN/m^2)	range (kN/m^2)	mean (kN/m^2)	range (kN/m^2)
Longitudinally-extruded smooth	30	3 to 113	85	10 to 155
Side-extruded (normal)	116	48 to 254	158	114 to 254
Side-extruded (stepped)	161	not reported	not reported	not reported
Side-extruded (embossed)	235	not reported	not reported	not reported

* denotes the mean stress of which relative movement begins
† denotes the maximum mean stress

H C Price carried out simulated tensioner tests using a set of four tracked tensioner pads equally spaced around the pipe circumference. The pipe sections were 300 to 350 millimetres (12 to 14 inches) long, and 273.1 and 323.9 millimetres (10.75 and 12.75 inches) diameter. The mean shear stresses at which the concrete slipped in the coating are reported in Table 7.2.4.2. H C Price caution that these results should be used to compare relative performance only and not to derive absolute values for design, because the test is not a faithful simulation of an actual tensioner and the actual loading is not symmetrical. They estimate the mean shear stress at which slippage may become a problem to be 15 kN/m^2 (2 lb/in^2) for concrete over smooth polyethylene, powder epoxy or Pritec, 30 kN/m^2 (4 lb/in^2) for concrete over enamel, and more than this for concrete over abraded polyethylene.

Table 7.2.4.2 Results of H C Price Tests on the Shear Stress between the Concrete and Corrosion Coatings

COATING	MEAN SHEAR STRESS RANGE (kN/m^2)
Concrete over asphalt enamel	100 to 170
Concrete over powder epoxy	55 to 100
Concrete over smooth polyethylene	55 to 00
Concrete over wire-brush abraded polyethylene	410 to 5800
Concrete over Pritec	55 to 100

A comparison between Tables 7.2.4.1 and 7.2.4.2 shows that there is agreement between the results of shear tests using smooth polyethylene. Furthermore, Table 7.2.4.2 shows that wire-brush abrasion increases significantly the adhesion between concrete and polyethylene, to a level well beyond that of other coating combinations which have proved satisfactory in practice.

7.3 Research and Development in Concrete Coating

The last decade has seen substantial progress in the technology of concrete coating and a marked improvement in quality. In the North Sea, concrete coating is no longer considered as a problem area. However, there remain significant gaps in the technology and in our understanding of the problems of concrete design.

Considering the design problem first, the present state of knowledge is unsatisfactory because it lacks a medium quantitative base. It can be contrasted with design for conventional structural concrete. When a designer writes the specification for a pre-stressed concrete beam, his selection of a concrete strength and a pre-stressing cable tension are founded on rational and repeatable calculation, not on 'experience'. Designers of concrete weight coating, on the other hand, generally proceed by a combination of experience, imitation and occasionally a few *ad hoc* tests. No professional engineer can accept this as a satisfactory position. Two steps need to be taken to resolve this problem. First the designer needs a sound basis for calculating the loads imposed on the coating. He now has this for trawl board impact, but not for other kinds of damage. Secondly, he needs a basis for calculating the concrete strength and reinforcement that are required. A more complete understanding has to be based on research programmes which include both systematic and carefully conducted tests and theoretical analysis of the fracture mechanics of reinforced concrete.

Only a fuller appreciation of the mechanisms of failure of concrete coatings can bring improved design. There is great potential for improvement, and ideas from fracture mechanics have already been applied to the design of better concrete. Unfortunately, as explained by Double (1981), the usefullness of concrete is limited by its low tensile strength and fracture toughness, which counteracts, to some extent, its advantages of cost, low thermal and electrical conductivity, resistance to corrosion and low energy content. Concrete is far weaker than natural materials which form in aqueous media at ordinary temperatures. For example, mother-of-pearl, which consists of plates of crystalline calcium carbonate linked by a protein polymer, has a work of fracture 100 times that of concrete. According to Birchall *et al* (1981), the tensile strength of conventional cement is limited to about 10 MN/m^2 by the presence of macro-defects about 1 millimetre across. However, these defects can be reduced to 15 microns by efficient mixing and using low water/cement ratio and additives to yield a tensile strength of 70 MN/m^2 (10,000 lb/in^2) and a compressive strength over 200 MN/m^2 (29,000 lb/in^2). In the context of submarine pipeline concrete, these remarkable increases in strength offer exciting possibilities of increased resistance to impact damage and perhaps elimination of the need for reinforcement.

New methods of construction also make demands on concrete technology. The reel method of pipe laying has great potential for large as well as small-diameter lines, but one of the factors limiting its application is the difficulty of devising a concrete coating that can be reeled successfully. This inability to use concrete-coated pipe involves using extra wall-thickness, (to give the pipeline the correct submerged weight), which in turn leads to welding problems and extra steel cost. Furthermore, without a concrete coating there exists a high risk of damage to the corrosion coating, both on the reel and in service. Development of a reelable concrete coating ought to be within the reach of technology, and would be a major breakthrough.

Bottom-tow construction, as described by Palmer (1981), makes different demands on concrete coating. Because accurate submerged-weight control is essential, water absorption by the concrete must be kept low and, in addition, the coating must resist abrasion on the seabed. Tows exceeding 400 kilometres in the North Sea have shown abrasion to be minimal where the seabed is mostly soft clay but also includes sand and boulder fields. In other seas however, the seabed is often rougher and rockier, so in these conditions it may be useful to add a hard and tenacious outer coating to the concrete. Here again there is scope for research and development.

Another area that may justify development work concerns the thermal conductivity of concrete. Most of the thermal resistance between the pipeline contents and the sea derives from the concrete coating. Usually, little attempt is made to control the thermal conductivity of the concrete coating but it is known that saturated concrete has a much higher conductivity than dry concrete. There is a demand for insulated pipelines, particularly if the insulation can be added at moderate cost, and much more can be done in this respect to explore the possibilities of more sophisticated concrete. This, indeed, is a general conclusion, and we can expect still more progress in the next decade than in the last.

7.4 Bibliography

Bergan, P. G. and Mollestad, E., 'Impact-Response Behaviour of Offshore Pipelines', Paper OTC 4065, *Proceedings, Thirteenth Annual Offshore Technology Conference*, 3, 9–16, (Houston, 1981).

Birchall, J. D., Howard, A. J. and Kendall, K., 'Flexural Strength and Porosity of Cements', *Nature*, 289, 388–390, (1981).

Double, D. D., 'Cement – a Respectable Material?' *Nature*, 289, 348–349, (1981).

De Groot, S. J. and Van Dongen, F., *Interactie Vistuigen en Pijpleidingen*, Report CA–79–09, Rijksinstituut voor Visserijonderzoek, (1979).

Kiernan, E. G., 'Treatment of Pipes for Submarine Pipelines', Paper Cl, *Proceedings, First International Conference on the Internal and External Protection of Pipes*, (Durham, 1975).

Kiernan, E. G. and Blake, F. E., 'Advances in Concrete Coating of Submarine Pipelines', Paper Cl, *'Proceedings, Third International Conference on the Internal and External Protection of Pipes*, (London, 1979).

Moshagen, H. and Kjeldsen, S. P., 'Fishing Gear Loads and Effects on Offshore Pipelines', Paper OTC 3782, *Proceedings, Twelfth Annual Offshore Technology Conference*, 383–392, (Houston, 1980).

Palmer, A. C., 'Bottom Tow of Submarine Pipelines', *Offshore Oil & Gas Pipeline Technology*, Oyez Scientific & Technical Services Ltd, (London, 1981).

White, P. A., *FRC-80 as a Protective Membrane for Thin-Film Anti-Corrosion Coatings.*

A C PALMER

8

CONTROLLING THE INTERNAL CORROSION OF SUBSEA PIPELINES

Several thousand miles of large diameter pipelines constitute the network now installed beneath the North Sea. Access to these pipelines is often limited because many are trenched beneath the seabed and rough weather prevents inspection, repair and maintenance for long periods. For corrosion engineers, these pipelines present a novel situation. Considerable experience exists regarding measures to combat the internal corrosion of land-based pipelines transporting hydrocarbons, but these measures are based on the principle of feed-back of information concerning corrosion to the engineer, guiding him towards a cost effective solution. Because of their inaccessibility, normal corrosion monitoring, measurement of chemical additive concentrations and discrete spot-testing cannot be done on offshore lines. By contrast, routine monitoring is restricted to their ends and, instead of injection at regular distances along the pipeline, chemical treatment can only be effected offshore using continuous dosing, the efficacy of which is often questionable.

Internal corrosion monitoring of offshore pipelines using instrumented pigs is feasible but this involves a considerable risk should the pig get stuck. Moreover, in their current state of development these pigs can only detect corrosion at relatively abrupt changes of pipewall section so more insidious general corrosion may escape detection.

8.1 Corrosion Mechanisms

Internal corrosion of pipelines is generally accepted to be caused by one or more of five corrosion mechanisms due to carbon dioxide, hydrogen sulphide, microbiological degradation, acids and erosion corrosion. In all cases, one major factor is the quantity of water present in the hydrocarbon.

Corrosion is an electrochemical phenomenon and can be separated into two distinct processes. Firstly, the anodic reaction involves the dissolution of the pipeline metal as positively charged ions known as cations. These ions react with material in solution to produce solid precipitates some of which may adhere to the metal surface and thereby reduce further dissolution. The electrons remaining in the metal are removed by the second corrosion process involving cathodic reactions. In pure or relatively pure water, these cathodic reactions lead to hydrogen evolution and oxygen reduction. Hydrogen evolution is especially rapid in acid solutions, but reduces in significance in neutral or alkaline environments. The two cathodic processes are illustrated in Figure 8.1.1.

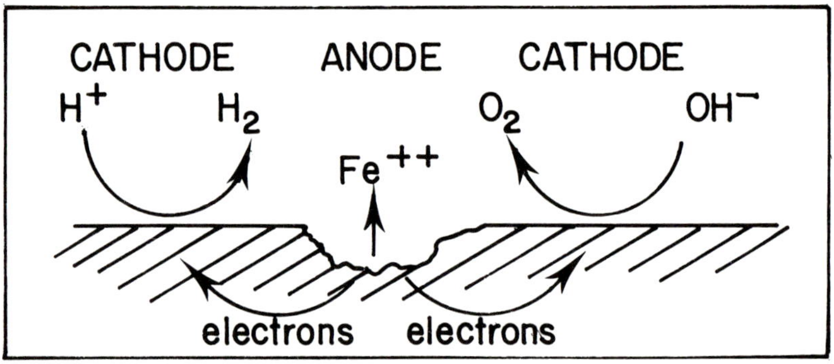

Figure 8.1.1 Corrosion of Ferrous Metals

8.1.1 Carbon Dioxide Corrosion

Carbon dioxide corrosion may occur simply or in combination with other gases, particularly hydrogen sulphide or oxygen. The wells of the North Sea are mainly 'sweet' and corrosion is by carbon dioxide alone or in combination with trace oxygen, which can enter entrained in low pressure systems or pumps or flood waters. Mechanical vacuum de-aeration normally removes all but the last traces of oxygen in the seawater used for well injection purposes but operational problems occur occasionally which can result in partially untreated water, in which case levels of oxygen as low as 50 ppb can result in rapid corrosion attack and pitting.

Carbon dioxide dissolves in water to produce carbonic acid which attacks the pipeline steel forming siderite, iron carbonate, which is not fully protective to the underlying metal. This form of corrosion usually results in deep pitting and the rate of attack depends on several factors such as the quantity and composition of the water, the concentration of carbon dioxide, the temperature, pressure and flow rates. Experience gained at the gasfields of Groningen, which have the longest production record in the North Sea, suggests that the partial pressure of carbon dioxide is the most critical factor. According to Geelen and Groenewond (1977) the following results pertain.

If the partial pressure of the carbon dioxide is:

(i) above two bars, then severe corrosion will ensue;
(ii) between one half-bar and two bars, then corrosion may occur; and.
(iii) below half-bar, then the system will usually be non-corrosive.

According to API (1958) a partial pressure of 0.2 bars is often considered the threshold of CO_2 corrosion.

Normally, the quantity of water in gas transmission is very low after gas drying, to avoid hydrate formation. Nevertheless, water does collect in sump areas in these lines so corrosion inhibitors are injected into the gas-flow to prevent their corrosion, combined with regular sphering or pigging of lines.

In oil transmission however, the percentage of water present is greater. Severe corrosion is usually regarded as initiating at water 'cuts' in excess of ten per cent and to become increasingly severe as the percentage increases. Furthermore, sump areas in pipelines always experience a high water content, despite pipeline pigging.

8.1.2 Sulphide Corrosion

Hydrogen sulphide is corrosive even in trace quantities and its corrosiveness is enhanced in the presence of carbon dioxide which reduces the pH. Its corrosive mechanism takes two forms. Firstly, it reacts with the metal surface, stimulating the anodic reaction. In the presence of carbon dioxide this leads to the production of smythite, a particular solid iron sulphide which is non-protective, whereas under high sulphide conditions alone sulphide films may be formed on the metal surface which can temporarily prevent further attack. Pitting occurs when these films break down. Continued sulphidation does occur beneath these adherent films, which may result in some inter-granular attack. Secondly, the reaction of the sulphide with dissolved cations produces solid sulphides which are cathodic to the pipeline steel and cause severe localised attack.

When temporarily protective sulphide films are formed, the cathodic hydrogen evolution continues but the hydrogen atoms diffuse into the steel. In high strength steels, or steels of hardness above R_c22 (VHN 240), this can lead to hydrogen embrittlement which makes the material susceptible to stress-corrosion cracking (SSC) when stressed. In controlled rolled pipeline steels, the recombination of adsorbed hydrogen atoms into molecular hydrogen at micro-flaws and sulphide stringers within the steel can result in hydrogen blistering or hydrogen pressure-induced cracking (HPIC).

Figures 8.1.2.1 and 8.1.2.2 (overleaf) illustrate the various effects of corrosion resulting from hydrogen sulphide attack.

8.1.3 Microbiological Corrosion

Microbiological corrosion is a particular form of hydrogen sulphide corrosion. Sulphate-reducing bacteria (SRB) are strict anaerobes which can flourish in hydrocarbon environments. These anaerobes obtain their energy for growth by oxidising trace organic acids in the hydrocarbons to carbon dioxide and acetates using the oxygen in the sulphate ion, and in the process they can produce massive quantities of hydrogen sulphide.

These organisms seem able to tolerate and/or adapt to the high

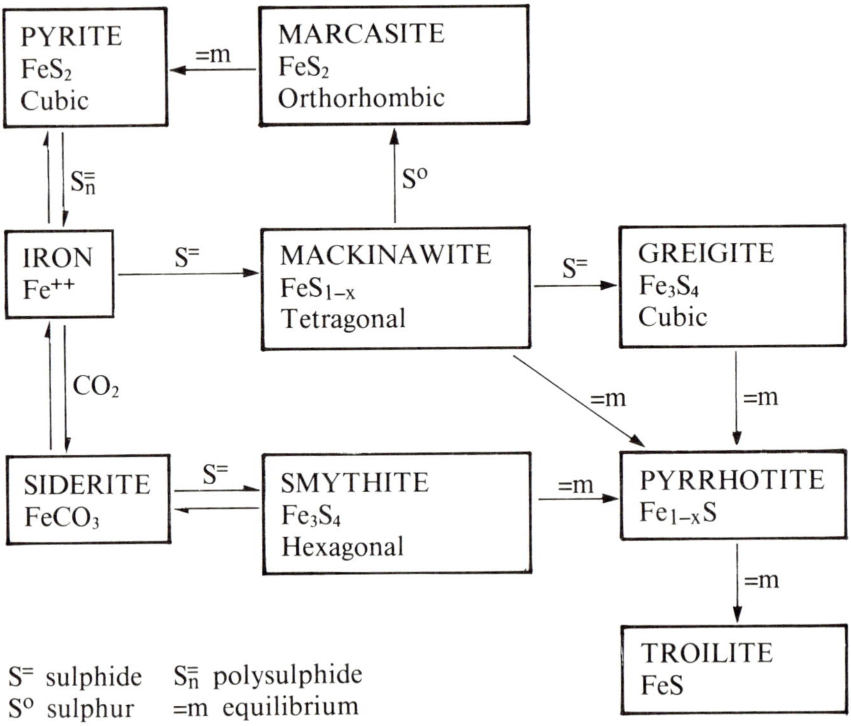

Figure 8.1.2.1 The Corrosive Iron Sulphides

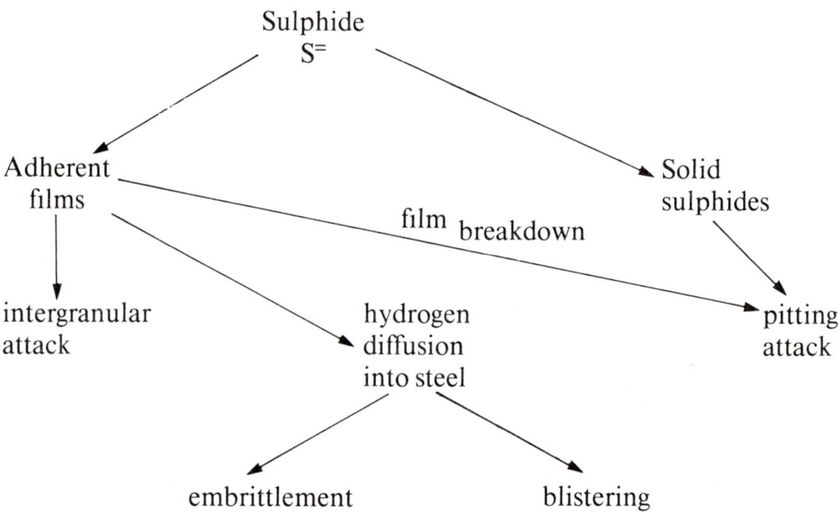

Figure 8.1.2.2 Corrosion by Sulphides

temperatures in production wells. In an attempt to combat this problem, downhole biocide treatment is increasingly practiced using organic material or heavy metal ions, but there are problems of thermal degradation of the organic bacterial inhibitors and also heavy metal precipitation by the biogenic sulphides, which may prevent the treatments from being totally effective. These bacteria may also enter a pipeline during its initial hydraulic testing. Unfortunately, biocide treatment of large quanities of seawater for hydraulic testing poses considerable technical difficulties and is increasingly problematical with the increasing restrictions on choice of biocide to avoid pollution. Moreover, many 'softer' biocides are inefficient against SRB in marine conditions.

Corrosion attack by SRB is made manifest in the form of rapid localised pitting in sumps and low flow areas. Growth is particularly likely under debris or loose scales. The presence of these bacteria in North Sea pipelines has been positively identified in several major lines and it is probably a general phenomenon. Furthermore, the giant concrete structures, which contain temporary oil storage tanks in their bases, exacerbate the situation further because their production will become increasingly sour over time, due to the fact that they are filled alternatively with oil and then seawater, thereby providing suitable conditions for the growth of SRB. As a result, pipelines taking oil from these tanks may be continuously reinfected.

8.1.4 Acid Corrosion

Some crudes contain trace amounts of oxidised acidic species which are corrosive to steel. These materials are only active in the aqueous phase and cause generalised etching or micro-pitting of the metal surface. The relationship between the acidity of the aqueous phase and the rate of corrosion under typical pipeline operating conditions is shown in Figure 8.1.4.1 (overleaf). In waxy oils, the surface is covered with a waxy film which prevents this attack. In most cases, other forms of corrosion are more significant and mask the effects of acid corrosion attack.

8.1.5 Erosion Corrosion

Particulate matter, especially sand, suspended in the oil or gas flow can result in massive attack at areas of impingement. In this respect, well casings, risers and offshore field lines are most at risk because of their very high flow rates of untreated fluids. Under these conditions enhanced corrosion occurs because of the mechanical stripping of the protective layers formed by either corrosion on the surface of the metal or by inhibitor films. Conventional inhibitor treatment can rarely combat erosion corrosion, so linings or more resistant materials must be used.

At present, the best linings for conventional pipe comprise epoxy-based paints, which require a very high quality of surface finish prior to their application. Uni-piece plastic liners have been less successful as exfoliation can result in the liner being stripped completely, leading to a flow blockage. Negative pressure surges are reported as a major reason for liner collapse. Furthermore, jointing problems occur with all coating/lining systems and remain a major disadvantage in their use. Alternative constructional materials would seem to offer a more practicable solution until a suitable coating/lining process has been developed. In this respect, high grade chromium steels have proved successful in many applications where erosion corrosion problems exist.

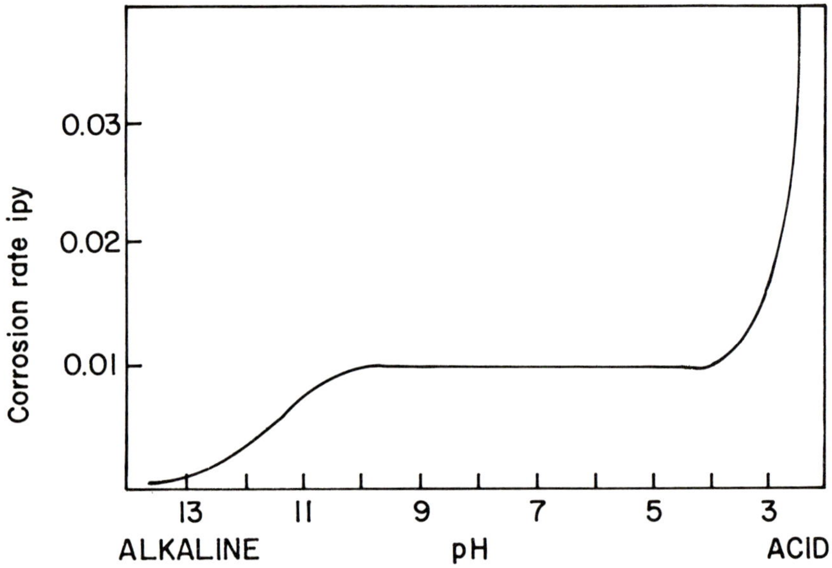

Figure 8.1.4.1 Effect of pH on Corrosion Rate

8.2 Corrosion Prevention Using Inhibitors

The most important method of reducing or preventing the internal corrosion of pipelines is to use organic corrosion inhibitors. These materials adsorb onto the pipewall to form a barrier to the corrosion processes. Most act to reduce the anodic dissolution reaction and the various cathodic reactions. The thickness and quality of the adsorbed layer depends on the characteristics of the particular chemical and its concentration in the bulk phase. This layer is continually desorbed/readsorbed and an equilibrium is established between the inhibitor in the adsorbed layer and that in the bulk phase.

Below a certain critical concentration, the whole of the metal surface is not protected and since the exposed areas of metal change continuously general corrosion, rather than localised pitting attack, occurs under these conditions. By contrast, most inorganic corrosion inhibitors, and some special organics, are chemi-sorbed onto the metal surface and remain fixed.

Since major corrosion only occurs in hydrocarbon lines in which water is present, the inhibitors must act in both the hydrocarbon and water phases or be transported in the bulk hydrocarbon phase and transferred to the corrosive water phase when required. Most inhibitors are surfactant containing active sulphur, nitrogen and phosphorous groups. Injection into the pipeline may be continuous or by batch treatment, although for long pipelines continuous dosage is most common.

Microbial inhibitors may be either biocides which kill the organisms or biostats which prevent their activity and growth. Quite often the difference in action of a particular chemical is merely in concentration. The activity of a biocide is affected markedly by several factors, such as:

(i) *pH*
Most biocides are only active over a restricted range of pH and may polymerise or precipitate outside that range.

(ii) *Particulate matter*
Surfactant biocides, especially quaternary compounds, act by adsorbing onto the wall of the bacterial cell and causing localised cell wall breakdown, thereby killing the organism. Unfortunately, these surfactants will also adsorb other finely divided particulate matter, thereby reducing the concentration available for killing the bacteria.

(iii) *Heavy metal ions and salts*
Certain heavy metals cause precipitation or inactivation of the biocides; salts can also lower the threshold concentration at which the biocide forms miscelles: once miscellisations occurs, the concentration of active material is reduced considerably.

(iv) *Temperature*
Several biocides are either thermolabile or polymerise at even modest temperatures. Injection into hot oil lines to prevent attack in cooler areas further down the line should therefore take account of this factor.

(v) *Compatibility*
Biocides and corrosion inhibitors may be cationic or anionic depending on their polar bonding. Mixing cationic materials with anionic or neutral materials may reduce their effectiveness whereas, in some cases, selected mixtures can exhibit synergy.

For North Sea pipelines, there is a growing awareness of the need to treat lines to prevent corrosion and microbial activity. To this end, blends of inhibitors and biocides are being formulated, based on laboratory tests such as the rotating bottle and bubble-test techniques. Though these tests have proved useful in the past, and will undoubtedly form a good basis for the initial selection of mixtures, it is important to realise that such tests have certain limitations. For example, these tests give overall effects only, and do not assess mixtures as well as they do single inhibitors. Furthermore, they do not give information on flow effects, temperature gradients or instantaneous corrosion rates.

8.3 Monitoring Corrosion in Pipelines

North Sea pipelines can only be monitored regularly for corrosion at their ends. Whereas corrosion engineers usually test land pipelines at many points to ensure that inhibition is adequate and that sump areas are protected, this is not economically viable with buried pipelines in deep waters, at present. Instead, monitoring methods are restricted to bulk analysis of flow material and pig scrapings. Typically, such analysis involves chemical determination of iron and manganese, corrosion testing using linear polarisation, electrical resistance probes and weight-loss samples, gas and oil analysis and microbiological analysis. Some lines have also been checked using instrumented pigs. The salient attributes of these various monitoring techniques are illustrated in Table 8.3.1 and Table 8.3.2.

Table 8.3.1 Non-continuous Techniques Used for Corrosion Monitoring in Oil Pipelines

TECHNIQUE	PRINCIPLE	ADVANTAGES	DISADVANTAGES
Coupons	Weight loss of exposed metal specimen.	Rate and type of corrosion by visual inspection.	Integrated corrosion rate. Time lapse before data obtained.
Iron counts	Analysis of iron level in the environment.	Fairly rapid results.	Need soluble corrosion product, not very reliable in sour systems. Iron in formation water must be known. No indication of pitting.
Ultrasonic radiographic caliper	Measurement of thickness changes.	Definitive information on metal loss.	Expensive, time consuming, not very sensitive to low corrosion rates. Local application only.
Magnetic pig	Magnetic flux leakage from pipewall irregularities.	General corrosion and pitting indication identifies area in which problems are present.	Difficult to interpret data quantitatively. Very expensive. Danger of pig sticking in line. Interferes with production.

In pipelines with low concentrations of water in oil, the problem of continuous in-flow monitoring of corrosiveness is difficult because conventional electrochemical methods of monitoring do not work effectively in such highly resistant environments. In these circumstances probe placement is critical to ensure that the measurements are made in the water phase. An alternating current impedance technique is being

Table 8.3.2 Continuous Techniques Used for Corrosion Monitoring in Oil Pipelines

TECHNIQUE	PRINCIPLE	ADVANTAGES	DISADVANTAGES
Electrical resistance probes	Change in resistance as metal thickness decreases.	Faster than coupons, good for seeing change in rate of corrosion. Useful in gases and non-conducting liquids.	Fouling of probe by conductive deposits cause irregularities, pitting not always detected.
Polarisation resistance	Current response to small potential impressed across electrodes.	Instantaneous, changes in rate easily monitored, pitting indication.	Fouling of probe can cause 'shorting out', needs conductive media.
Polarisation curves	Current response to dc potential change.	Type and rate of corrosion.	Conductive media only, complex interpretation, alters electrodes irreversibly.
ac impedance	Current response to small ac voltage perturbation at several ac frequencies.	Information on rates and mechanisms in low and high resistivity systems.	Complicated analysis required. No proven site equipment.
Hydrogen probes/sleeves	Monitor build up of pressure due to migration of atomic hydrogen through the probe or pipeline wall.	Instantaneous reading, changes easily monitored. Patch probes need not breach pipewall.	Does not give corrosion rate, only useful where atomic hydrogen produced as in sour (H_2S) corrosion.
Thin-film activation	Irradiated spool piece, or probe, monitor radiation in process fluid or remaining radiation on probe.	Continuous monitoring, indication of pitting.	Costly and complex. Must be soluble/ detached corrosion product. No service experience.
Electrochemical noise	Analysis of microvolt noise from corrosion process.	Measures corrosion on pipeline.	Development stage only.

developed to obviate this problem. In this technique, a frequency varying ac voltage signal is impressed between two electrodes. The current response of the electrodes is monitored and from calculations based on the perturbation voltage signal and current response at each frequency, the corrosion rate can be calculated. The use of an ac signal thus enables electrochemical monitoring to be extended to high resistance environments.

Other potential corrosion monitoring techniques involve the use of spool pieces with irradiated inner surfaces, whereby corrosion is continuously monitored by a radiation check of the flow-stream. By layering the radioactive layers, general and pitting attack can be distinguished.

Microbiological testing to determine the efficacy of biocide treatment can only be done over a period by quantifying the concentration of organisms in the flow or pig scrapings. An increasing population would indicate that treatment was ineffective. Existing methods of microbiological analysis are not standardised and need refinement, particularly for field testing for the SRB. A particular count of bacteria in the flow may indicate a low, even population generally spread over the pipeline interior (which is unlikely to be a high corrosion risk situation) or a massive localised concentration of organisms (a high corrosion risk situation).

Because of the limitations imposed on corrosion monitoring along a buried offshore line, the use of inhibitors and biocides does require very careful consideration. Flow rig simulations of conditions and a better understanding of the filming properties of inhibitors is required to ensure line integrity.

The development of more sensitive pigs to monitor internal pipeline conditions will be a major advance. It is important to note that the number of magnetic pigs being improved or developed is increasing due to the efforts of: AMF Tuboscope, Vetco, General Descaling (Trans Canada) and British Gas. Ultrasonic pigs are also being developed by DnV and British Gas. Furthermore, microprocessors programmed to read and interpret data from sensors and record only the problem areas, will markedly increase the range and usefulness of such pigs.

8.4 Bibliography

API, *Corrosion of Oil and Gas Well Equipment*, API Publication, (Dallas, 1958).

Geelen, P. M. and Groenewond, H., *Proceedings, 6th European Congress Metallic Corrosion*, 403–412, (1977).

R A KING
D GEARY

9

CONTROLLING THE EXTERNAL CORROSION OF SUBSEA PIPELINES USING CATHODIC PROTECTION

As Cathodic Protection (CP) is a well established technology, we shall not review it here. Instead this chapter deals with the modern, computerised design of CP systems for submarine pipelines and offshore structures and it incorporates experience gained from the North Sea over the past few years. It also includes data on the CP performance of submarine pipelines as obtained recently using new and advanced monitoring techniques.

9.1 Introduction

Submarine pipelines for oil and gas are usually protected against external corrosion by a coating of electrically insulating material and cathodic protection. Typically, coating thicknesses up to 20 millimetres for bitumen or 0.4 millimetres for epoxy resin are used beneath a concrete weight coating approximating 25–100 millimetres in thickness.

In addition to its primary purpose of preventing corrosion, the insulation coating reduces the current and the number of sacrificial anodes needed to protect the pipeline. Regularly spaced anode bracelets are usually attached to the pipeline at every twelfth pipe joint. These anodes are designed to supply the current required to protect exposed steel on the pipeline where the corrosion protection is damaged.

9.2 Traditional CP Design

In excess of 4,000 kilometres of submarine pipelines have now been laid in the North Sea but although pipelaying started in this area over ten years ago, detailed information on the performance of cathodic protection has been scarce.

A survey conducted by Strømmen (1979) into the design practices adopted for the cathodic protection of pipelines in the North Sea revealed considerable changes in the design parameters for new pipelines compared with earlier ones. These changes comprised reductions in the anode bracelet spacing on the pipelines from about 500 metres down to

100–150 metres: an increase in anode to cathode area ratios from 1:632 to 1:222. These changes represent a strong up-grading of the protective capacity of the anodes, to meet the needs of significant coating defects increasing the current drainage of the anodes.

Depending on the influence of salinity and according to NACE Recommended Practice (1975), steel is protected against corrosion in seawater when polarised to –800 mV or less when compared with the Ag/AgCl/seawater reference electrode. If sulphate reducing bacteria are active at the steel surface, as can happen when pipelines run through mud areas in the North Sea, then the requisite protective potential can be less than –900 mV. In the case of high strength steels however, caution has to be exercised to avoid their over-protection, lest they suffer from hydrogen embrittlement.

The current required to protect a pipeline is the product of the total area of exposed steel and the current density required for polarisation of the steel to the protective level. Current leakage through the corrosion coating also has to be accommodated. Current density requirements depend on environmental factors such as: the salinity and oxygen content of the seawater; the skin temperature of the pipe; the flow-rates of tides and currents and the presence of marine growths on the pipeline. As a result, considerable variations are found in the current requirements along a protected pipeline, which can also be exacerbated by the break-away or scouring of scale and rust.

Typical cathodic protection for North Sea pipelines is based on an average current requirement of 5 mA/m^2 of total pipe surface. This allows effectively for the exposure of five per cent of the steel surface of the pipeline with a current density of 100 mA/m^2. Evidence suggests that this level of protection is satisfactory for the major proportion of North Sea pipelines.

Three basic metals can be used to fabricate sacrificial anodes namely: magnesium, zinc and aluminium.

Magnesium alloys have limited application in submarine situations. The high driving potential of magnesium anodes makes them suitable only for high resistivity applications, such as freshwater and estuarine waters, where inadequate driving potentials are available from zinc or aluminium, and retrofit situations, where a small number of larger anodes having a restricted life can be of technical and economic advantage.

Zinc alloys complying with US Mil Spec 18001H have been used throughout the world in the successful cathodic protection of submarine pipelines. The material is suitable for either buried or open seawater applications and yields a capacity of 780 ampere hours per kilogramme of material consumed in environments of resistivity up to 500 ohm cms. With an open circuit potential of 1050 mV (*versus* a silver/silver chloride reference electrode) a driving potential of 250 mV against protected steel is available in aerobic conditions and 150 mV against protected steel in anaerobic conditions. The material has virtually no tendency to passivate at even the lowest current densities and can therefore be used with considerable reliability in the majority of projects.

Normally the anodes are operated at a current density of 0.1–1.5 A/m^2 at a specified potential of –1030 mV for zinc.

The use of zinc on higher temperature pipelines is, however, restricted. Under heat transfer conditions, zinc exhibits inter-granular corrosion and a reduction in capacity results from sections of anodes being lost without cathodic protection current being generated. A commensurate reduction in capacity occurs. Also, under heat transfer conditions, the anode potential is affected and an erratic corrosion potential is obtained, sometimes significantly cathodic to the corrosion potential in the absence of heat transfer conditions. This renders the material unsuitable for cathodic protection purposes. Alternative compositions for anodes, based on higher purity zinc alloys, have to a limited extent overcome the difficulties at intermediate temperature, but for higher temperature conditions zinc anode alloys must be regarded as unsuitable.

Aluminium anodes have found extensive application for the protection of immersed steelwork in marine applications. Aluminium, whilst appreciably more costly in terms of price per kilogramme, yields a considerably higher capacity, up to 2,830 ampere hours per kilogramme as against 780 ampere hours per kilogramme for zinc. One ampere hour delivered by zinc costs approximately one third more than one ampere hour delivered by aluminium. There are therefore considerable economic benefits to be gained from the use of aluminium alloys. However, the corrosion engineer must be fully aware of the limitations of the alloys at his disposal before making a final choice of anode material.

Under normal conditions in the protection of bare steel in free-flowing seawater, aluminium operates at a surface current density of 2 amperes per square metre for the protection of a coated surface. However, this current density is reduced and, in the early years of operation of a pipeline, the operating current density of an aluminium anode may be so low as to permit the formation of an adherent corrosion product which restricts further activity. Anode alloys, which operate at high efficiency in open seawater at high current density, may not operate at all at low current densities particularly if the anodes are buried in the sea mud.

One aluminium anode alloy, developed by Dow Chemical specifically for operation at low current density in seabed type environments, is Galvalum II, an aluminium zinc mercury alloy which yields approximately 2,700 ampere hours per kilogramme at an open circuit potential of 1,050 mV *versus* silver/silver chloride reference electrode.

Under heat transfer conditions many aluminium alloys, like the zinc alloys, exhibit erratic and unreliable corrosion potentials and low efficiencies arising from gross loss of material or passivation. One indium-containing aluminium anode material has shown exceptional performance in elevated temperature applications. This material, Galvalum III, also developed by Dow Chemical, contains zinc and indium and has shown itself to be uniquely well behaved amongst aluminium anode alloys. Extensive testing at a variety of locations has shown that Galvalum III can give reliable and predictable performance under the most severe conditions, typical capacities being:

Open seawater	2,550 ampere hours per kilogramme.
15% NaCl oilfield brine at 75°C	2,095 ampere hours per kilogramme.
Seabed mud	1,650 ampere hours per kilogramme.
Estuarine water (200 ohm cms)	2,425 ampere hours per kilogramme.

As a result of these considerations, zinc is generally the preferred material of anode construction for the majority of pipeline situations.

9.3 Anode Design

The current output for each anode can be calculated from Equation 9.3.1:

$$I_a = \frac{E_c - E_a}{R_a} = \frac{\Delta E}{R_a} \qquad (9.3.1)$$

where:

I_a = anode current output per anode
E_a = closed circuit potential of anode (Spec)
E_c = closed circuit potential of protected structure, normally set equal to the protection potential
R_a = anode (contact) resistance (resistance of electrolyte), reflecting anode geometry and specific resistance of the surrounding electrolyte.

The integrated current output for all the anodes has to match the total current consumption/current requirements for the pipeline throughout the design life.

Additional requirements include:

(i) the total and the individual anode weights have to match the design life ie, the net weight divided by anode material capacity (ampere hours/kilogramme) has to match the total number of ampere years required for protection;
(ii) the anodes have to be located so that the spread of protection is satisfactory.

The formulae used to calculate anode resistance have been questioned with regard to reliability and accuracy. Two such formulae commonly in use for anode bracelets are:

$$R_a = 0.315\,\rho/\sqrt{A} \qquad (9.3.2)$$

and:

$$R_a = \rho/2S \qquad (9.3.3)$$

where:

ρ = electrolyte resistivity
A = surface area of anode
S = mean length of sides of anode.

9.4 Computerised CP Design

During the last few years, the author has developed computer programs which can simulate the three-dimensional performance of CP systems including sacrificial anodes, impressed current and hybrid systems, for various offshore structures. These programs are based on the use of finite element analysis using the algebra of finite differences. These programs are described by Strømmen (1979, 1980 and 1981). Considerable experience has now been gained through the use of this technique especially with North Sea structures and pipelines.

These programs include the solution of Laplace's equation for proper boundary conditions, which allows for the geometrical, physical and electrochemical properties of the cathodic protection system. The finite difference technique involves the replacement of the seawater/mud environment of the pipeline by a grid of elements, as shown in Figure 9.4.1. The grid represents conducting bars connected at junctions, each bar having an electrical resistance equal to that of the electrolyte, from the centre of one element to another.

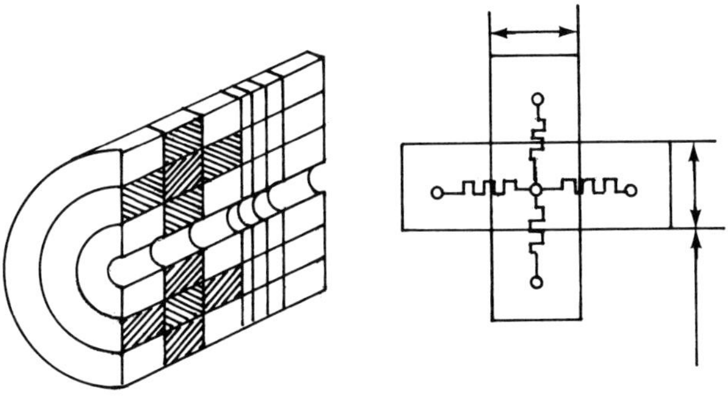

a b

Figure 9.4.1 The Electrolyte Surrounding the Tubular Body is Divided into Elements Connected by Conducting Bars

Following Ohm's and Kirchof's laws, a set of equations can be established, one for each element, describing the system. By these means, large systems involving more than 3,000 equations (and elements) can be solved on large computers, to model the distribution of current and potential and hence the spread of protection on a structure.

The application of this technique to pipelines has included the evaluation of the effects of number, size and distribution of coating defects and current density requirements on the protective level of the pipeline. By way of an example the results from one analysis are shown in Figure 9.4.2 for a 32-inch diameter pipe in seawater of 30 ohm cm conductivity. Anode bracelets of 1.63 m^2 surface area were attached every 170 metres. One large, two medium and three small circumferential coating defects were modelled, corresponding to six per cent of the pipe steel surface exposed to the seawater. The average cathodic current density was 130 mA/m^2, corresponding to 2,080 mA/m^2 anodic current density on the bracelets. An examination of Figure 9.4.2 shows that while most of the pipeline was polarised to –820 mV, a potential of –780 mV locally resulted in unsatisfactory protection at the large coating defect.

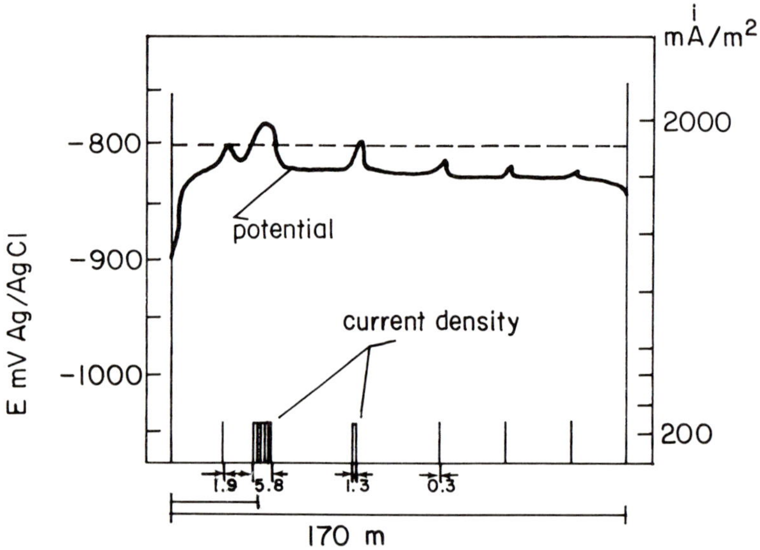

Figure 9.4.2 Calculated Potential Variations along a 32-inch Pipeline with Coating Defects of 0.3 Metres to 5.8 Metres Length, Corresponding to 6 per cent Bare Steel. (i_a = 2.08 A/m^2 –i_c = 130 mA/m^2 average. Seawater resistivity = 30 ohm cm)

A simulation of a similar pipe section, buried to a depth of 3 metres in mud of 90 ohm cm resistivity, is shown in Figure 9.4.3. Anode bracelets of 1.26 m^2 surface area supply 628 mA each, corresponding to 500 mA/m^2. Five defects in the coating, corresponding to 2.5 per cent exposed steel, have been simulated with evenly distributed cathodic current of 54 mA/m^2. If SRB are present the protection potential is specified to

–900 mV, which is obtained for all the pipeline, except at the largest coating defect.

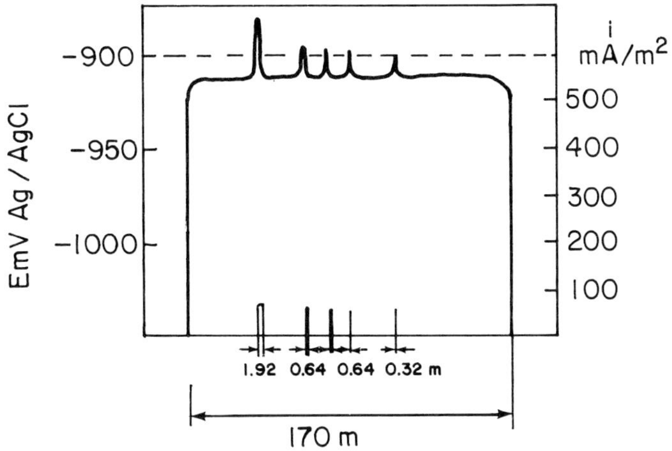

Figure 9.4.3 Potential Variations along a Cathodically Protected Pipeline Buried to a Depth of 3 Metres in Mud of 90 ohm cm Resistivity. (Length of anode bracelets is 0.4 metres)

As a last example, the local I·R drop at an anode bracelet of 0.4 metres length and 1 metre diameter has been calculated for various depths in mud, including one with the pipe lying on the bottom, as shown in Figure 9.4.4. As shown by Figure 9.4.5, the I·R drop is gradually increased for increasing depth of burial.

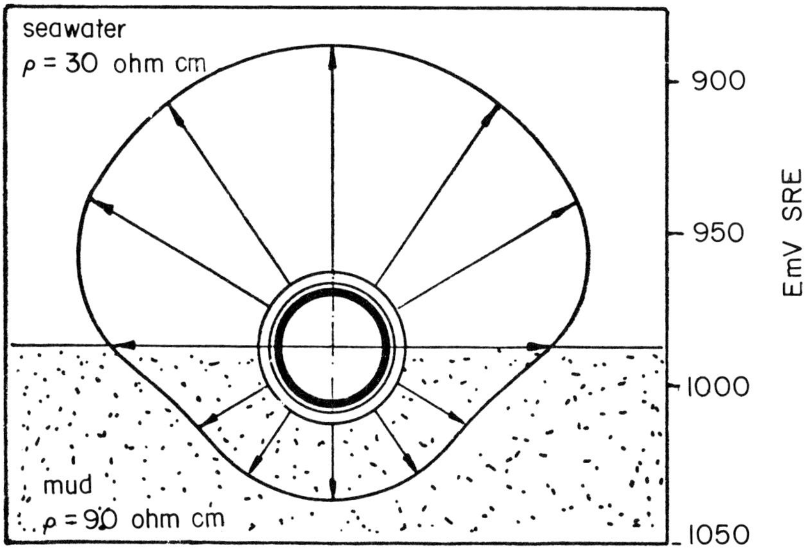

Figure 9.4.4 Current Density Variations on the Cirumference of a Bracelet on the Bottom

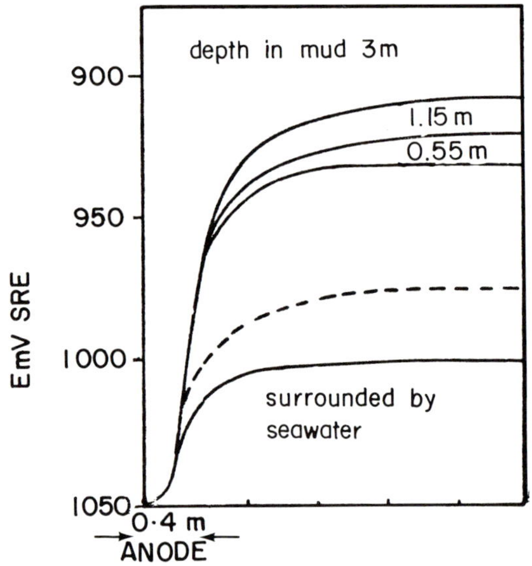

Figure 9.4.5 Potential Variations at Anode Bracelet for Various Conditions. ($i_a = 0.5$ A/m^2)

9.5 A Comparison between Computerised and Traditional Designs

A considerable number of examples similar to those shown above have been examined and the results have been used to check traditional CP design methods. Of particular interest has been the evaluation of the formula for assessing anode resistance and the effects of coating defects on the level of protection.

A principle potential diagram for a cathodically protected submarine pipeline is shown in Figure 9.5.1. This reveals that the total potential

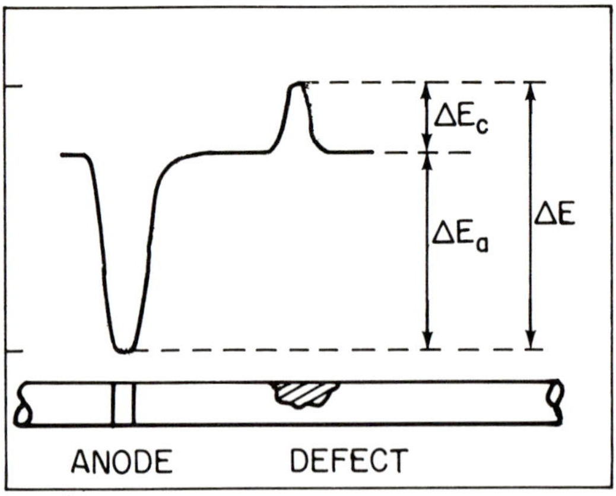

Figure 9.5.1 Principle Potential Diagram for a Cathodically Protected Submarine Pipeline

drop ($\triangle E$) from the anode surface to a remote coating defect, is made up of two contributions, as shown by Equation 9.5.1:

$$\triangle E = \triangle E_a + \triangle E_c \qquad (9.5.1)$$

where:

$\triangle E_a$ = voltage drop from anode to 'remote' locations in the sea
$\triangle E_c$ = voltage drop from exposed steel to 'remote' locations.

As already observed from Figure 9.4.1, the peak of the potential at exposed steel, $\triangle E_c$, increases with the size of the defect. Figures 9.5.2 and 9.5.3 (overleaf) show how $\triangle E_c$ depends on the defect size for a 32-inch line, for unburied and buried conditions respectively. A correct value for the anode's driving force can now be obtained from:

$$\triangle E_a = \triangle E - \triangle E_c \qquad (9.5.2)$$

where:

$\triangle E_c$ is obtained from Figure 9.5.2 and Figure 9.5.3.

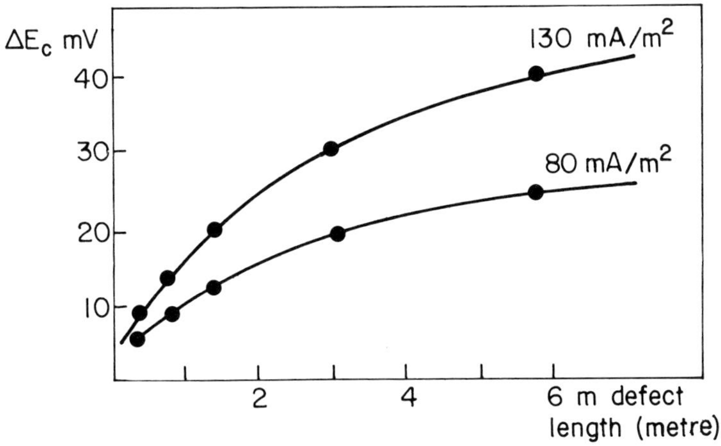

Figure 9.5.2 Error in Potential Readings at Remote Locations from Coating Defects on 32-inch Line. (Circumferential defect in 30 ohm cm seawater)

The CP system would be designed for the most serious condition which might reasonably be expected to occur between routine inspections. $\triangle E_a$ = 30 mV might, for example, correspond to a 3-metre long defect with a cathodic current density, on the seawater-exposed steel, of 130 mA/m². If the protective potential was –900 mV and the closed circuit potential of the anode was –1050 mV leaving $\triangle E$ = 150 mV, the driving force for the anode would be found according to the following relationship:

$$\triangle E_a = 150 - 30 = 120 \text{ mV} \qquad (9.5.3)$$

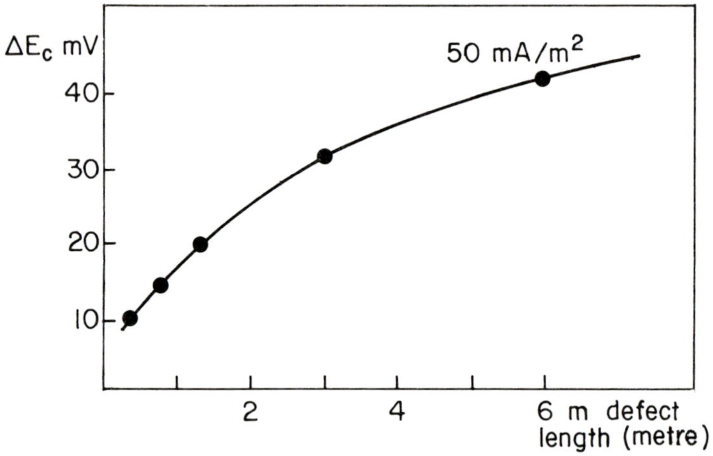

Figure 9.5.3 Error in Potential Reading at Remote Locations from Coating Defects on 32-inch Line. (Defects cover half the circumference. Pipeline buried to 2-metres depth in mud of 90 ohm cm resistivity)

Provided that ΔE_a rather than ΔE is used in Equation 9.3.1 to calculate the current output from the bracelets, the numerical analysis reveals that the anode resistance (R_a) can be calculated using Equation 9.3.2 above. However, the constant depends on the anode bracelet's length to diameter (L/D) ratio. Changing the L/D ratio from 0.2 to 1.4 changes the constant from 0.26 to 0.30. (A constant of 0.282 corresponds to a spherical anode and a homogenous spherical field.)

In conclusion therefore, traditional design, based on the Equations 9.3.1 and 9.3.2 above, normally overestimates anode current output compared with the real output, for protected pipelines with coating defects exceeding a certain size.

9.6 CP Experience in the North Sea

CP design methods have steadily improved over the years, as a result of experience offshore. However, much of the data obtained offshore has been limited to potential readings, with very little critical evaluation of such data. In more recent times however, there has been a considerable development in, and application of, equipment for electric field strength/current density recording on structures subsea. Such equipment is now used extensively during the inspection of submarine pipelines, buried and exposed. The information contained in such readings includes:

(i) protective level of the pipeline (based on a combination of potential and field strength readings);
(ii) the positions of active and inactive anodes;

(iii) current output from sacrificial anodes from which estimates of remaining life are obtained;
(iv) the position and size of coating defects.

An example of a computer plot of readings obtained for a 100-metre long section of a submarine pipeline is shown in Figure 9.6.1. This section had very minute coating defects. The largest positive peak on the curve (corresponding to 130 μV/cm) reflects an anode bracelet, while the rest of the anodic peaks (10 – 30 μV/cm) reflect the pipe joints. The computerised technique is used in the interpretation of such readings, according to the methods described by Strømmen and Rødland (1981).

Out of the large number of pipelines which have been surveyed, it was discovered that the majority were very well protected, although marginal protection has been observed in a few cases for minor sections close to platforms and risers.

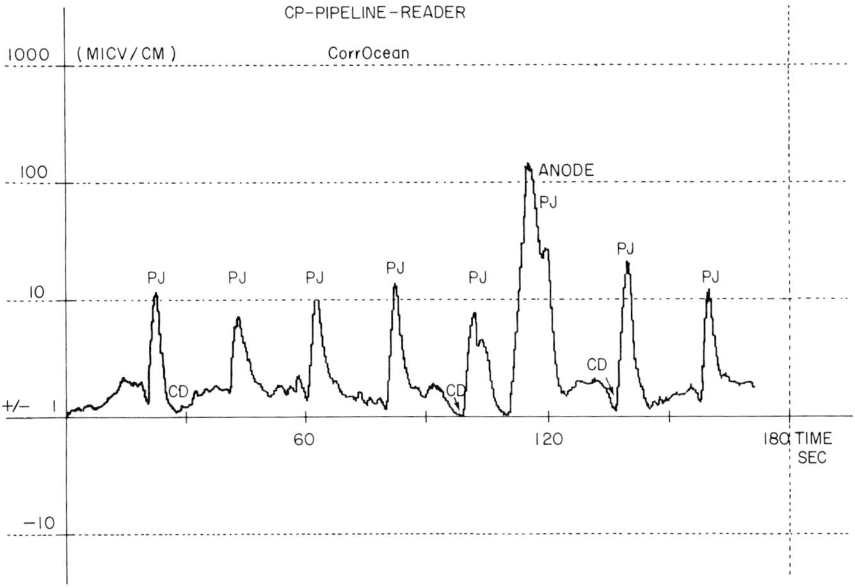

Figure 9.6.1 Typical Raw Data Print-out from Current Density/Field Strength Monitoring of Submarine Pipeline, Showing Anode, Pipe Joints (PJ) and Minor Coating Defects (CD)

However, despite this adequacy, it was discovered that anode operation was not always optimal. The strength of this monitoring method is its capability of simultaneously obtaining both anode current output and potential. It was thus revealed that the zinc bracelet anodes normally operated at a potential less negative than specified, despite a fairly low current output. This is therefore an additional reason for reducing the driving force in anode design/current output calculations.

To take this fact into account, as well as the effect of coating defects, as already discussed in the previous section, the zinc anode bracelets should be designed with a somewhat larger surface area compared with conventional design.

9.7 Bibliography

King, R. A., 'Prediction of Corrosiveness of Seabed Sediments', *Materials Performance*, 19, No 1, 39, (1980).

NACE Recommended Practise, NACE Standard RP–06–75, (Houston, 1975).

Smith, S. N., Reding, J. T. and Riley, R. L., 'Development of a Broad Application Saline Water Al-Anode-Galvalum III', Paper No 180, *Proceedings, Corrosion/76, NACE Conference,* (Houston, 1976).

Strømmen, R., 'Current and Potential Distribution on Cathodically Protected Submarine Pipelines', *Trans I Mar E,* 91, (Edinburgh, 1979).

Strømmen, R., 'Aspects Related to the Cathodic Protection of Offshore Facilities', Thesis for the Dr Ing degree at NTH – The Norwegian Institute of Technology, (Trondheim, December 1978).

Strømmen, R., 'Numerical Methods for Calculation of Current and Potential Distribution in Galvanic Cells and Cathodic Protection Systems', *Proceedings, Second Israel-Norwegian Technical & Scientific Symposium, Electrochemistry-Corrosion,* (NTH, Norway, June 1978).

Strømmen, R. and Rødland, A., 'Computerised Numerical Techniques Applied in Design of Offshore Cathodic Protection Systems', Paper No 241, *Proceedings, NACE Conference Corrosion/80,* (Chicago, March 1980).

Strømmen, R. and Rødland, A., 'Corrosion Surveillance of Submarine Pipelines by Electric Field Strength Monitoring', Paper No 111, *Proceedings, NACE Conference Corrosion/81,* (Toronto, April 1981).

R STRØMMEN

10
3-D WAVE THEORY APPLIED TO PIPELINE SPANNING PROBLEMS

The overall aim of this chapter is to present the application of three-dimensional (3-D) wave theory to pipeline spanning problems, especially those involving vortex shedding. The main body of the chapter deals with the wave theory and the methodology for calculating the induced velocity of the wave action on a submarine pipeline. It then applies this method to pipeline spanning problems involving vortex shedding. A comprehensive account of the theory of vortex shedding is provided in Appendix C.

10.1 The Justification for a 3-D Wave Model

Studies of pipeline stability require reasonably accurate data concerning the velocities of currents along the prospective pipeline route. Unfortunately these data are seldom available in the detail which is normally required, so it becomes necessary to develop models of the specific current conditions, which can then be verified by sampling actual conditions. This approach is now commonly used in offshore pipeline design, which is facilitated by the theory contained herein.

For a number of engineering applications, a three-dimensional deterministic description of irregular waves would be of great value, and definitely preferable to many commonly used wave theories. With this in mind, a research project on three-dimensional waves was carried out in Denmark by the Danish Hydraulic Institute, the Institute of Hydrodynamics and Hydraulics and the Technical University of Denmark. The theoretical work was done by Professor H Lundgren and Dr S E Sand and is described by Gravesen (1979).

In the interests of script economy and to ensure that we focus our attention on the transient current velocities, as they apply to pipeline spanning problems, only extracts of the work by Lundgren *et al* (1979) are presented here.

The work done by Forristall, Ward, Cardone, and Borgman (1978) pointed the way to the need for a three-dimensional wave theory. During the tropical storm Delia in 1973 in the Gulf of Mexico, they recorded simultaneously the surface elevation and two horizontal components of the orbital velocity at three levels (4, 10 and 18 metres beneath the surface). The simplest indication of the directional spreading of energy is obtained through a plot of the horizontal velocity vector shown in Figure 10.1.1. Furthermore, comparisons of the measured velocities with those predicted by uni-directional theories commonly used in design practice showed considerable scatter. Figure 10.1.2 demonstrates that, in the case of Stokes' fifth-order theory, the wave-induced velocities are overpredicted by about 40 per cent. Both linear/nonlinear and regular/irregular wave theories were considered. The errors, however, were found to be similar to those in Figure 10.1.2. Consequently, there is a real need for a three-dimensional description of waves.

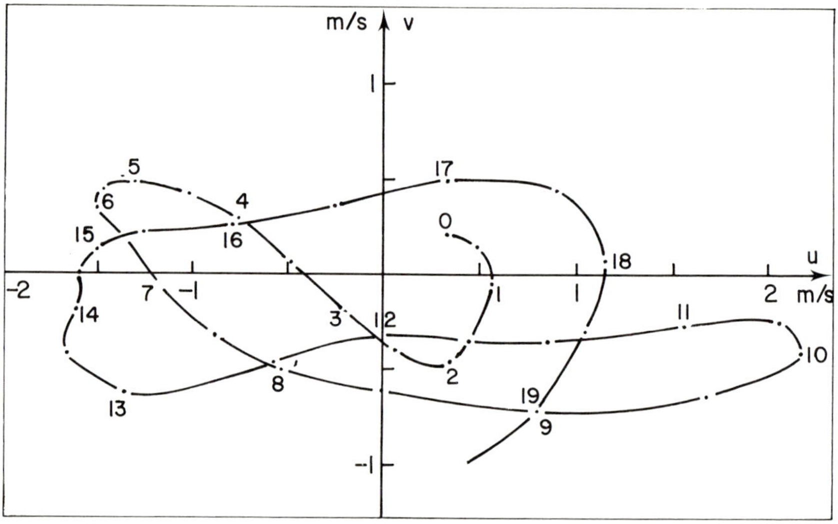

Figure 10.1.1 Time History of the Horizontal Velocity Vector Measured at the Top Current Meter during the Largest Wave. *[Forristall et al (1978)]*

Traditionally, the natural sea is represented by a directional spectrum, although it must be remembered that a spectrum cannot distinguish between wave crests and wave troughs. The main interest, however, is often attached to the actual shapes of the highest waves and the distribution of their orbital velocities as illustrated above. These parameters are of vital importance for the design of offshore structures, where both the rising wave crests, the maximum wave loads, and the frequency of wave loadings must be considered.

Thus, it would be advantageous to replace the stochastic description of three-dimensional waves by a deterministic one. Such a description is presented below as an extension of a previous work by Lundgren and Sand (1979) dealing with two-dimensional irregular waves.

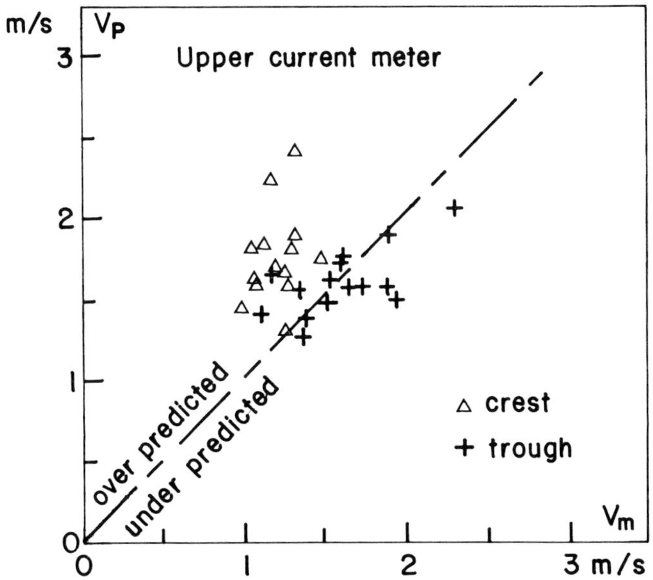

Figure 10.1.2 Comparison of Stokes' Fifth-Order to Measured Maximum Velocities during the Largest Waves for Top Current Meter. *[Forristall et al (1978)]*

10.2 3-D Wave Theory

This theory assumes linearity, and is based on the concept that the sea is composed of many, mutually independent, small, regular wavelets, each with its individual amplitude, direction of travel and phase. The basic equations are derived for simultaneous records of the surface elevation, η_l, and two horizontal orbital velocities, u_l and v_l at one level, all taken at one vertical, P_l. These three time series are decomposed into Fourier series, which will be shown later, and mathematical optimisation principles are then applied in order to obtain the least possible error when η, u, v at P_l are reconstructed as the sum of all wavelets. The directional information is a function of the basic frequency interval, ie, the longer the records, the more directional information.

To test the accuracy of this model, the Danish Hydraulic Institute recorded the three time series (η_l, u_l, v_l) at one point, P_l, during a storm in the Baltic Sea in May 1977. The water depth was 13.4 metres and the electromagnetic (u, v) current meter, as well as an upwards η echo sounder, was mounted 9.25 metres above the seabed.

The three-dimensional approach was applied to data recorded during a 35-minute experiment. The energy was computed for each wavelet in certain frequency bands, each comprising 75 basic frequency intervals, and plotted as a vector, in order to indicate the direction of travel.

In Figure 10.2.1, which corresponds to the band with maximum energy, the x-axis is the mean energy direction for all frequencies, and each cross represents a vector (from the origin) with length proportional to the energy.

The velocities u_2, v_2 were recorded also at another point, P_2, at a distance of 1.3 L from P_1, L being the mean wave length. The three-dimensional theory was tested by calculating the velocities u_2 and v_2, which one should expect from the knowledge of η_1, u_1, v_1. Figure 10.2.2 gives a comparison for the larger of the two components, u_2. The line from P_1 to P_2 formed an angle of 22 degrees with the mean energy direction.

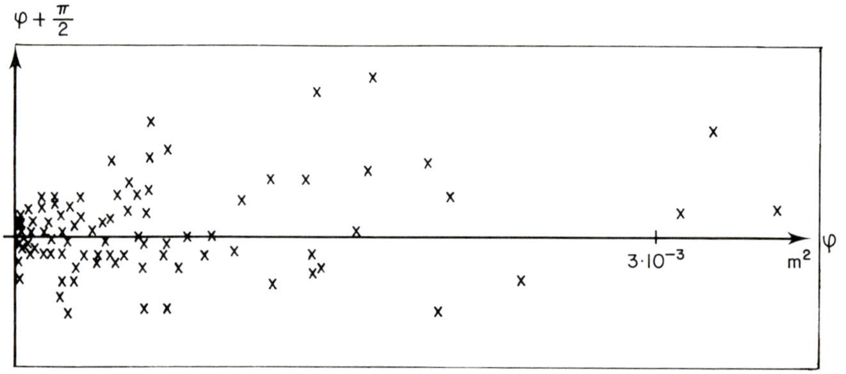

Figure 10.2.1 Energy Distribution in Frequency Band 0.1995 Hz to 0.2265 Hz

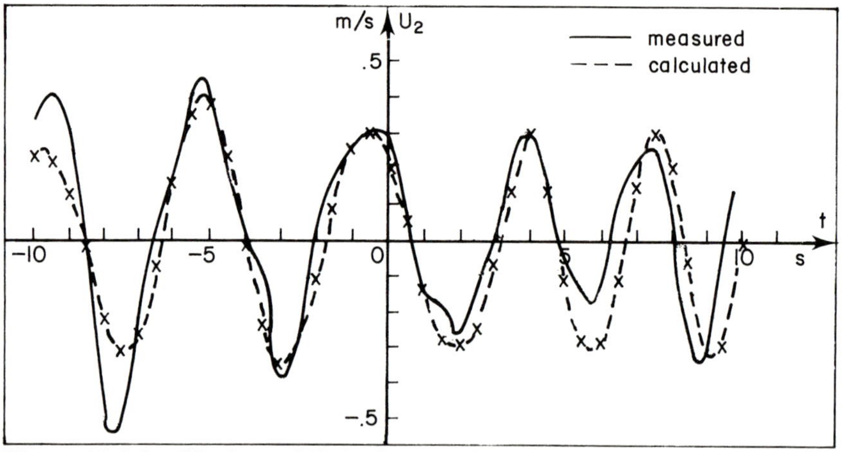

Figure 10.2.2 Horizontal Velocity u_2 at Point P_2

Considering the linearisation used in the deterministic description, the nonlinear interaction over the relatively large distance 1.3 L, the directional spread of energy, and the accuracy of the recorders, the agreement in Figure 10.2.2 seems satisfactory.

In many stochastic approaches concerning the concept of directional spectra, the spreading function s(f,θ) is assumed to be proportional to:

$$\cos^{2s}(\theta - \theta_o)/2$$

where the spreading parameter s is a function of frequency f, while θ is the angle from the x-axis and θ_o the mean energy direction for the frequency considered.

In order to determine the functional form of s, energies in each frequency band (as Figure 10.2.1) have been fitted to the above functional form. s is plotted in Figure 10.2.3 with the frequency normalised by:

$$f_{1,-1} = m_1/m_{-1} \qquad (10.2.1)$$

where m_n is the n'th moment of the one-dimensional spectrum, s(f). It is seen that s is large for low frequencies, corresponding to a very narrow directional distribution, while the energy of the higher frequencies is more widely spread, and s is smaller.

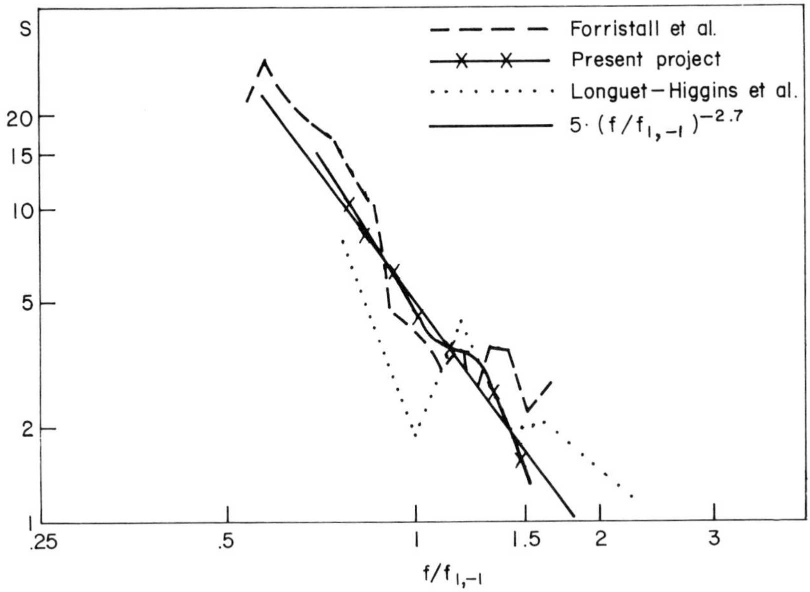

Figure 10.2.3 Spreading Parameter S as a Function of Frequency

Although the other data analysed and plotted in Figure 10.2.3 have been obtained under conditions unlike those of the present study, the various curves seem to be reasonably close to one functional form which empirically may be written as:

$$s = 5(f/f_{1,-1})^{-2.7} \qquad (10.2.2)$$

Accordingly, the directional spectrum, s(f, θ), is given by the equation (Equation 10.2.3) shown overleaf:

$$s(f,\theta) = \frac{2^{2s-1}\, r^2(s+1)}{r(2s+1)} \cos^{2s}(\theta-\theta_o)/2\; s(f) \qquad (10.2.3)$$

where r(n) = (n–1)! is the gamma function, which is added for normalisation purposes, such that the total integral equals unity.

10.3 Application of the 3-D Wave Theory

The three-dimensional theory presented in Section 10.2 can be used to generate physically consistent velocity design conditions for a marine pipeline. By the deterministic description in the time domain, it is possible to obtain a consistent description not only at one point but also of the simultaneous conditions in neighbouring points along the pipeline. This gives the necessary input to either a stability analysis or an analysis of the resonance development for a spanning pipe section. This information is particularly important to spanning calculations since vortex shedding is dependent on the peak current velocity and the duration of this peak current. This aspect is covered in Section 10.4 and a detailed account of the phenomenon of vortex shedding is given in Appendix C.

The procedure to determine wave-generated current velocities is presented in the following example, as illustrated in Figure 10.3.1.

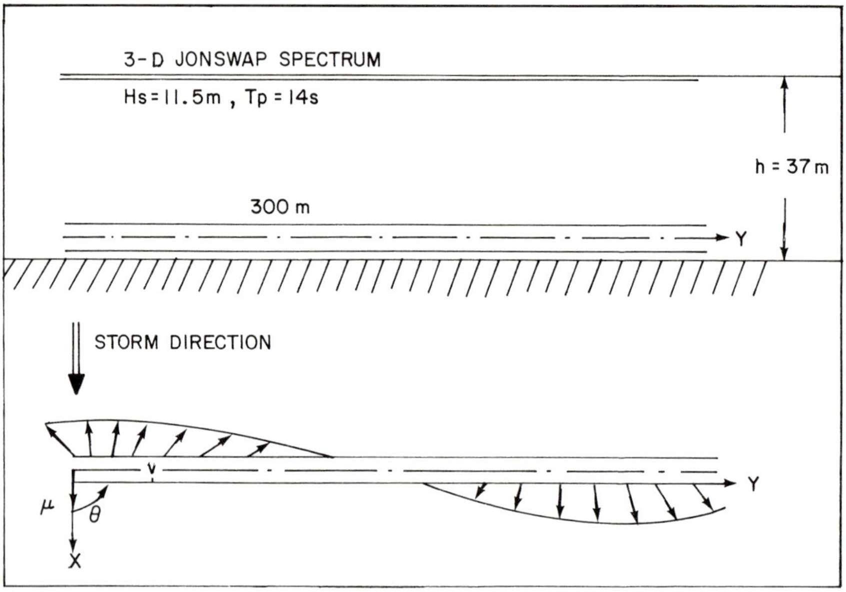

Figure 10.3.1 A Schematic Presentation of the Wave-generated Currents about a Submarine Pipeline

The storm direction has been selected as perpendicular to the pipeline. The time series of the velocity vector (u(y,t), v(y,t)) and the corresponding acceleration vector (acc_x(y,t), acc_x(x,t)) have been calculated for steps of y = 10 metres over a time period of 1,000 seconds and the following Jonswap spectrum characteristics have been chosen:

$$f_m = 0.0714 \text{ Hz}$$

$$\tau = 4.8 \qquad\qquad H_s = 11.5 \text{ m}$$

$$\sigma = \begin{cases} 0.07 & f < f_m \\ 0.09 & f > f_m \end{cases} \Longrightarrow \begin{array}{l} T_p = 14 \text{ s} \\ f_{1,-1} = 0.080 \text{ Hz} \end{array}$$

$$\alpha = 0.009$$

This is based on the Rødvig three-dimensional compilation of a_n, θ_a, b_n, θ_b for 500 frequencies, [after Sand (1979)].

This results in the spectrum shown in Figure 10.3.2.

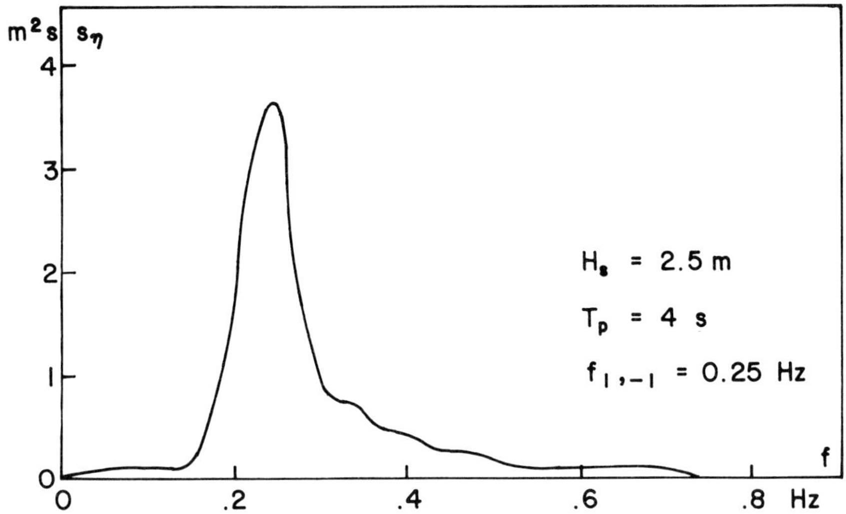

Figure 10.3.2 Spectrum of n(t), Rødvig

For additional discussion on the directional spreading, see Lundgren (1979).

The scaling procedure has been made in accordance with Sand (1979). The rotation of the mean direction $\theta.(f)$ is taken directly from the Rødvig data as 35 degrees. The spreading conditions correspond to those shown in Figure 10.3.3.

$$\text{Frequency – factor: } M_f = f^{I}_{1,-1}/f^{R}_{1,-1}$$

$$\text{Amplitude – factor: } M_a^2 = M_f . S_g(f)/S_R(f)$$

The distance y = 10 metres is a reasonable distance for wave calculations and it is furthermore checked that:

$$\int_o^\infty S_{u,v} df = \overline{\vec{uv}}^2 = \int_o^\infty S(f)_{Jonsw} \cdot \left(\frac{\omega}{\text{Sinh } kh}\right)^2 df = 1$$

For reasons of avoiding aliasing the t-step, t, has been designated as unity.

The equations for deterimining the velocity and acceleration are as follows:

$$\text{Velocity, } u_b = \eta \cdot \frac{\omega}{\sinh kh} \quad \text{(one component)} \tag{10.3.1}$$

$$k_f = \frac{\omega}{\sinh kh} \tag{10.3.2}$$

$$u(o,y,t) = \sum_f (a_n \cos \omega t - ky\sin\theta_a) \,.\, k_f \,.\, \cos\theta_a + b_n \sin(\omega t - ky\sin\theta_b) \,.\, k_f \,.\, \cos\theta_b) \tag{10.3.3}$$

$$v(o,y,t) = \sum_f (a_n \cos(\omega t - ky\sin\theta_a) \,.\, k_f \,.\, \sin\theta_a + b_n \sin(\omega t - ky\sin\theta_b) . k_f \,.\, \sin\theta_b) \tag{10.3.4}$$

$$acc_x(o,y,t) = \sum_f (a_n \sin(\omega t - ky\sin\theta_a) \,.\, k_f \,.\, \cos\theta_a - b_n \cos(\omega t - ky\sin\theta_b) \,.\, k_f \,.\, \cos\theta_b) \tag{10.3.5}$$

$$acc_y(o,y,t) = \sum_f (a_n \sin(\omega t - ky\sin\theta_a) \,.\, k_f \,.\, \sin\theta_a - b_n \cos(\omega t - ky\sin\theta_b) \,.\, k_f \cdot \sin\theta_b) \tag{10.3.6}$$

Figure 10.3.3 shows four examples (stepped in 200-second intervals) of the simultaneous velocity field over the 300 metre length of the pipeline. These results can be used to determine the effective velocity for stability design. This analysis is important for shallow water pipelines, where the wave-generated velocities dominate the design, since it converts the surface conditions to bottom velocities in a correct manner. This is also the case when there are substantial local variations in velocity, due to wave refraction and other bottom effects. At first this is taken into account in the wave height and wave spreading on the surface, and then converted to bottom velocities.

It must be stressed that there exists no unique design factor between surface waves and bottom velocities, as described by H_S and T_Z. The actual factor will depend on wave spreading (which is much influenced by refraction) and the relative water depth which directly changes the 'effective period'.

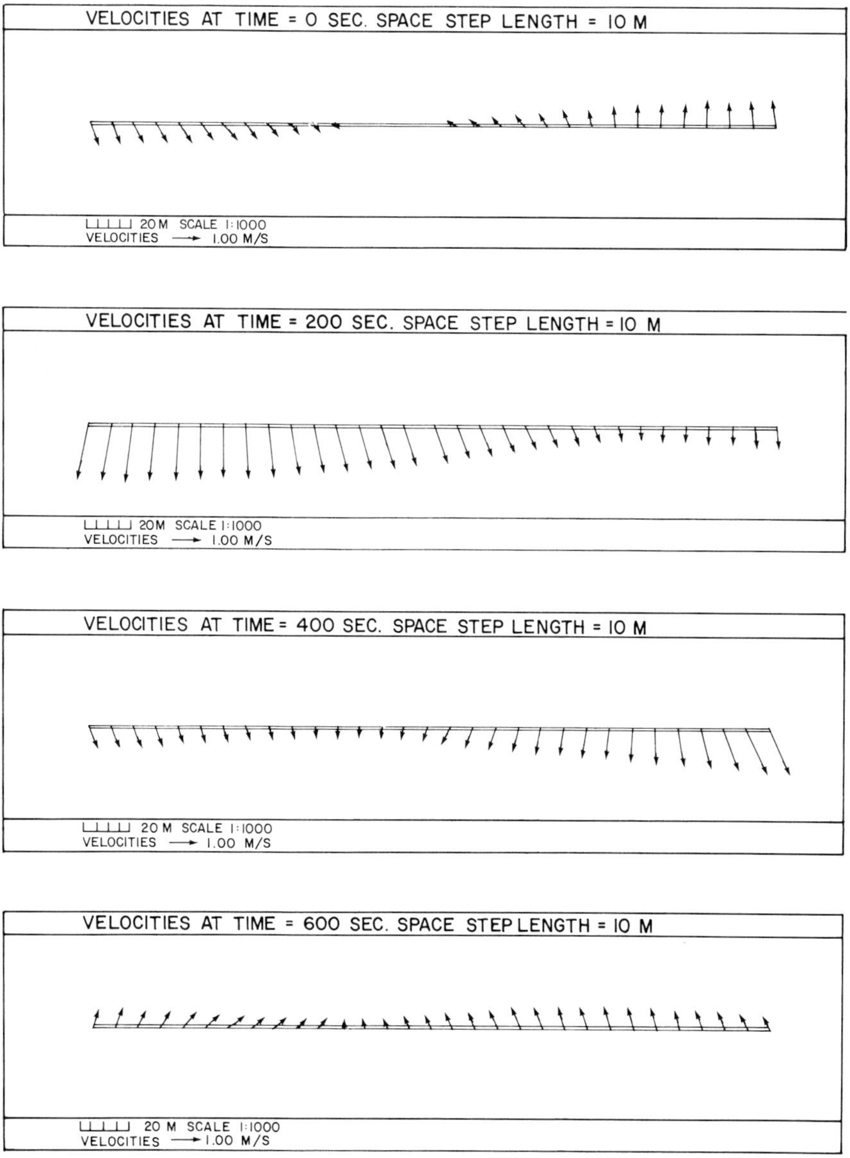

Figure 10.3.3 Simultaneous Bottom Velocities over a Pipeline Length of 300 Metres

Figure 10.3.4 shows the time history of the horizontal velocities at the origin of the pipeline. It is these results which are the most interesting for pipeline spanning problems and for natural backfilling predictions. It should be pointed out that the pipeline designer is faced with determining the velocity envelope as a maximum for stability calculations and as a minimum for natural backfilling calculations.

The spanning calculation depends on the combination of two aspects; namely the peak values and the frequency of application.

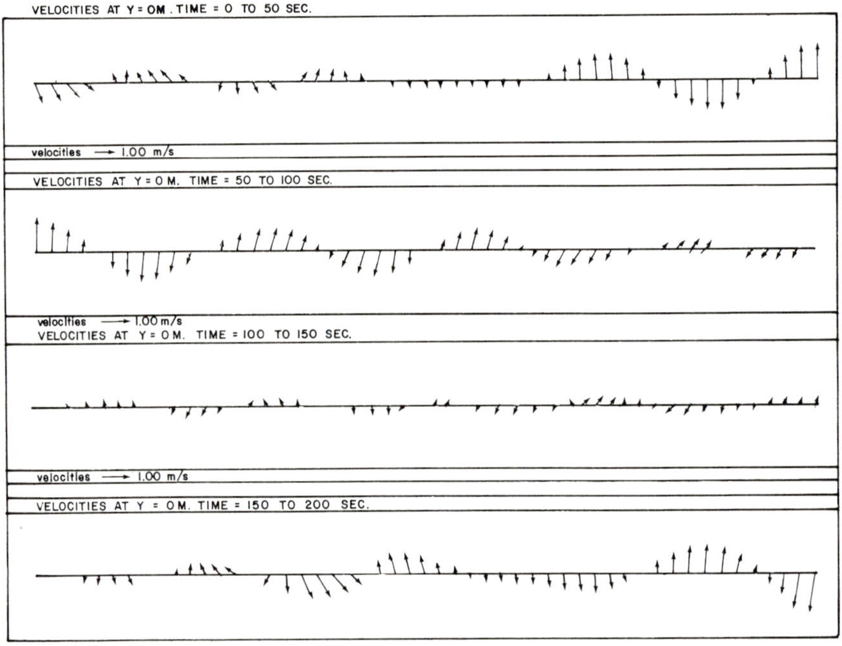

Figure 10.3.4 Bottom Velocities in One Point Generated by 3-D Wave Description

10.4 Application of 3-D Wave Theory to Pipeline Spanning Problems

The oscillations generated in a spanning pipeline are determined by the pipe characteristics, the existing local current velocity, the history of the preceding local velocities and the stresses in the pipe. Figure 10.4.1 illustrates the physical conditions which determine the frequency of oscillation.

In this model it is assumed that the pipe oscillation corresponds to the first note of oscillation only.

The energy input and the damping which occurs during the resonance period may be determined for different points, depending on the local reduced velocity and the local amplitude of oscillation. Using energy balances, the new amplitude of oscillation may then be determined.

This makes it possible to convert the velocity time series for a prescribed pipeline condition into a time series of stresses.

As discussed in Appendix C, pipeline spans will have a damping parameter (δ) approximating 0.12–0.5 which means that the time scale $T_L = T_O/\delta$ for the oscillation (following a linear description of the oscillation) lies between two and eight times the resonance period T_O. Here the ratio $T_L/T_O = 1/\delta$ describes how many forced oscillations of a certain fixed amplitude are necessary to obtain an oscillation amplitude equal to the asymptotic one times the factor (1–1/e), according to the relationship shown below in Figure 10.4.2.

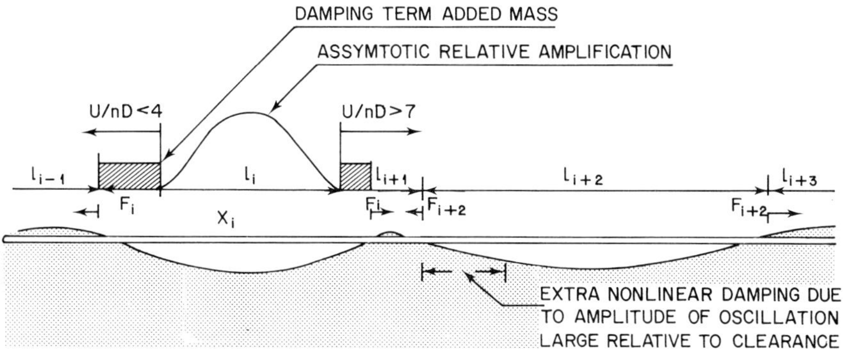

Figure 10.4.1 The Physical Conditions which Govern the Oscillation of a Spanning Pipeline

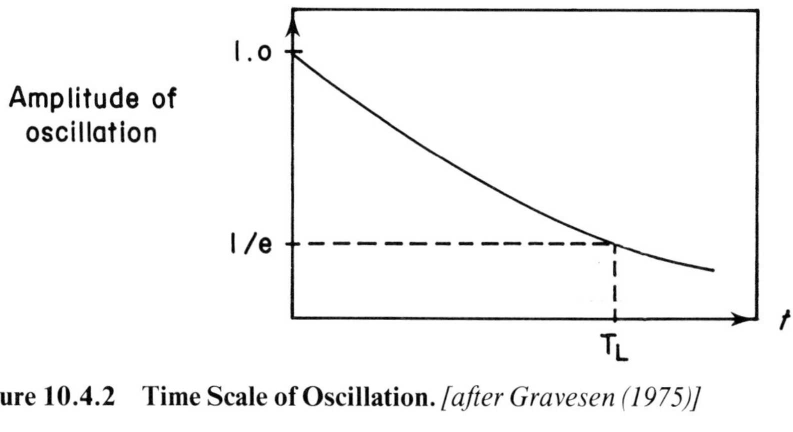

Figure 10.4.2 Time Scale of Oscillation. *[after Gravesen (1975)]*

We have thereby an approximate method of determining the number of oscillations needed to develop an asymptotic resonance amplitude in a pipeline. The length of time required for the oscillatory currents to make a significant contribution to the steady-state current, without diminishing the vortex shedding regime, approximates T_L.

Referring to Figure 10.4.3 this means that, for the same wave height (H), the short duration T_1 may be less effective than the longer duration T_2 depending on the damping ratio δ.

Figure 10.4.4 shows the estimated development of the amplitude of oscillation. It is assumed that it takes three oscillations of the pipe to generate an approach to the asymptotic maximum.

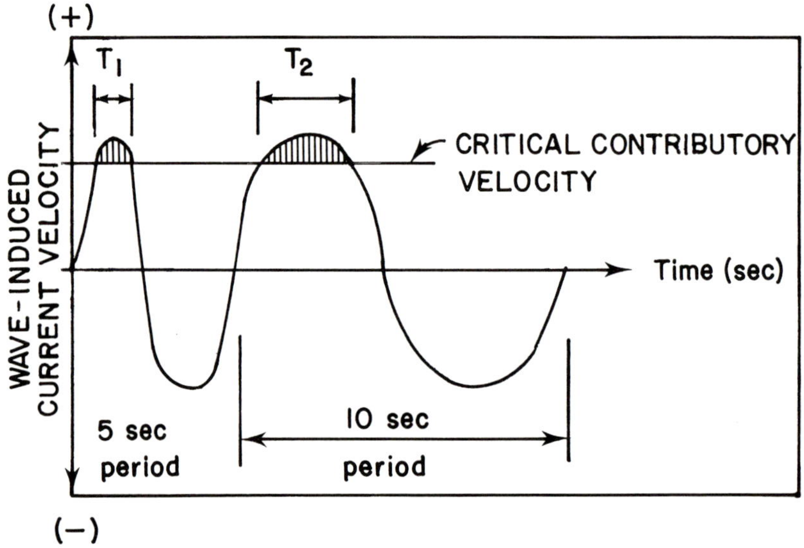

Figure 10.4.3 The Effect of Wave-induced Currents on Pipeline Oscillations

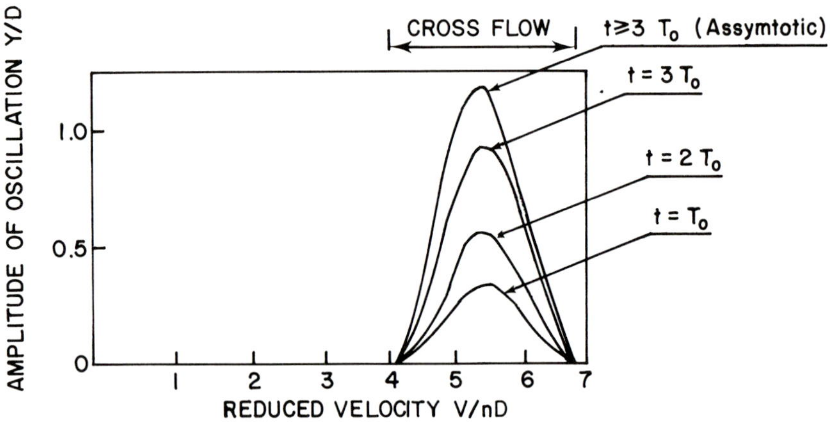

Figure 10.4.4 Estimated Characteristic Development of Amplitude of Oscillation (in principle)

This may be expressed in a simple way as follows. The criteria to endure an oscillation of more than one pipe diameter are:

(i) $5 < V/nD < 5.7$

(ii) the critical velocity range is only effective after $T_L = 3T_O$.

To illustrate the usefulness of this approach, consider the example, due to Nielsen (1979), where a pipeline has the following characteristics:

Outside metal diameter	24 inches
Metal thickness	0.625 inches
Concrete density	150 lbs/ft
Concrete thickness	2 inches
Pipeline atmospheric weight	338 lbs/ft
Pipeline submerged weight	57 lbs/ft

This means that for a span length of 50 feet, which has a resonance period of 0.3 seconds, the effective T_L is one second and this would require a critical velocity range of 11.3 to 12.9 metres per second.

For a span length of 150 feet the resonance period would be two seconds which would require an effective T_L equal to six seconds and a critical velocity range between 1.7 and 2.0 metres per second.

This means that for this pipeline storm waves never create an oscillation with an amplitude of one pipe diameter for a span length of 150 feet.

10.5 Conclusion

Although limitations of space prevent a thorough description of 3-D wave theory here, nevertheless the value of the three-dimensional wave and current model has been demonstrated as an aid in determining the critical spans on a pipeline. It gives realistic bottom velocities which are less than the velocities calculated by two-dimensional wave theories. With a wave time series generation it is now possible to generate a dynamic simulation of the vibrating pipeline, as Sarpkaya (1978) has demonstrated for highly damped short simple beams.

In this it is characteristic that more than one oscillation is required to build up a maximum response in the pipeline. Since it requires a certain duration with critical velocities, the maximum oscillation may often be significantly less than one pipe diameter with a corresponding reduction for stresses and fatigue.

10.6 Bibliography

Forristall, G. L., *et al,* 'The Directional Spectra and Kinematics of Surface Gravity Waves in Tropic Storm Delia', *J. of Physical Ocean.*, 8, 5, (1978).

Gravesen, H., *Analysis and Reproduction of a Natural Three-Dimensional Wave Field*, Report to the Danish Council of Technology. Danish Hydraulics Institute, (1979, not published).

Longuet-Higgins, *et al,* 'Observations of the Directional Spectrum of Sea Waves Using the Motion of a Floating Buoy', *Proceedings, Conference on Ocean Wave Spectra,* National Academy of Science, (1963).

Lundgren, H. *et al,* 'Improved Physical Basis of Wave Forces', *Proceedings, Conference, Civil Engineering in the Oceans,* (San Francisco, 1979).

Sand, S. E., 'Three-Dimensional Deterministic Structure of Ocean Waves', Unpublished PhD Thesis, Institute of Hydrodynamics and Hydraulic Engineering, Technical University of Denmark, (1979).

Sarpkaya, T., 'Lateral Oscillations of Smooth and Rough Cylinders in Harmonic Flow', *Proceedings, Symposium on Wave-induced Forces on Cylinders,* (Bristol, 1978).

R NIELSEN
H GRAVESEN

11

PIPELINE SPAN DETECTION, ASSESSMENT AND CORRECTION

In recent years, the installation of oil and gas pipelines has taken place in increasingly deeper and rougher waters. In such adverse environments, high stresses are imposed on the pipelines when being lowered from the laybarge to the seabed. For some of the lines, this has resulted in damage to their concrete weight coating, with consequent reduction in their stability on the seafloor. Also, the onset of severe storms, especially in the northern North Sea, results in higher than normal bottom current velocities – which in turn lead to scouring around the pipelines. Under these various conditions of reduced pipe stability, spans are liable to develop in the lines, allowing oscillations to occur which may lead to loss of weight coating. Free spans in marine pipelines, whether caused by irregularities in the seabed or by erosion of soil from beneath a line, have proved to be a costly problem for offshore oil and gas operators – not least of all in the North Sea.

The stages of initially detecting a span, making an accurate assessment and recommending appropriate corrective action, require both a detailed survey of the span configuration and an understanding of the behaviour of the pipeline before, during and after correction. Early detection and analysis of spans with regard to the existing and future safety of a pipeline, is of the utmost importance. The risks and cost effectiveness of proposed remedial work must therefore be fully evaluated when determining corrective action.

11.1 Pipline Span Detection Using Acoustic Inspection

The combination of installational and environmental hazards to pipelines, plus the threat of damage by ships' anchors or fishing trawl boards, make it essential to monitor the condition of pipelines on a regular basis, preferably at least once a year, at the end of the winter storm season. Sonar techniques are a most cost-effective means to accomplish this. To ensure that routine surveys do actually detect spans in, or damage to, a pipeline, it is important to perform an acoustic inspection along its full length. Based on the results of this inspection, a more detailed inspection of selected sections of line should then be made using a manned, or unmanned submersible.

A pipeline acoustic inspection is normally carried out by a survey support vessel equipped with an accurate positioning system, and acoustic sounding gear which includes an echo sounder and side-scan sonar. The results of an acoustic survey will determine the extent of pipeline burial and indicate any changes in the bottom condition, such as the trail of a dragging anchor in the vicinity of the pipeline. Figure 11.1.1 is an example of a typical echo sounder recording showing (left) a pipeline exposed in its trench, (centre) a pipeline spanned above the bottom, and (right) a buried pipeline. Where a pipeline is found to be spanned, an inspection by submersible is necessary to determine the exact length of the span and/or the extent and nature of any damage.

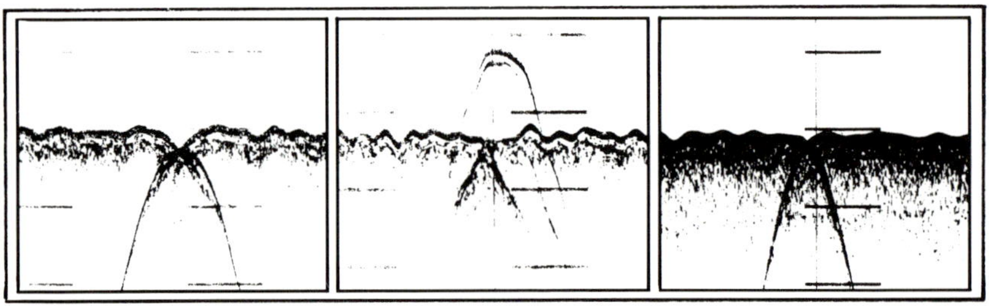

Figure 11.1.1 An Example of a Typical Echo Sounding Recording of a Submarine Pipeline

11.2 Pipeline Submersible Inspection

Specialised equipment has been developed in recent years which enables a submersible to obtain pipe span information with the degree of accuracy needed for detailed analysis and planning of remedial action. One such submersible is illustrated in Figure 11.2.1. The main items of equipment required on board a submersible are:

(i) a long-base positioning system for locating the submersible relative to a pre-positioned transponder adjacent to the pipeline;
(ii) a short-base navigation system to provide for accurate positioning of the submersible relative to the support vessel;
(iii) an accurate pressure-sensitive depth measuring device;
(iv) a pipe tracker to enable the submersible to follow a buried pipeline, even with 3 metres of cover;
(v) a continuous seabed profiler;
(vi) a sector scanning trench profiler to provide cross-sections of the pipe and trench; and
(vii) optical equipment, such as video systems and still-photography to facilitate the provision of permanent records, and hence an assessment of overall span condition.

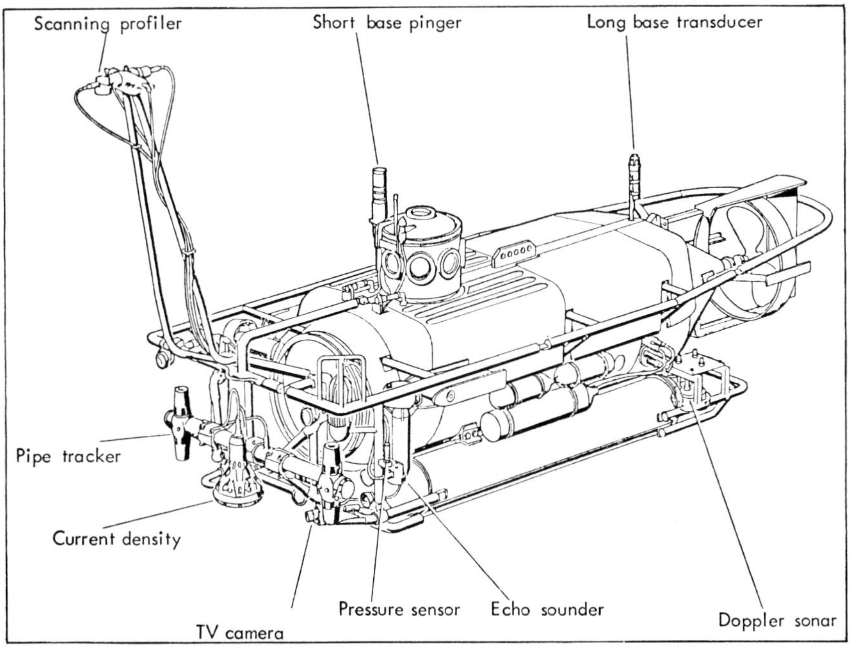

Figure 11.2.1 An Illustration of a Submersible

A representative pipe tracker system is shown in Figure 11.2.2 and a typical sector scanning trench profiler is illustrated in Figure 11.2.3 (both overleaf).

To make the pipe span assessment and plan the remedial action, a survey should provide the following information:

(i) *The position of the span along the pipeline*
The accuracy of the position reading should be adequate to facilitate relocating the span. The position information is also important when establishing the environmental conditions prevailing in the vicinity of the span.

(ii) *The depth of water*
The water depth is required when determining the magnitude of the wave-induced velocity around the pipeline. The depth also has to be taken into consideration when deciding on the appropriate correction method.

(iii) *The length of span*
The length of the pipe span is an important parameter in determining the static stresses in, and dynamic behaviour of a spanned section.

(iv) *The clearance under span*
The height of the span above the bottom is also a factor which influences the dynamic response of the pipeline to currents and

waves. The clearance under the span can normally only be determined by visual inspection.

(v) *The condition of the concrete coating*
The combination of a long span and deteriorated concrete coating represents an adverse situation in which the spanned section can grow to excessive length.

(vi) *The pipe position in trench*
The position of the pipe, horizontally as well as vertically, relative to the trench is important in the planning of remedial action.

(vii) *Any supplementary information*
This can include information on such matters as the presence of debris, anchor scour, sandbags, etc.

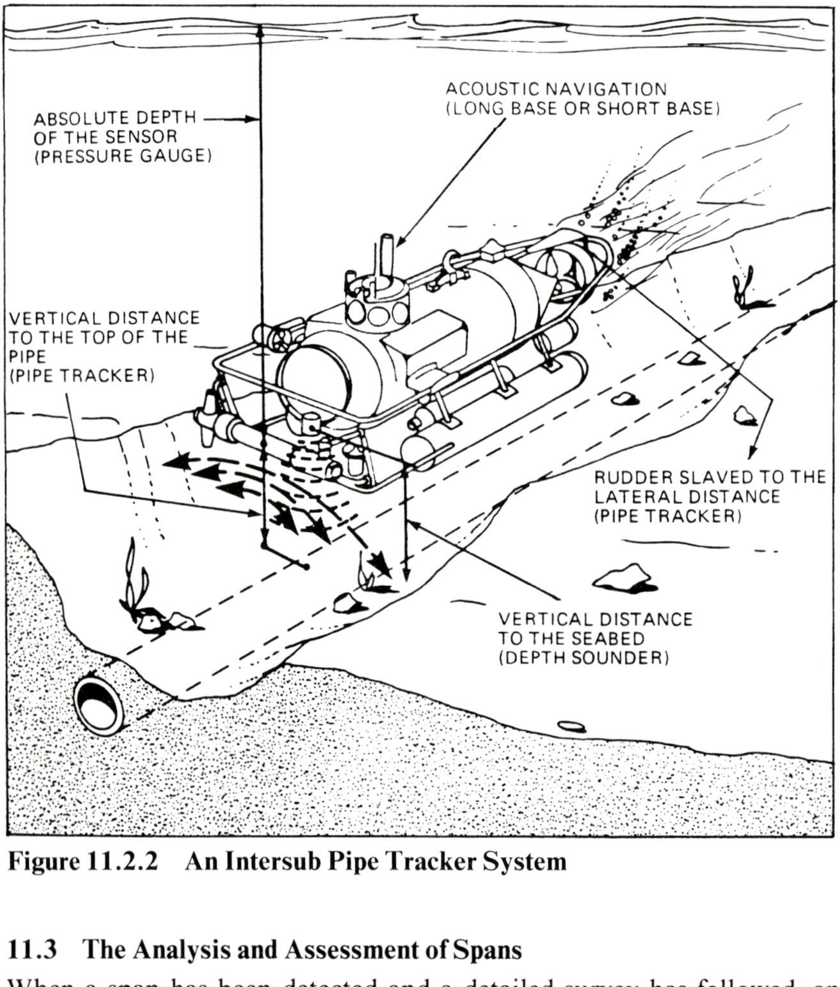

Figure 11.2.2 An Intersub Pipe Tracker System

11.3 The Analysis and Assessment of Spans

When a span has been detected and a detailed survey has followed, an analysis must then be carried out to determine whether the span needs

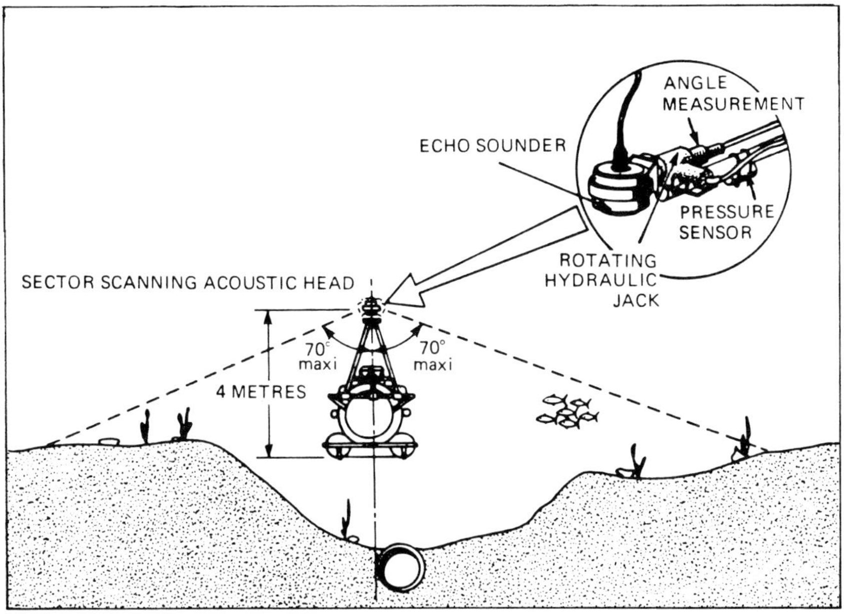

Figure 11.2.3 An Intersub Sector Scanning Trench Profiler

immediate correction, or whether it can be left uncorrected until the next inspection. Such an analysis must include a static stress assessment and an evaluation of the pipeline's dynamic behaviour.

11.3.1 The Static Analysis

A pipeline span is subject to bending moments imposed by the weight of the spanned section of line, by currents and waves, and by dragging anchors or fishing gear. The magnitude of these moments depends on the size of the forces, the length and configuration of the pipe span, and such properties of the pipe as its weight and flexural rigidity. The bending stress can be determined from the bending moment, and can be combined with the induced temperature and pressure stresses to provide a comparison with the specified minimum yield stress of the pipe material. The margin of safety can then be determined. Unfortunately, this process of analysis is complicated by the following considerations:

(i) What is the relevant theoretical model for a given span configuration?

(ii) Do axial forces exist which influence the bending moments?

(iii) Does an existing curve in the pipeline caused by bottom irregularities require to be taken into account?

For practical reasons, the effects of possible adjacent spans are usually ignored, and the span is considered initially as a single beam. Alternatively, all adjacent spans may be taken into account up to the point where the pipeline section is fully restrained. However, this latter approach would almost certainly involve an excessive increase in the amount of work entailed in the initial analysis. In view of this, and where other factors may further complicate the analysis, more sophisticated methods, such as finite difference techniques, may be used.

Where a single beam model is adopted, the maximum bending moment depends on the manner in which the beam is supported. The choice of end constraints, whether hinged, fixed or otherwise, should reflect the results of the detailed survey of the span. A hinged end constraint gives a conservative result; whereas a fixed or encasted constraint leads to a slightly more optimistic assessment. A more realistic solution is sometimes obtained by considering the reactionary moments to be carried beyond the end supports.

Another influence which is normally disregarded is the elastic behaviour of the soil. If this factor is to be taken into consideration, then for a meaningful result, considerably more information than is provided by a routine survey is required on the nature of the soils near the end supports.

In addition to other factors, the magnitude of the bending moment, and hence of the longitudinal stress, in the spanned section of pipeline, depends on the presence of an axial force. A tensile axial force tends to reduce the bending moments, whereas a compressive force has the opposite effect. The following sources of axial force can be identified:

(i) longitudinal component in the hoop stress induced in the pipewall. When considering a span as axially constrained, this component is equal to Poisson's ratio times the hoop stress. A positive hoop stress gives a tensile axial force;

(ii) temperature-induced axial stress, which in most circumstances tends to induce a compressive axial force;

(iii) residual pipelaying forces, usually considered as tensile. The magnitude of these forces is, however, difficult to determine;

(iv) tensile force resulting from any sag in the pipeline.

(v) transverse loading due to internal pressure in zones of pipe curvature. This loading accentuates the curvature and bending moment. The transverse loading per unit length is equal to pressure times pipe cross-section divided by pipe radius of curvature at point in question.

For the majority of cases, the resultant of these various axial forces is tensile.

Occasionally, a span or spans may occur in areas where the pipeline has assumed a curvature corresponding in general to the topography of the seabed. This combination of spanning and natural curvature also occurs at pipeline cross-overs. One analytical approach to this situation is to take the bending moment caused by the curvature into account, and to add it to the bending moment resulting from the pipeline weight. In determining the allowable span length, this approach may result in a conservative assessment.

An alternative approach is to regard the curvature as having existed during the hydrostatic test so that the combined stress exceeded the yield strength of the pipe steel. In this case, the effects of this curvature should not be taken into account when determining the bending moments.

11.3.2 Dynamic Analysis

A further consideration in the appraisal of a pipe span is its dynamic response due to the phenomenon known as 'vortex shedding'. This aspect will not be discussed here since it receives a profound treatment in Appendix C.

11.4 Pipeline Span Correction

Analysis of a pipeline span with respect to static and dynamic stress criteria may show that corrective action is needed to avoid further deterioration of the line or extensive damage. The remedial action depends on the nature of the stresses and the cause of spanning. In some instances, spanning may be caused by water turbulence around the pipe. This condition is frequently aggravated by the presence of a nearby structure or irregularities in the bottom profile. Spans are also liable to occur where a pipeline is laid on a hard, irregular bottom in which jetting is unlikely to provide a regular trench.

A number of methods are currently available for correcting pipe spans. One is to deposit gravel or sand under, around and on top of the span so that the pipe is supported and oscillation cannot occur. Specially designed dumping vessels can be used to minimise the dispersion of material and enhance the accuracy of the dumping operation. The choice of gravel is a compromise between being light enough not to damage the pipeline concrete coating during dumping, and being sufficiently heavy to ensure its stability on the seabed when subject to current and wave action. Unfortunately, gravel dumping sometimes leads to local scour of the seabed due to hydrodynamic effects.

When the static stresses in a span are found to be high, supports must be placed beneath the pipeline prior to dumping. This is especially important where a bund or depth of gravel is required on top of the line to provide stability. The weight of this additional gravel increases the static stresses in the pipe, especially where compressive yielding of unconsolidated soil occurs beneath the line.

A second, very common method of correction is to support the span with sandbags. Carefully positioned piles of bags can both provide some support to the pipe and suppress any oscillations.

A more recent development has been the use of grout bags, which are placed empty, in suitable positions on the seabed, under the pipeline and then pumped full with a suitable grouting mixture from a supply vessel *via* flexible hoses.

Where a span is very long and has shed a considerable amount of weight coating, concrete saddles can be used to restore stability to the line. Special heavy mattresses offer an alternative method of applying weight.

Unfortunately, a disadvantage with the use of bags, saddles or mattresses is the secondary scour which takes place in their vicinity, which can reduce their stability and limit their effectiveness.

An alternative method of span correction which has been tested by some pipeline operators, is the use of artificial seaweed to suppress scour and induce natural backfilling. The seaweed consists of polypropylene strands locked into synthetic mats. The strands float upwards to form a curtain which decreases the flow velocity locally and causes suspended particles of sand to be deposited on the seabed. This action eventually results in the pipeline being buried. The efficacy of this method is very much dependent on the prevailing sand transport, and in any case takes time before becoming effective.

11.5 A Case History in Pipe Span Correction

A span of at least 550 metres was discovered during a routine survey of a well-known 36-inch gas transmission line. The size of the span was such that the line was floating near the surface. An immediate investigation of the span was made by submersible, followed by diver inspection, depth sounding and side-scan sonar surveys, the chronological results of which are shown in Figure 11.5.1

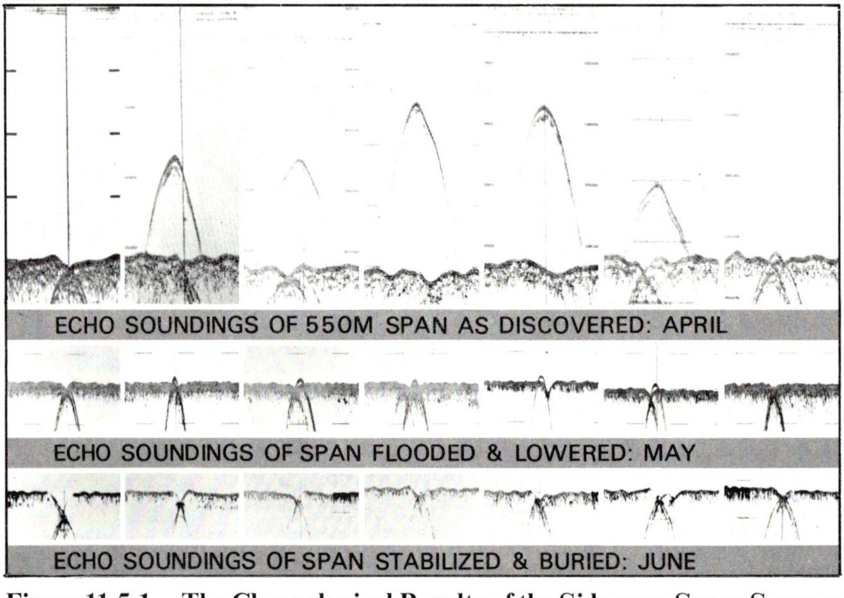

Figure 11.5.1 The Chronological Results of the Side-scan Sonar Surveys

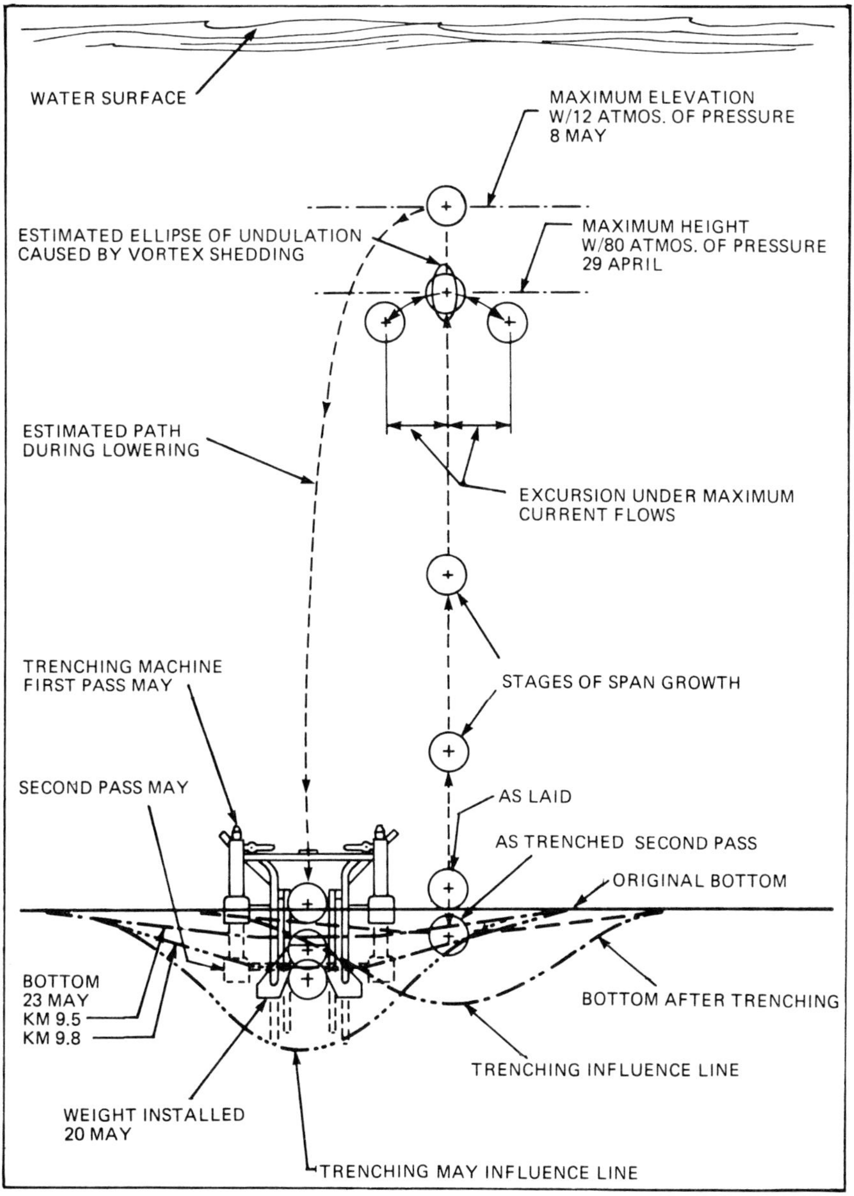

Figure 11.5.2 The Chronology of Events to Restore the Integrity of the Pipeline

The initiating cause of the span was reckoned to have been an anchor, which a work boat had dragged over the pipeline, resulting in a local loss of concrete weight coating. The state of the line was thought to have deteriorated during a storm and subsequent current-induced oscillations of the pipe, which resulted in further concrete loss and partial flotation.

After a check had been made to ensure that the line had not been overstressed during the development of the span, corrective measures were engineered, which enabled the line to be safely lowered to the seabed, reburied, and the spanned section restabilised. The various steps involved are illustrated in Figure 11.5.2. The completion of this work through to the recommissioning of the line took six weeks, and an acoustic survey was used to confirm the integrity of the pipeline.

T J CORBISHLEY
R W LUYNENBURG

12

RECENT DEVELOPMENTS IN PIPELINE INSPECTION USING SIDE-SCAN SONAR

Commercial side-scan sonar equipment has been used in the North Sea since the late 1960s. Typical applications have involved seabed geological reconnaissance, bathymetric surveys and surveillance operations for a variety of purposes.

This chapter describes recent developments to the side-scan sonar method which have been accomplished over the past four years with the active support of Shell and Esso, and it presents ideas for the integration of the sonar-derived data with information obtained by direct visual means.

These developments represent significant improvements in the acquisition and presentation of sonar data, which are likely to yield much greater data quality and reliability in future survey results. Ultimately it is hoped that a system of computer-based information storage will evolve in which pipeline data, obtained by sonar, may be readily compared with observations by diver or submersible television; all referenced to a common distance measurement.

12.1 Basic Side-scan Sonar Technology

By definition, a sonar system is one which uses acoustic energy for remote sensing, observation or communication underwater. Thus sonar is a branch of applied acoustics which uses water as the propagating medium. Sonar systems comprise a source of acoustic energy, a path over which the energy propagates, a receiver and a display unit. An 'active' side-scan sonar is one where a well defined signal pulse is transmitted from the transducer. This signal insonifies a swept volume of water, and any acoustic scattering medium within the volume will reflect part of the incident acoustic energy back to the towed transducer. The range of the 'target' is then determined by the time delay between the transmitted and received signal using the following formula:

$$R = \frac{t}{2C}$$

$$R = \frac{t}{2C}$$

where t = time delay
C = speed of sound in the water.

The reflected acoustic signals are detected by the transducer, processed and displayed as amplitudes on some form of graphic recorder. By moving from point to point, a form of relief map or sonograph of the target area can be formed.

Since its inception, many different types of sonars have been developed, such as the single side-scan, dual side-scan, transit side-scan and a scanning sonar which rotates on an axle and gives radial information from a point source.

The most common systems currently employed on pipeline inspection are the dual channel side-scan sonars produced by EG & G, OAK, Klein and ORE, which operate at frequencies close to 100 kHz; and the UDI system which works at 50 kHz.

12.2 Pipeline Inspection by Side-scan Sonar

Pipeline inspection surveys by side-scan sonar are used for the following reasons:

(i) to obtain a rapid overview of the pipeline and the immediate emplacement zone or corridor;
(ii) to identify and locate any significant changes in the seabed geomorphology, such as evidence of current action and active erosion or scouring along the pipeline; and
(iii) to identify and locate evidence of local shipping activity such as seabed marks or scars caused by trawl boards or anchor drag.

Surveys by experienced personnel using modern analogue equipment can normally produce sonar records of sufficient quality, resolution and positioning reliability to fulfil these objectives. The method is fast, with surveying speeds of 4–5 knots in sea conditions up to Force 5–7. It is also comparatively inexpensive in that a survey costs about one-third to one-half the daily cost of a manned submersible spread. Unfortunately the sonar method does have some disadvantages. It is an indirect remote sensing technique, producing no meaningful evidence of itself. It records data which must be interpreted, and it lacks the high resolution and positioning accuracy to discriminate and pinpoint important fine scale detail such as broken weight coatings or incipient small spans in areas of scour. It is also unable to produce any useful information about the state of a pipeline lying in a deep trench.

12.3 New Developments in Side-scan Sonar Technology

Despite these shortcomings, side-scan sonar has achieved a wide measure of acceptance by the offshore petroleum industry in the North Sea area, and most major operators commission annual or bi-annual inspection surveys of their pipelines using this method. One such company is Shell-Esso, which currently operates about 780 kilometres of oil and gas pipelines in the northern and central North Sea. Much of this complex system of trunk lines, intrafield lines and infield lines is surveyed annually by side-scan sonar, with inspection of varying proportions or certain routes by manned submersible or RCV mounted television, as well.

In recent years, Shell-Esso has sponsored two important technical developments designed to jointly improve the performance and accuracy of the side-scan sonar for pipeline surveillance work. The first of these is the Colour Video Sonar System (CVSS) designed to make better use of sonar data and improve the resolution of the system as a whole. The second is the Sonar Navigation Control Package (Nav Pac) developed to provide the necessary accuracy and control which the CVSS requires to perform to its full capability. Both systems have been designed and manufactured by UDI Group Ltd, Aberdeen over the period 1977–1980 and are described briefly below.

12.4 The UDI Colour Video Sonar System

The Colour Video Sonar System (CVSS) enables side-scan sonar data, previously recorded on monochrome analogue paper charts, to be presented as rolling or still map display on a standard high quality television monitor, as illustrated by Figure 12.4.1 (overleaf).

Reflected sonar signal amplitude levels are represented within a scale or spectrum of 16 colours to express variations in the reflectivity of the seabed and any objects lying thereon.

The principal component of the CVSS is a sophisticated microprocessor which performs three functions:

(i) it digitises the analogue input signals held on magnetic tape, according to a control input signal and other instructions from the operator;
(ii) it stores the processed digital data; and,
(iii) it prints out the stored data as standard composite PAL encoded television signals.

The CVSS offers a number of advantages which are not available with orthodox analogue systems. It makes more effective use of the reflected sonar signals and thus improves detail discrimination and resolution. Within limits, it permits the colour coding of certain typical seabed returns, which assist with rapid identification. It enables the automatic selection on demand of any desired coverage which can then be displayed in single or dual mode, with single or double expansion in the lateral direction. Finally, it allows direct interfacing with the navigation computer so that accurate position data, in alpha-numeric format, can be superimposed on the screen.

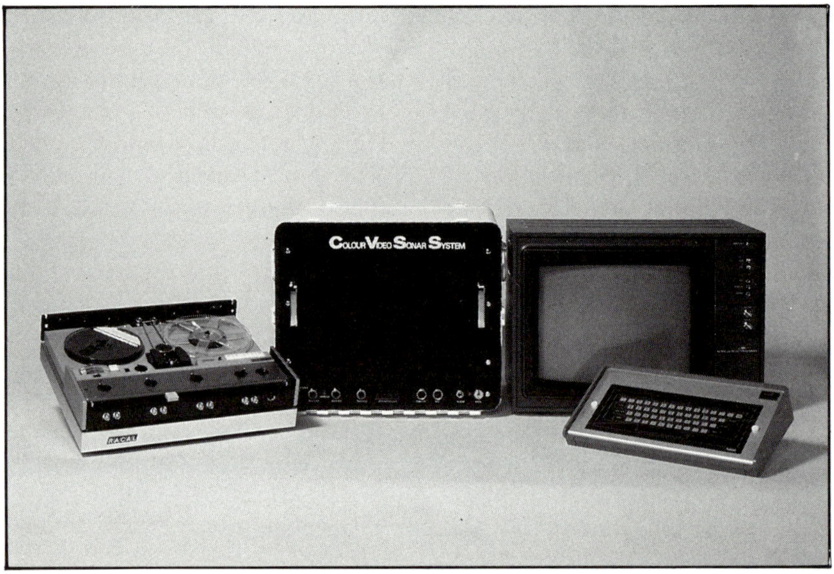

Figure 12.4.1 The CVSS System Complete with Racal Tape Deck and Television Monitor

12.5 The UDI Navigation Package System

To realise the full potential of side-scan sonar with CVSS as an inspection tool, it is necessary to consider the role which the associated navigation system plays during a pipeline investigation. As all information is derived from a towfish, it is essential to know the exact position and heading of the fish at all times, within the limits of accuracy imposed by surface navigational systems. A method is also required in which the fix numbers are generated at the correct intervals by the navigation computer and presented with the sonar data, so that any pipeline incident can easily be referenced to the accurate towfish position. During the study, which is reported herein, the system also had to meet Shell's requirement to accept the digitised position data of their pipelines using the Kilometre Post (KP) method as the main interval measurement along a pipeline.

The Navigation Package (Nav Pac) was developed to meet all of these requirements. It is based on a Hewlett-Packard HP 9825 desk top computer and peripherals, and a sophisticated suite of programs which were developed by UDI.

12.6 Practical Experience with the New System

The CVSS was first used in 1980 and the integrated package was fully operational for the inspection surveys carried out in April 1981. These covered some 745 kilometres of pipelines, including the 36-inch Sullom

Voe main oil-line, the 36-inch FLAGS gaslines and the Cormorant and Brent intrafield and infield lines.

The 1981 operations, carried out as a joint venture by Gardline Surveys and UDI, began with a preliminary two-month period of office-based familiarisation with the CVSS, which included a desk study of the sonar and visual data from earlier surveys. The proper survey comprised an initial five day sea trial of equipment and operational routines, followed by 19 days of data acquisition, during which a total of 1,130 kilometres of survey lines was completed. Data interpretation began by stages as tapes and charts were brought ashore, and the results were communicated progressively to Shell Aberdeen, so that the follow-on submersible inspection programme could be planned in a timely fashion.

The survey was carried out using the MV *Gardline Locater* navigated and positioned by Decca Pulse 8 and it required a party of 13 surveyors, geophysicists and engineers, who were provided jointly by Gardline and UDI. The party was organised into four teams: three teams stood normal watches and the fourth team provided day-time quality control. This unusual staffing arrangement probably contributed as much as anything to the high rate of acceptable data acquisition and to the relatively small amount of re-surveying which was necessary; actually less than 180 kilometres of survey line.

The basic on-board equipment consisted of the following:

(i) twin Racal-Decca Pulse 8 receiver Mk 4 interfaced to a Nav Pac HP 9845A computer for on-line control of the sonar system, and Gardline GA computers for station keeping and general navigation;

(ii) one single and one dual channel UDI side-scan sonar model AS 350A: the former mounted within a gate valve on the survey vessel to assist with maintaining the correct ship and towfish position relative to the pipeline; the latter was used with the towed transducer and operated in single channel mode to acquire the basic pipeline route sonar data; and

(iii) two UDI colour video sonar systems, complete with television monitors and four track fm Racal Store 4DS magnetic tape recorders: one for on-line quality control and the other for off-line data review and interpretation.

From work carried out in the preliminary desk study phase it was determined that the towfish should be streamed so as to maintain an optimum strike angle (towfish to pipeline) of between 10 and 25 degrees, using as short a cable length as possible. In practice this configuration was usually achieved thanks to the excellent towing characteristics of the catamaran towfish. Typically, at depths of 150 metres the towfish would run at a lay-back astern of 300 metres, with a cable length of 450–500 metres. The off-set distance from the pipe was kept at about 50 metres with the unit working on a 100 metre range. The on-board data management and quality control were organised so that all forms of data, both navigational and sonar, could be checked, reviewed, catalogued and stored within 12 hours of acquisition. This ensured that any re-survey could be planned with the maximum economy of effort.

The techniques for interpreting the CVSS data presentation were developed in the 1980 surveys, and then much improved during the preliminary phase of the 1981 project. They were based largely on experience of interpreting analogue (paper) sonar records, and the knowledge gained in comparing representative 1980 CVSS records with companion video taped films obtained from a submersible.

The pipeline zone incidents to be identified, measured and fully reported during the 1981 Shell surveys included damage or displacement of the pipeline, debris, leaking gas, local fishing, shipping or construction activity, the condition of the trench, signs of scouring, and free spanning.

All but the last of these incidents produce characteristic and recognisable sonar contact images on CVSS. However, the identification of free spanning is not so straightforward. This is partly due to the fine scale of the sonar image generated by these incidents, and partly to the fact that other pipeline zone features can produce very similar and thus ambiguous sonar display images. This is especially the case with sand ripples aligned parallel to the pipeline.

The identification and measurement of spans above a certain minimum size was a specific contract requirement. Considerable efforts, therefore, were made to develop reliable criteria for their interpretation from the sonar evidence. In most cases one or more of the following proved effective:

(i) the absence of sonar reflection from the area close to the near-side of the pipeline, due to the presence of a hollow or depression along the pipe itself, suggested the presence of scouring or possible spanning;

(ii) reflections from the seabed immediately beneath and slightly to the far-side of the pipe were usually observed as a broadening of the bright trace of the pipeline on the CVSS display – the underpipe gleam;

(iii) a bulge on the extreme edge of the pipeline's far-side sonar shadow, caused by elongation in a hollow or depression in the seabed, suggested the possibility of scouring or spanning;

(iv) other general sonar evidence of seabed irregularities or undulations, which might cause depressions within the pipeline zone and the possibility of spanning.

The pipeline incidents were measured in all significant cases. Longitudinal dimensions and KP positions were estimated by interpolation using multi-point dividers, scaling between fiducial lines. Transverse dimensions were measured directly using a screen scale constructed for the lateral range to be displayed.

The interpretation, location and measurement of the sonar evidence called for great care and concentration. This work was undertaken throughout by the same three-man team of marine geophysicists to ensure maximum familiarity with the data. Later, selected sections of the pipeline system were inspected by a submersible. A comparison of the CVSS-based interpretation with the visual evidence on video tape suggested that the correct identification of incidents by sonar has been achieved in better than 70 per cent of cases, and that their measurement was subject to a maximum error of ten per cent. Overall, the survey control gave a position repeatability of $\pm$ 5 metres.

12.7 Future Developments of Side-scan Sonar for Pipeline Work

Work is already in hand at UDI to increase the accuracy of towfish positioning, and control over both towfish and survey ship by means of more sophisticated software for the Nav Pac system. This should improve the ability to maintain the crucially important towfish-pipeline geometry.

Work is also underway which should lead to the automatic slant range correction of the sonar image, with a reduction of the present scale distortion, associated with the CVSS. Other ongoing research and development relates to improvements in the quality and versatility of the alphanumeric information and plans have been made to create an integrated hardware-software package for a storage and retrieval system in which pipeline zone incident information from all sources are entered with relevant KP values and archive reference. This is only now feasible because of the much improved quality, reliability and accuracy associated with sonar data obtained on surveys using the new systems mentioned above.

12.8 Acknowledgement

The authors would like to thank Shell Expro UK Ltd for permission to present the material contained herein. Their thanks also to the Managements of both UDI Group Ltd, Aberdeen and of Gardline Surveys, Great Yarmouth for their active support and encouragement.

M W RAINBOW
I McKENZIE
N S WIJESINGHE

13

EMERGENCY PIPELINE REPAIR

Total Oil Marine plc, as operator of the twin 32-inch Frigg Field to St Fergus gas pipelines, has always been mindful of the fact that it transports about one-third of the UK's national gas supplies. Given that gas is difficult to stock anywhere outside a reservoir, it means that any incidents on the 220-mile (360-kilometre) trunk lines can have a rapid and far-reaching impact on British Gas customers. Having noted that the North Sea had had its fair share of pipeline incidents, Total Oil Marine carried out an evaluation to see what measures should be taken to reduce the risks, and in the unfortunate event of an incident occurring, what could be done to minimise resulting downtime.

The study led to the decision to establish an Emergency Pipeline Repair Scheme (EPRS) with a diving contractor. The EPRS contract provides continuous stand-by of a full underwater hyperbaric welding habitat along with short notice availability of qualified diver/welder personnel.

It is recognised that the actual welding repair time only represents a small portion of the total repair cycle. For example, the effect of having to flood the gaslines for repair requires many days of de-watering and recommissioning of pipelines. A research project was therefore launched to devise a means of plugging the lines prior to a repair and minimise the ingress of water. This gave rise to the 'Cold Tapping Project' launched by Total Marine Norsk A/S.

This chapter reviews considerations behind the policies adopted by the operator and reports on the progress made on both projects.

13.1 Hazard Evaluation

The initial gas flow down Line 1 commenced in September 1977 with Line 2 starting the following year. Production levels built up rapidly until Frigg was providing over a third of the UK's gas supply. During the late seventies a number of major pipeline incidents had occurred in the North Sea. In this somewhat nervous climate, Total Oil Marine commissioned a hazard evaluation using data available for the UK sector of the North Sea.

Whilst evidence now shows that pipeline risks have come down due to the diminishing amount of offshore work by large semisubmersibles, the conclusions of the report published early in 1981 provided much food for thought. Of the 83 notified incidents covered by the survey period, the largest part involved major trunk lines transporting hydrocarbons from central collection points to onshore terminals. Much of the remainder concerned infield/flow-lines.

By far the larger part of the incidents took place relatively close to platform installations, largely due to the ongoing development activities round installations after the pipelines had been laid. However, the overall risk assessment from the study was that it was likely that one could expect an incident every three or four years on the Frigg pipelines.

Following the hazard assessment, an economic study was then carried out to establish how best to reduce the financial impact of a major pipeline incident. Neither Elf Aquitaine Norge who operate the five platforms on Frigg Field nor Total Oil Marine operating the pipelines and the intermediate compression platform have a permanantly chartered marine support vessel. Thus there is no immediate availability of intervention by a large vessel complete with diving equipment, etc.

However, pipeline repairs can be achieved very satisfactorily by using small diving support vessels rather than large semisubmersibles. Since such vessels have a reasonable market availability, it was felt that the best course of action would be to have a fully operational repair system which could be deployed easily from a normal diving support vessel.

13.2 The EPRS System

From the very beginning of the Frigg project, Total Oil Marine had made a firm commitment to have all welded pipelines with no mechanical connectors. This was felt necessary to retain the high integrity of the 150 bar design pressure pipelines. Whilst there have been improvements in mechanical connector systems over the years, Total still consider that a hyperbaric weld is the most satisfactory repair for high pressure gas pipelines.

In addition to the hyperbaric weld repair, it was also recognised that minor mechanical damage could be 'patched' by means of specially designed mechanical split sleeves.

Following a market survey and a call for bids, a contract was eventually awarded to Comex Houlder Diving Limited of Aberdeen for the long term hire of the spread of equipment which became known as the Emergency Pipeline Repair Service (EPRS). Both Comex Houlder and Total Oil Marine had worked closely together during the pipelaying of the original Frigg project and had built up extensive experience of hyperbaric welding procedures in the North Sea.

The complete set of equipment comprises over 300 items, the main ones being as follows:

(i) hyperbaric welding spread consisting of two H frames and one Sea Horse IV welding habitat fitted with hydraulic clamps for final pipeline alignment;

(ii) all the miscellaneous equipment required for hyperbaric welding bearing in mind that the system must be as independent as possible from the surface support vessel. (Figure 13.2.1 is page one of a seven-page inventory of equipment and gives an idea of the comprehensive nature of the equipment lists.);

FAMILY	GEMAT	CONTAINER MINREQ	BACKUP	DESCRIPTION	LOCATION
0010	06106	1		Seahorse 4	ASCO PHEAD
0020	06206	1		CSSM 6	ASCO PHEAD
0030	13063	1		ECV2 Air Tugger (Handling Frame)	ASCO PHEAD
0035	13344	1		Lucas (2002) Air Tugger (Handling Frame)	ASCO PHEAD
0040		2	1	32 inch MS Doors	ASCO PHEAD
0070		1		Set of Shackles and Slings for Seahorse	ASCO PHEAD
0080		4		5/8 diameter Turnbuckles	ASCO PHEAD
0090		1	1	Hydraulic Umbilicals	ASCO PHEAD
0100		2		Sets Adaptors for Seahorse Claws (2 inches thick)	ASCO PHEAD
0110		2		Sets Seahorse Handling Clamps (41 1/4 in)	ASCO PHEAD
0120		2		Sets Seahorse Handling Clamps (39 in)	ASCO PHEAD
0130		2		Sets Seahorse Handling Clamps (37 in)	ASCO PHEAD
0140		2		Sets Seahorse Handling Clamps (35 3/4 in)	ASCO PHEAD
0150	77428	1		30–32 inch Dearman Clamps (long)	ASCO PHEAD
0151	77486	1		30–32 inch Dearman Clamps (long)	ASCO PHEAD
0180		1		Set of Through Water Communication Equipment	CHDL BASE
0185		1		Set HP Water Jetting Equipment	BOUGHT IN
0190	06301	1		Miss 01	ASCO PHEAD
0195	34345	1		THV 970 Television Camera Thompson 144 (Miss Docking)	ASCO PHEAD
0200		1		Miss Deck Receptacle	ASCO PHEAD
0210	80003	1		Miss Storage and Transport Receptacle	ASCO PHEAD
0220		1	1	Miss Umbilical	ASCO PHEAD
0230		1	1	Miss Umbilical Baskets (1 per basket)	ASCO PHEAD
0240	13502	1		10T Air Tugger (Miss) (Popeye Winch)	ASCO PHEAD
0250		1		Welding Basket and Umbilibloc Mounting Gantry	ASCO PHEAD
0260	90120	1		Umbilibloc 5CV 2000 mm diameter	ASCO PHEAD
0270	50009	1		7T Portique Frame	ASCO PHEAD
0300	90061	1		Umbilibloc 8CV 970 diameter	ASCO PHEAD
0310	13335/7	2		(16016) Lucas Winch	ASCO PHEAD
0320	38011	1		Aurenty Winch	ASCO PHEAD
0330	06407/8	2		H Frames (Numbers 7 and 8)	ASCO PHEAD
0335		1		Set of Shackles and Slings for H-Frames	ASCO PHEAD
0340		1		Set of Adaptors for H Frame Clamps (2 1/2 in) (Fitted)	ASCO PHEAD
0350		2		Sets Adaptors for H Frame Clamps (1 7/8 in)	ASCO PHEAD
0360	76484	1		Dredge Pump	ASCO PHEAD
0370		1		Dredge Pump Cable	ASCO PHEAD
0380	90711	1		Dredge Pump Cable Winch	ASCO PHEAD
0390	45084	1		Variable Flow Tool Hydraulic Power Pack	ASCO PHEAD
0400		1		Plidco Repair Clamp Handling Frame	ASCO PHEAD
0410		1	1	Plidco Repair Clamp (Cast 5.5T/FAB 3.5T)	ASCO PHEAD
0415		2		Spare Hinges for Plidco	ASCO PHEAD
0420		12	3	Plidco Studbolts (41, 47, 51 and 57 inch lengths)	ASCO PHEAD
0430		23		Nuts and Locknuts for Plidco Studbolts	ASCO PHEAD
0440		1		Sealant Injection Ram (including Clamps and Mounting)	ASCO PHEAD
0460		1	1	Ovality Gauges (32 inch pipe)	BOUGHT IN
0470		1		Set of Acoustic Means of Relocation	CHDL BASE
0480		1		Gammagraph Container complete	CHDL BASE
0490		1		Pocket Radioactivity Detector	CHDL BASE
0500		1		Negatoscope	CHDL BASE
0510		1	1	Radioactivity Detector Light	CHDL BASE

Figure 13.2.1 Part of the Comex/Total EPRS Equipment Listing

(iii) a gantry frame with torquing equipment for handling and installing the Plidco split sleeve mechanical connectors;
(iv) two Plidco split sleeves for 32-inch diameter pipe;
(v) unburying equipment comprising a dredge pump modified to ease handling by divers;
(vi) pipe cutting and bevel machining equipment.

All of the above equipment can be mobilised within 24 hours.

The personnel to be mobilised for the pipeline repair with hyperbaric welding would comprise:

(i) a 30-man diving team;
(ii) a 13-man welding team;
(iii) a three-man project management team based onshore.

The contract allows for all personnel to be mobilised in less than three days with the exception of the diver/welders who have to be on site ready to work seven days after the spread has been mobilised. This time scale allows the contractor to have 'Frigg pipe qualified' welders working in other geographical locations on a regular basis.

The welding procedure which would be used in such repairs was evolved after much study of the several pipe steel/hyperbaric weld combinations used at construction.

Figures 13.2.2 to 13.2.6 (pages 164 to 168) show a typical sequence of a pipeline repair using the hyperbaric spread. It is obvious from the illustrations that a repair would have to be preceeded by extensive preparation works on the seabed. Since both Frigg pipelines are trenched over their entire length, it would be inevitable that substantial unburying

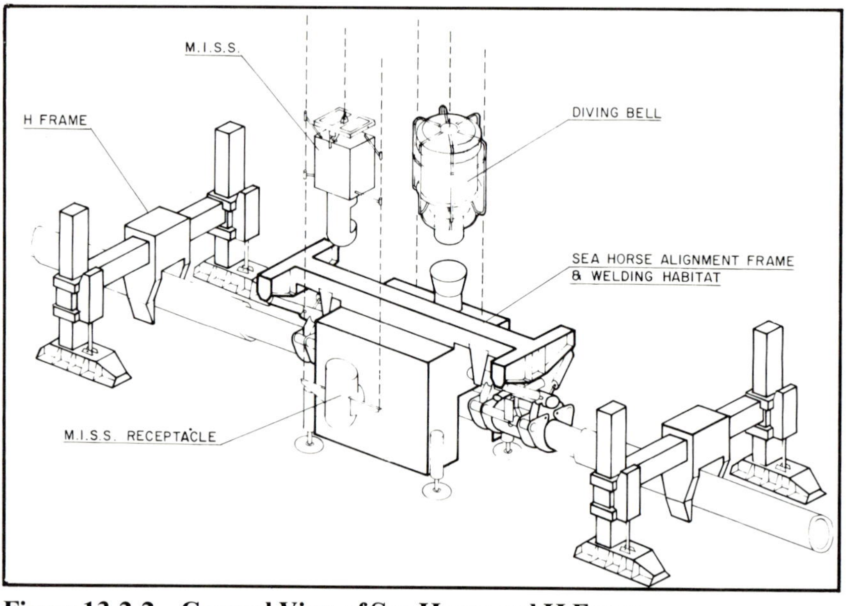

Figure 13.2.2 General View of Sea Horse and H Frame

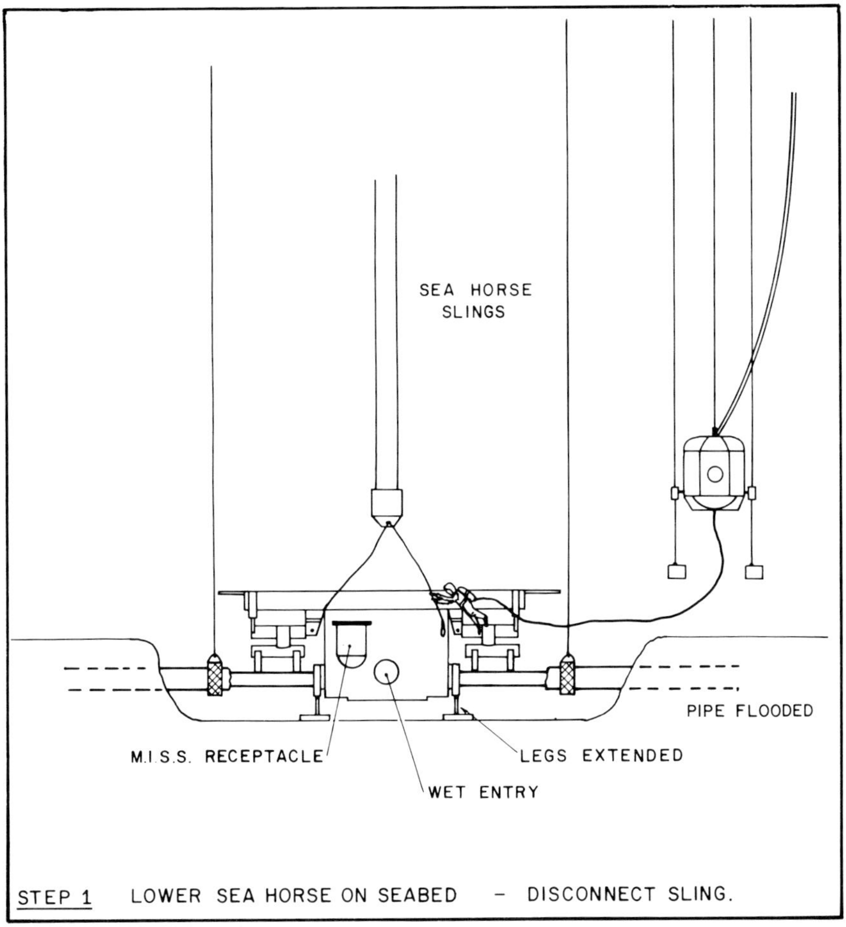

Figure 13.2.3 Step 1 of the Pipeline Repair Sequence

works would be required. Similarly, even for the installation of a Plidco split sleeve (Figures 13.2.7 to 13.2.9, pages 169 to 171) extensive preparation works would be required. In the case of the Plidco repairs, concrete and mastic removal has to be carried out, followed by grinding to a bare metal surface condition.

13.3 Reducing the Repair Cycle Time

Figure 13.3.1 (page 172) shows a notional bar chart for a complete repair cycle including all of the preparatory works and the recommissioning phases. The durations given must be regarded with extreme caution. One has no means of predicting the variable time required to carry out the unburying work, for example. Also many of the time-consuming tasks are due to the fact that the gas pipeline has been flooded in the first place.

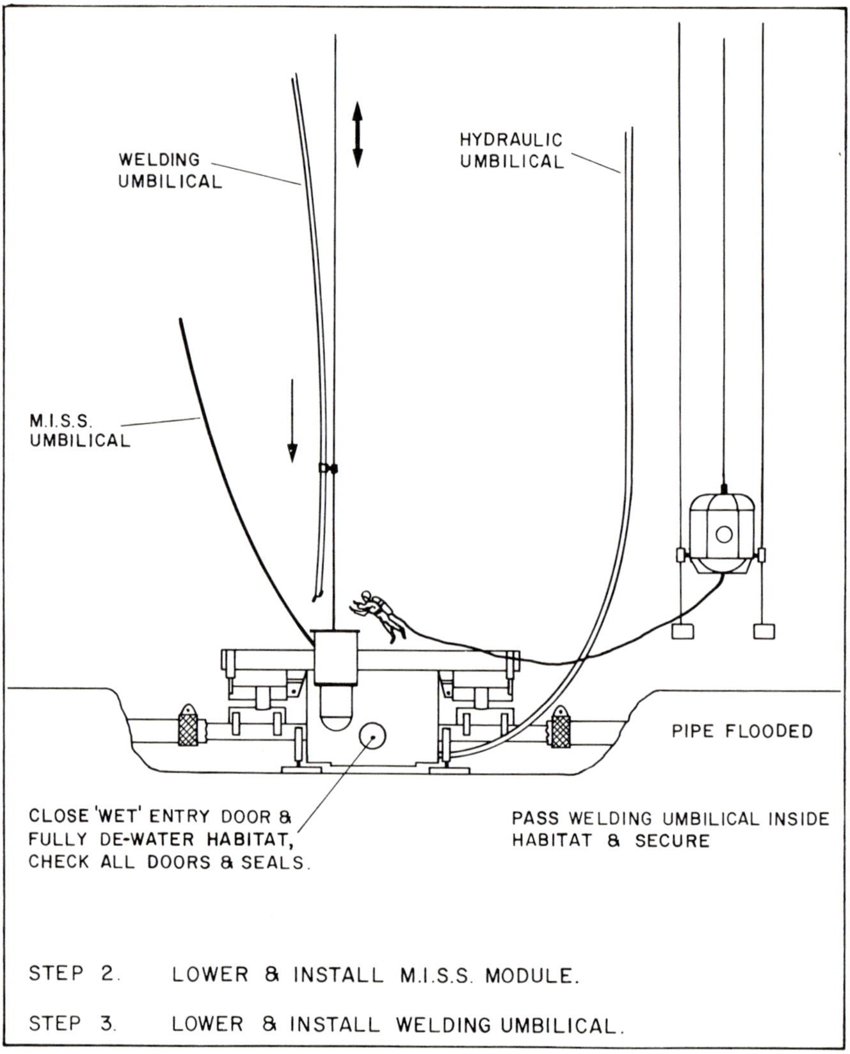

Figure 13.2.4 Steps 2 and 3 of the Pipeline Repair Sequence

Therefore any technique which could be developed either to prevent flooding, or at least minimise the extent of the flooding, could have significant benefits all round. Ignoring the effects of waiting on weather, which could be quite variable depending on the activity in progress at any time, there is probably little saving that can be made on the initial reconnaissance activities in locating the leak. The problems of unburying the lines have already been mentioned and in this respect the dredging equipment already provided within the EPRS is considered a reasonably economic option for the time being.

The removal of concrete and the coal-tar enamel coating is a time con-

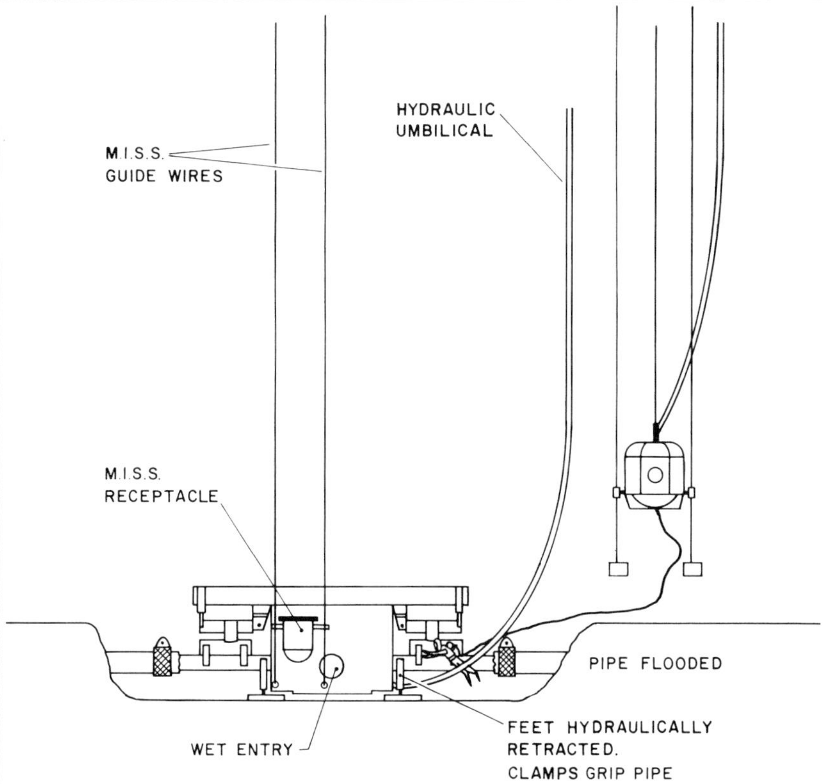

STEP

4 HYDRAULIC UMBILICAL LOWERED & CONNECTED.

5 LOWER SEA HORSE 2 ON THE PIPE – CLOSE CLAMPS.

6 CUT EITHER SIDE OF THE TEMPORARY CLAMP TO PUP-PIECE LENGTH, EVACUATE SCRAP PIECE.

7 POSITION & INFLATE SPHERES, INSTALL HABITAT DOOR & SEAL, CONNECT M.I.S.S. GUIDE WIRES ON CONSTANT TENSION.

Figure 13.2.5 Further Steps in the Pipeline Repair Sequence

suming operation which must nevertheless be done carefully. By the same token, there is little that can be done to speed up the hyperbaric repair phase, given that all of the equipment has been on stand-by and that fully qualified divers are available. However, when one considers the five or six days required to flood the pipeline followed by perhaps 25 days after the repair weld to de-water, clean and dry the lines then there is obviously merit in eliminating the flooding requirement or at least minimising its effects.

If a gas pipeline is ruptured, there is an obvious loss of gas through leak-

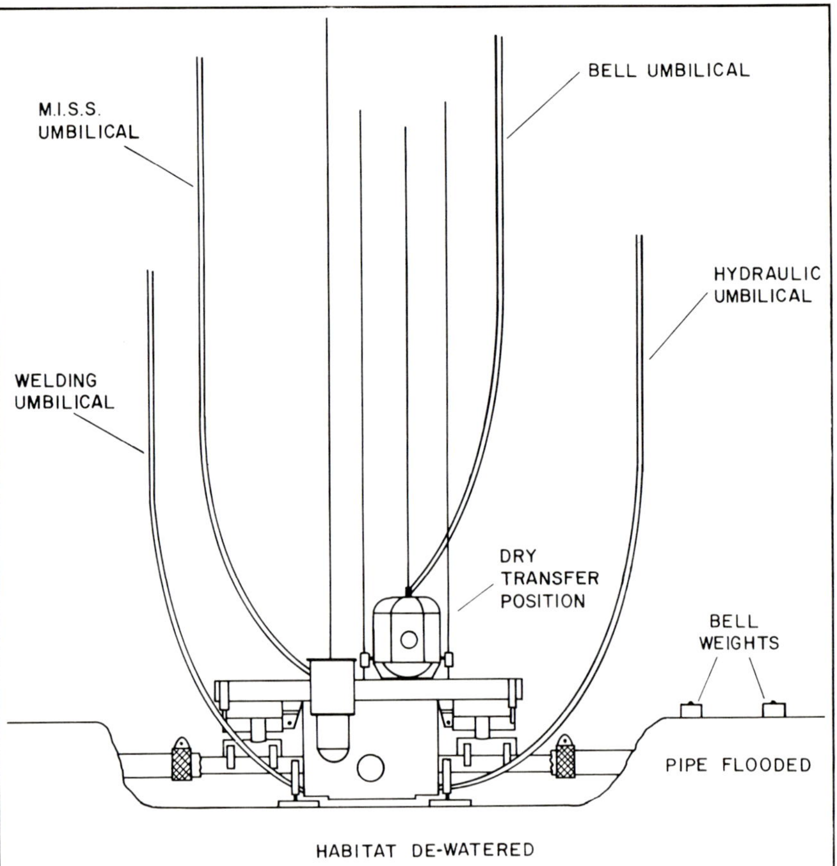

STEP

8	BELL GUIDE WIRES TRANSFERRED TO HABITAT.
9	WELDERS & FITTERS DRY TRANSFER TO HABITAT.
10	PUP-PIECE FITTED & CLAMPED FOR WELDING.
11	WELDING IS COMPLETED, GAMMAGRAPHY IS COMPLETED.
12	PIPE COATING IS MADE GOOD.

Figure 13.2.6 Remaining Steps in the Pipeline Repair Sequence

age. With operating pressures up to 150 bars, even the immediate closure of isolating valves at each end of a 180-kilometre pipeline section means that many hours will elapse before the gas pressure decays to seabed ambient. With this in mind, one could make use of the elasticity of the gas by launching pigs from both ends of the pipeline and allowing them to converge on the site of the rupture. Provided a means could be found to stop the pigs at an appropriate location and they could lock themselves in

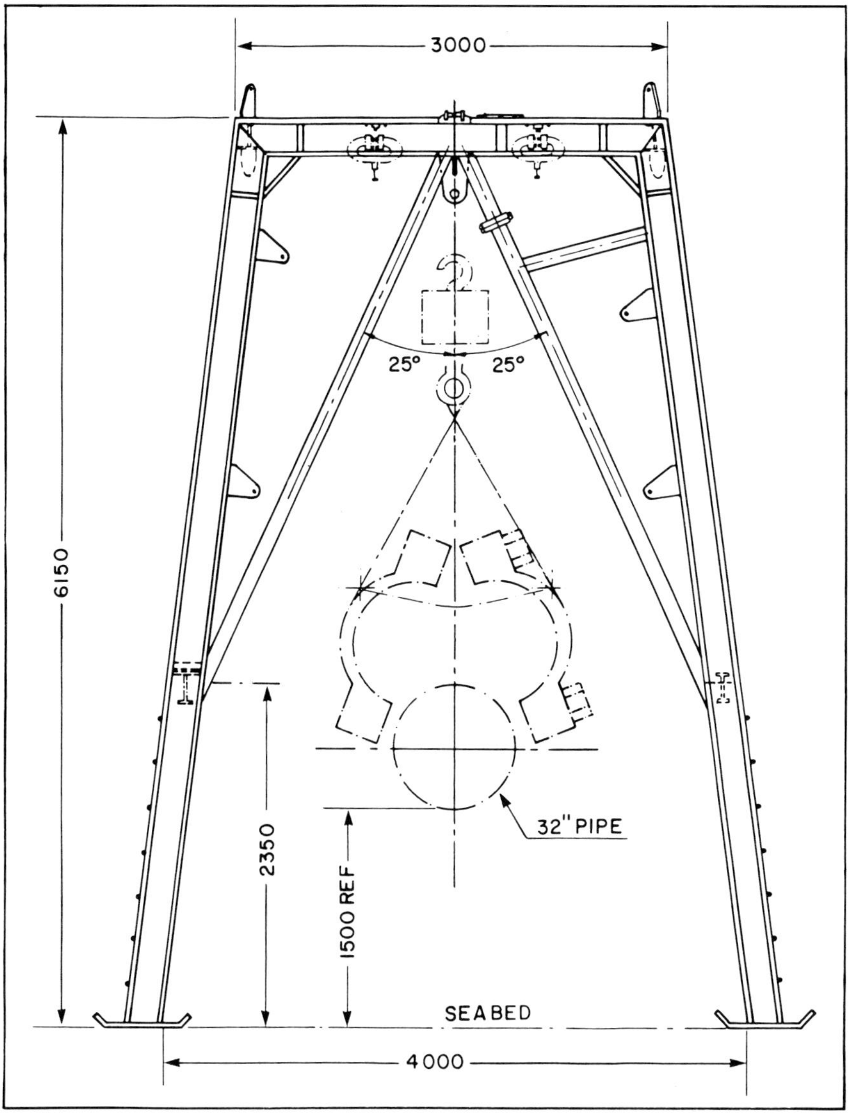

Figure 13.2.7 Gantry Frame Used to Install a Plidco Split Sleeve

a sealing position, a major time saving could be made on the overall repair cycle.

A few years ago there were strong doubts as to the feasibility of devising a pig to carry out these tasks. Therefore attention was focused on the concept of tapping into the line adjacent to the rupture, by so-called 'conventional methods' such as the well-established T D Williamson systems. Instead of cutting into the pipe 'hot', the operation could be carried out 'cold' at virtually seabed ambient conditions.

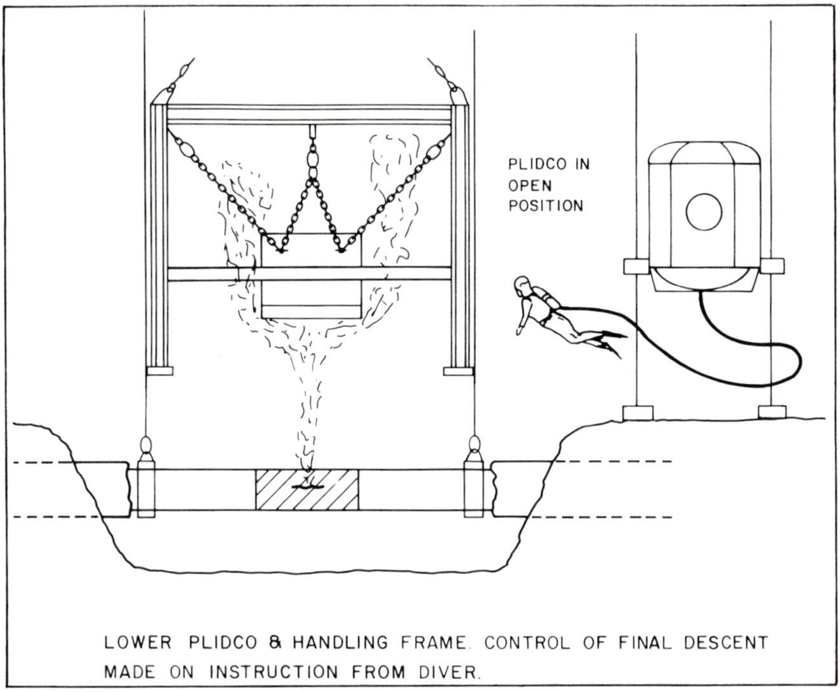

Figure 13.2.8 Steps in the Sequence of Repairing a Pipeline Using a Plidco Split Sleeve

13.4 Cold Tapping Project

Total Marine Norsk A/S started work on the project in October 1980 with a joint venture between Kongsberg Vapenfabrikk and Comex Services as main contractors. The cold tapping operation is performed in two separate stages:

(i) the oblique boring of the pipe by a trepanning cutter;
(ii) the installation of an inflatable plug in the pipeline.

Figure 13.4.1 (overleaf) shows the sequence of the operation.

The main handling frame supports the saddle which is clamped around the pipe. The saddle is provided with a 45-degree branched connection on top of which is mounted a sandwich valve. The sandwich valve is part of the tapping equipment and is permanently bolted to the saddle. The tapping machine, which cuts through the pipewall at 45 degrees, is a specially adapted version of the 1200 T D Williamson tapping machine. Weighing some 4.2 tons, the assembly includes the hydraulic drive systems, an adapter, and a specially designed cutter/pilot assembly.

When the cutting operation has been completed the steel coupon is withdrawn and the sandwich valve is closed to isolate the pipeline. The tapping equipment is then disconnected from the sandwich valve and replaced by the plug launching box.

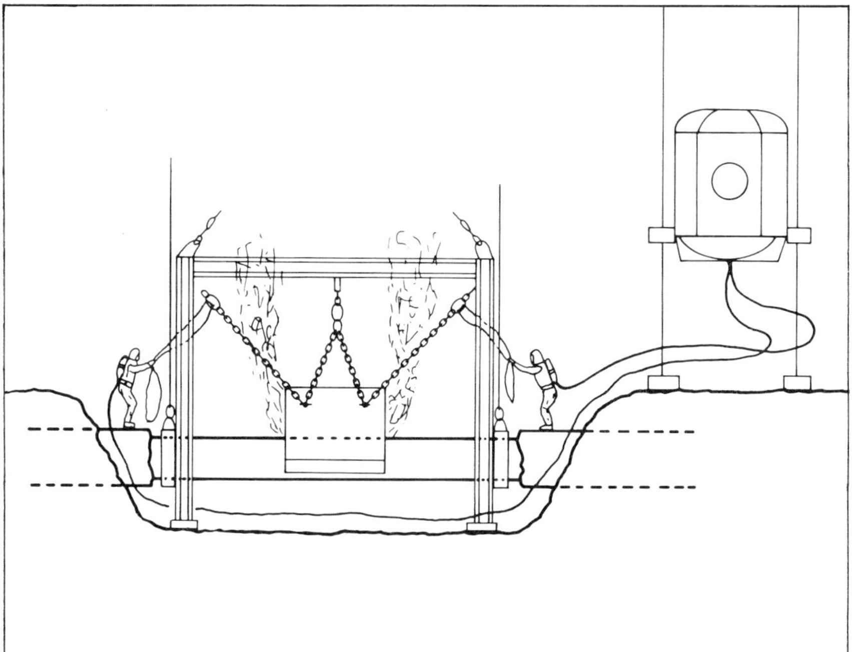

PRIOR TO FINAL POSITIONING – PIPE TO BE CLEANED BY H.P. WATER JETTING. PLIDCO CAN BE MANOEUVRED INTO FINAL POSITION BY THE USE OF CHAIN BLOCKS BY THE DIVERS.

Figure 13.2.9 Remaining Steps in the Sequence of Repairing a Pipeline Using a Plidco Split Sleeve

The launching box is designed to install the inflatable plug inside the pipeline. It comprises a pressure vessel section containing guides for the pig and space for the pressurising hoses. The launching mechanism is housed inside the pressure vessel and consists principally of hydraulic ram and linkage mechanisms. The inflatable plugs have been the subject of much development work by Kleber Industries. They have to be flexible enough in the uninflated condition to traverse the 45-degree insertion geometry and yet able to withstand an inflation pressure of 15 bars. Tests have shown that the plug is able to seal the pipe with a 20 bar differential pressure.

Once the plugs have been inserted on each side of the damaged pipeline section, the entire damaged section including the two complete sets of cold tapping equipment is removed. The replacement spool piece is than brought into position and welded into the line by conventional hyperbaric welding methods. It is obvious that mechanical coupling methods could be used instead of hyperbaric welding.

The extensive research and development work culminated in a successful trial at 141 metres water depth near Bergen on 9th, 10th and 11th December, 1983. The entire operation took only 32 hours including recovery of the handling structure by the diving support vessel.

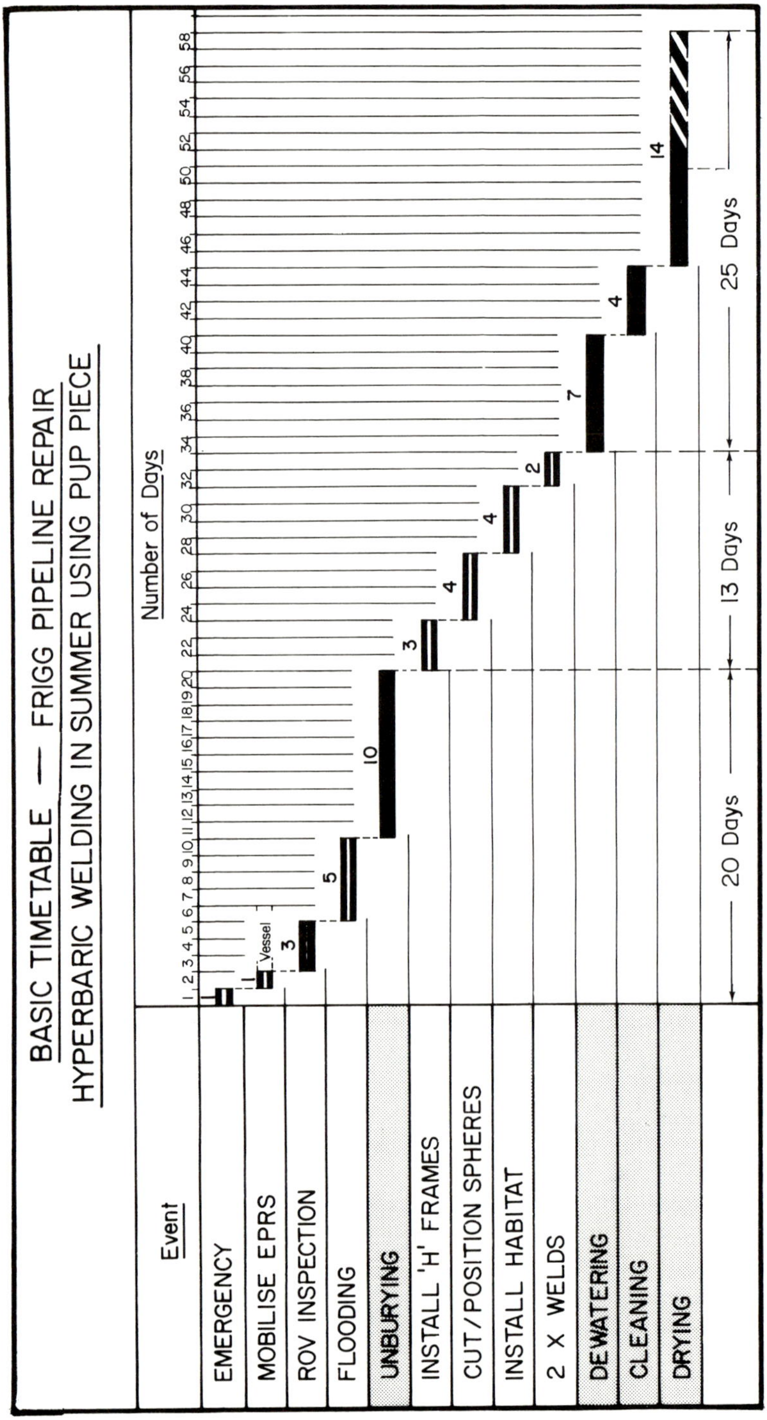

(VENTING, HYDROTEST, MISCELLANEOUS + 20%)

Figure 13.3.1 Notional Bar Chart for the Time to Repair a Frigg Pipeline Failure

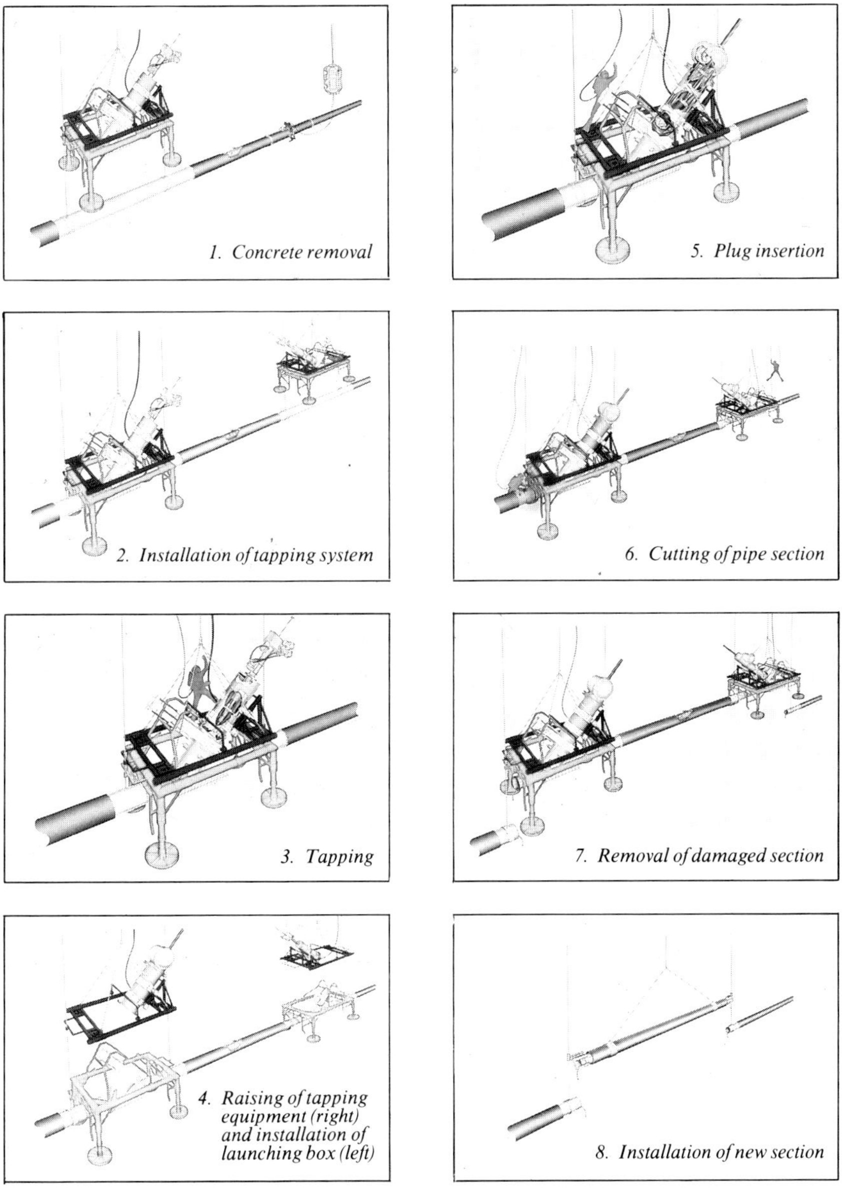

1. Concrete removal
2. Installation of tapping system
3. Tapping
4. Raising of tapping equipment (right) and installation of launching box (left)
5. Plug insertion
6. Cutting of pipe section
7. Removal of damaged section
8. Installation of new section

Figure 13.4.1 Sequence of Events Involved with a Pipeline Repair Using Cold Tapping

F J ALEXANDER
R QUIN

14

PIPELINE SAFETY ANALYSIS

In this chapter we concentrate our attention on the problems of pipeline safety and protection and address, *inter alia*, the conflicting issues which arise from trenching and backfilling, which also have major cost implications. To make the safety assessment as realistic as possible, we have used parts of the risk analysis undertaken recently for the new Danish Offshore Gasline, which is described elsewhere in this book. Later in this chapter, we shall have occasion to compare the results of this study with those obtained from a more global study of pipeline reliability, which follows in the next chapter. Throughout this chapter we refer to the trenched depth of a pipeline meaning the depth from the seabed to the top of the pipeline, whether exposed or buried.

14.1 The Importance of Pipeline Safety

The 30-inch gas pipeline, running from the Tyra field in the Danish sector of the North Sea, across Jutland, Funen and Zealand to Copenhagen and then to Dragør in Sweden, will be the only contractual means of supplying natural gas to Denmark and Sweden, under normal circumstances, for the foreseeable future. The 24-inch line running from Egtved, in mid-Jutland, down to the German system at Frøslev, will be used to supply the Danish system prior to commissioning the Danish North Sea gas pipeline in 1984, and will represent technically an alternative source of supply thereafter. However, this alternative source will only be available to the extent that there exists a German gas surplus to draw on. In effect, therefore, this means of supply cannot be guaranteed in the event of a winter breakdown of the Tyra-Egtved system. Nor does it provide any guarantee to consumers located east of the Danish belts in the event of failure of one of the belt links, across Funen and Zealand.

One of the fundamental differences between the failures of land and submarine pipelines is that a land pipeline can be repaired within approximately 48 hours, whereas a submarine pipeline may take one to two months to repair and recommission, even under favourable weather conditions. For these reasons both the Danish Energy Agency and the Danish Oil and Natural Gas Corporation (DONG) have given much consideration to the question of security of supply and have concentrated in particular on the safety of the marine pipelines.

Much attention has also been given to evaluating the hazards to persons and property and the general environmental impact in the event that a pipeline rupture should occur.

14.2 Shipping Damage to the Gasline Crossing the Great Belt (Storebælt)

Early in 1980, DONG, in consultation with the Danish Energy Agency, commissioned Det norske Veritas (DnV) to perform a risk analysis of the gas transmission marine pipelines as envisaged in the Project Concept (Revision 1) which was submitted to the Energy Agency as a part of the application for Approval in Principle of the system. That analysis covered the North Sea pipeline, the Danish Little Belt (Lillebælt) and Great Belt (Storebælt) crossings, which are described in Chapter 17. The Project Concept predicted a trenching requirement in the North Sea and Great Belt of 1 metre generally with no artificial backfilling, but deeper trenching and artificial backfilling were prescribed in the shore approaches and in the Little Belt crossing.

In March 1980, DnV reported the level of risk from various causes for each discrete section of the marine pipeline system and made recommendations for a number of design adjustments, with a view to risk reduction and mitigation. These recommendations are noted in DnV Report Number 80–0129 (1980). The Danish Hydraulics Institute Marine Pipelines Division, henceforth known as DHI, accepted this report as their starting point for detailed risk evaluation and pipeline protection design and most of its recommendations were adopted. Some of the risk figures, however, because of their implications, became the subject of more detailed investigations as described herein.

One of the conclusions of the DnV report was that ship accident risks, due to anchoring, grounding, and sinking, were a dominant factor in the overall risk appraisal, so the DHI risk analysis programme concentrated on this aspect. This was particularly the case in the Great Belt (Storebælt), which accommodates Transit Route T, the main shipping route between the North Sea and the Baltic, where shipping risks were a very important consideration for decisions on trenching depth, route location, the number of pipeline crossings and the spacing between them. As a result, the modelling techniques, which are referred to below, were originally formulated to assess the frequency of events causing damage to the pipeline crossing the Great Belt, though it should be noted that the same modelling techniques were applied later to the Little Belt crossing and the transmission line in the North Sea.

Because of the complexity of the models, they have been confined to the Appendices. Originally it was hoped to develop stochastic models, but it was discovered that a dearth of statistical information exists on shipping accidents involving pipelines. As a result, DHI found it necessary to develop deterministic models to interpret the incidence of ship aberration events in terms of the risk of damage to pipelines, using the two papers by Fujii (1971) and MacDuff (1974) as the base line for their study.

Ships anchor when their speed has fallen to a level where they have either lost or are about to lose steerage way. This loss of speed might result from a collision with another vessel or machinery failure. In either event, there exists a good chance that the vessel can be steered outside the fairway prior to dropping anchor or possibly running aground. Herein lies the potential danger to any pipeline which crosses a busy sea lane.

The calculations of the frequency of events leading to pipeline damage in the Great Belt resulting from ships anchoring, following machinery failure, are given in Appendix D and those due to ships anchoring, following collision, are given in Appendix E. These calculations are specific to a concrete-coated 30-inch pipe in the Danish Great Belt: calculations for other sizes of pipeline in other soils at different locations must be modified accordingly. In the interests of script economy, the corresponding calculations for the Little Belt crossing and the North Sea transmission line are not provided here but are incorporated in the DHI references provided later in the bibliography.

The development of a rational procedure for evaluating the risk of pipeline damage due to ships grounding proved even more difficult but had a direct bearing on the trenching depths in certain critical areas. This procedure is somewhat more complicated than the methods mentioned previously and it is presented in Appendix F.

The results of all three risk models are presented in Table 14.2.1 (overleaf). As a word of caution, however, it is thought that these models cannot assimilate accurately the real hazards of shipping. As such, the results of Table 14.2.1 are no more than order of magnitude indicators of the risks involved. Any conclusions derived from these results must be interpreted in this light and compared with results obtained by other means. A comparison of these results with those from the earlier DnV study is given in Table 14.2.2 (see over).

These risks are not distributed evenly along the length of the pipeline. The risk from anchor dragging following machine failure is concentrated in the Route T shipping channel, with an effective width of 4 kilometres. Ships which still have some steerage way will aim to anchor outside this main lane. By constrast, a study of the chart indicates that, for the alternative Route 4, the risk will be spread over a total lateral distance of 9 kilometres, yielding a damage event frequency of $2 \times 10^{-3}/9$ or 2.22×10^{-4} per kilometre per year.

The corresponding risk from anchor dragging following ship collision is $4.5 \times 10^{-5}/9$ or 5.0×10^{-6} per kilometre per year.

The risk from ship grounding is concentrated in the first shoaling zones outside the main shipping channel. The critical areas are between the 14 metre and 8 metre depth contours. The total length between these contours, excluding the zones in Musholm Bugt which are protected from the main traffic by Slettings Grund, is effectively 2.5 kilometres; 1.5 kilometres on the shoal east of Route T and 1 kilometre on the steeper shoal towards the Fyn shore. This yields a damage event frequency of $2.74 \times 10^{-4}/2.5$ or 1.1×10^{-4} per kilometre per year.

Table 14.2.1 Ship Accident Risks for the Great Belt Pipeline

	Anchoring event following engine failure	Anchoring event following collision of 2 ships	Ship grounding event	Total ship accident events
Annual probability of an accident event in the Storebælt passage (Røsnæs-Omø) for all ship sizes	4.7	0.025	2.5	7.23
Expected number of events in a 10-year period	47	0.25	25	72
Actual observed frequency of events per 10-year period	No information available	Only one event reported involving ships over 5,000 dwt	10 ships of over 5,000 dwt corresponding to 20 events for all ship sizes > 500 brt	
Annual frequency and return period of damage (including rupture) to a single 30-inch pipeline in the Project Concept Route	3.9×10^{-3} 256 years	$>4.5 \times 10^{-5}$ <22,222 years	2.74×10^{-4} 3,650 years	4.22×10^{-3} 237 years
Annual frequency and return period of damage (including rupture) to a single 30-inch pipeline in Route 4	$<2 \times 10^{-3}$ >500 years	$<4.5 \times 10^{-5}$ >22,222 years	2.74×10^{-4} 3,650 years	2.32×10^{-3} 431 years
Probability of damage (including rupture) event during 30-year design life. Project Concept Route Route 4	0.12 0.06	0.001 0.001	0.01 0.01	0.13 0.07

Table 14.2.2 A Comparison of the DHI and DnV Pipeline Failure Frequencies

STOREBÆLT ROUTE 4 ANNUAL FREQUENCY[1] OF RUPTURE OR SERIOUS DAMAGE FOR THE CROSSING AS A WHOLE		
Damage[2] caused by:	DHI	DnV
Anchoring after machine failure	2.0×10^{-3}	1.4×10^{-2}
Anchoring after collision	4.5×10^{-5}	
Sinking following collision	~ 0	5.9×10^{-3}
Grounding	2.7×10^{-4}	2.6×10^{-3}
	2.3×10^{-3}	2.3×10^{-2}

1. Frequencies are defined as the number of events per year per kilometre.
2. The DHI analysis makes the conservative assumption that all pipeline damage events deriving from ship accidents result in rupture. Since the ship accident risk is the predominant one in Tables 14.2.3 and 14.2.4, the corollary is that serious damage events, requiring shutdown for repair, not involving rupture, have a probability which is small by comparison with the rupture event probability. The DHI results therefore simplify the picture by showing the two combined and treating them as a predominantly rupture type risk. It should be noted however, that the DnV calculations maintain a separation of the rupture risk and the serious damage risk.

These risk levels are listed for Route 4 in Table 14.2.3, together with the other general risks, whose values have been adopted from the DnV report mentioned previously, and the corresponding risk values for the Project Concept Route are listed in Table 14.2.4 (overleaf).

For Route 4, the total risk of serious damage, including rupture, can be seen from Table 14.2.3 to be 4.3×10^{-3} per annum, which is equivalent to a return period of 233 years. The probability of such an event occurring within the 30-year design life of the pipeline therefore approximates 13 per cent. The highest level of risk per unit length of pipeline is seen to be 2.9×10^{-4} per kilometre per year. This compares favourably with typical landline figures of 2.3×10^{-4} per kilometre per year, reported by ASEA – ATOM (1979).

It should be noted that the risk level in the main channel, which is predominantly derived from anchor damage following ship machinery failure, is not sensitive to trenching depth, whereas the risk level on the shoals where grounding can occur is highly sensitive to the trenching depth assumption. If the pipeline lies untrenched on the seabed, the risk level in the critical ship grounding zones is substantially greater than the risk level for land pipelines.

The criteria used in the final selection between the Project Concept Route and Route 4 are summarised in Table 14.2.5 (see page 182). According to these results the superiority of Route 4 is obvious.

Table 14.2.3 Distribution of Risk Level along Route 4 Pipeline with Trenching Depth of 1 Metre from Seabed to Top of Pipe

Chainage (10)	KP 0.5–1.5	1.5–8	8–12	12–17	17–21.0	21.0–23.5	23.5–28.4	28.4–29.4	Total risk per year
Offshore activities	~ 10^{-5}	~ 10^{-5}	~ 10^{-5}	~ 10^{-5}	~ 10^{-5}	~ 10^{-5}	~ 10^{-5}	~ 10^{-5}	2.9×10^{-4}
Anchoring after machine failure	~ 0	~ 0	2.2×10^{-4}	2.2×10^{-4}	~ 0	~ 0	~ 0	~ 0	2.0×10^{-3}
Anchoring after collision	~ 0	~ 0	5.0×10^{-6}	5.0×10^{-6}	~ 0	~ 0	~ 0	~ 0	4.5×10^{-5}
Ship grounding	1.1×10^{-4}	0	0	0	0	1.1×10^{-4}	0	~ 0	2.7×10^{-4}
Anchoring intentional but position erroneous	~ 10^{-6}	~ 10^{-6}	~ 0	~ 0	~ 10^{-6}	~ 10^{-6}	~ 10^{-6}	~ 10^{-6}	2.0×10^{-5}
Trawling	~ 0	~ 0	~ 0	~ 0	~ 0	~ 0	~ 0	~ 0	~ 0
Dropping of heavy objects	~ 10^{-6}	~ 10^{-6}	~ 10^{-6}	~ 10^{-6}	~ 10^{-6}	~ 0	~ 10^{-6}	~ 10^{-6}	2.7×10^{-5}
External corrosion	4.5×10^{-5}	2.3×10^{-5}	2.3×10^{-5}	2.3×10^{-5}	2.3×10^{-5}	4.5×10^{-5}	2.3×10^{-5}	4.5×10^{-5}	7.4×10^{-4}
Other external loadings	—	—	—	—	—	—	—	—	—
Total external loadings	1.7×10^{-4}	3.5×10^{-5}	2.6×10^{-4}	2.6×10^{-4}	3.5×10^{-5}	1.7×10^{-4}	3.5×10^{-5}	5.7×10^{-5}	3.4×10^{-3}
Internal corrosion	3.0×10^{-5}	3.0×10^{-5}	3.0×10^{-5}	3.0×10^{-5}	3.0×10^{-5}	3.0×10^{-5}	3.0×10^{-5}	3.0×10^{-5}	8.8×10^{-4}
Other internal loadings	—	—	—	—	—	—	—	—	—
Total risk per km per year	†2.0×10^{-4}	6.5×10^{-5}	*2.9×10^{-4}	*2.9×10^{-4}	6.5×10^{-5}	†2.0×10^{-4}	6.5×10^{-5}	8.7×10^{-5}	4.3×10^{-3}

* These figures are unaffected by trenching depth
† These figures are highly sensitive to the local trenching depth

Table 14.2.4 Distribution of Risk Level along the Project Concept Route Pipeline with Trenching Depth of 1 Metre from Seabed to Top of Pipe

10

Chainage KP 0.5 1.5 11 17 18 19.5 21

							Total risk per year
Offshore activities	$\sim 10^{-5}$	$\sim 10^{-5}$	$\sim 10^{-5}$	$\sim 10^{-5}$	$\sim 10^{-5}$	$\sim 10^{-5}$	2.1×10^{-4}
Anchoring after machine failure	~ 0	~ 0	6.5×10^{-4}	~ 0	~ 0	~ 0	3.9×10^{-3}
Anchoring after collision	~ 0	~ 0	7.5×10^{-6}	5.0×10^{-6}	~ 0	~ 0	5.0×10^{-5}
Ship grounding	1.1×10^{-4}	0	0	0	1.1×10^{-4}	0	2.8×10^{-4}
Anchoring intentional but position erroneous	$\sim 10^{-6}$	$\sim 10^{-6}$	~ 0	~ 0	$\sim 10^{-6}$	$\sim 10^{-6}$	1.3×10^{-5}
Trawling	~ 0	~ 0	~ 0	~ 0	~ 0	~ 0	~ 0
Dropping of heavy objects	$\sim 10^{-6}$	$\sim 10^{-6}$	$\sim 10^{-6}$	$\sim 10^{-6}$	$\sim 10^{-6}$	~ 0	1.9×10^{-5}
External corrosion	4.5×10^{-5}	2.3×10^{-5}	2.3×10^{-5}	2.3×10^{-5}	2.3×10^{-5}	4.5×10^{-5}	5.3×10^{-4}
Other external loadings	—	—	—	—	—	—	—
Total external loadings	1.7×10^{-4}	0.4×10^{-4}	6.9×10^{-4}	0.4×10^{-4}	1.5×10^{-4}	0.6×10^{-4}	5.0×10^{-3}
Internal corrosion	3.0×10^{-5}	3.0×10^{-5}	3.0×10^{-5}	3.0×10^{-5}	3.0×10^{-5}	3.0×10^{-5}	6.3×10^{-4}
Other internal loadings	—	—	—	—	—	—	—
Total risk per km per year	†2.0×10^{-4}	0.7×10^{-4}	*7.2×10^{-4}	0.7×10^{-4}	1.8×10^{-4}	†0.9×10^{-4}	5.6×10^{-3}

* These figures are unaffected by trenching depth
† These figures are highly sensitive to the local trenching depth

Table 14.2.5 Parameters for Route Selection

		Project Concept Route	Route 4	Preference
ENVIRONMENTAL CRITERIA				
1. Route configuration	Length	21 km	30 km	PC Route
	Water depth, max.	39 m	28 m	Route 4
	Topography	Irregular	Smoother	Route 4
2. Landfall conditions	Soil condition	Location of a moor on Fynside	Location on Sjælland is ideal	Route 4
	Access condition	Bad on Sjælland Side	Fairly good	Route 4
3. Soil condition	Ratio boulder clay/mud	0.5	0.3 (clay mainly at Shore Approach)	Route 4
	frequency of boulders	Higher	Lower	Route 4
4. Hydrographic conditions	Waves	Same level	Same level	Equal
	Current	Jet effect from Eastern Channel	Presumably reduced jet effect	Route 4
5. Navigational aspects	Interim location of Route T	Bad possibilities	Fairly good	Route 4
	Pipeline construction	Reduced installation rates	Normal installation rate	Route 4
6. Obstructions	Cables	50 m from pipeline route	None	Route 4
	Sand reservoirs	None	None	Equal
	Great Belt Bridge	Close to pipeline route	Far from proposed bridge	Route 4
	Power plants	None	6 km to a proposed plant	PC Route
	Sewerage outlet	None	Feasibility study for one just N.	PC Route
SAFETY ASPECTS				
1. Anchor damage 2. Damage due to ship groundings 3. Population density		Accidental impact damage once /237 years. Repair operations could lead to winter obstruction of Route T shipping	Accidental impact damage once /431 years. Obstruction from repair operations less critical	Route 4 Route 4 Route 4
4. Risk during construction (collision)		Higher than for Route 4	Lower than for PC Route	Route 4
AUTHORITIES APPROVAL STATUS				
1. Status		Approved conditionally Farvandsdirektoratet call for anchor safe trenching depth	Not yet approved Farvandsdirektoratets requirements more flexible	PC Route
CONSTRUCTION COSTS (Dual Crossing) at April 1978 price. Trenching depth generally 1 metre from seabed to top of pipe. No engineered backfilling		Dkr 424 million	Dkr 400 million	Route 4
CONSTRUCTION TIMING				
1. Flexibility		June to August	All year	Route 4

14.3 The Choice of Trenching Depth in the Great Belt

When these previous risk levels are compared with those generally prevailing for land and marine pipelines, as cited by de la Mare (1980), which appear in the next chapter, it can be concluded that, with a general trenching depth of 1 metre, the level of risk associated with ship accidents as assessed by DHI is within the acceptable range. However, the question remains as to whether there are any reasonable steps which could be taken to further reduce or eliminate the ship accident risks.

The trenching depth required to eliminate all risk of pipeline damage due to anchoring is indicated below in Table 14.3.1.

Table 14.3.1 Anchor Penetration Depths

Ship size (dwt)	15,000 tons	60,000 tons
Anchor weight (tonnes)	5 t	10 t
Penetration in moraine clay (metres)	2 m	3 m
Penetration in mud (metres)	5 m	7 m

From these results it becomes apparent immediately that trenching to ‘anchor safe depths’ under these circumstances would not only be prohibitively expensive but impossible to achieve with ordinary construction methods using the present state-of-the-art.

However, in response to the continued concern of the Danish shipping authorities over the ship grounding risk immediately east of Transit Route T, at a location called Slettings Bank, DHI undertook a supplementary study which concentrated on the hazard to the ship’s crew and the environmental pollution problem in the event of an oil tanker running aground on the pipeline, it being known that oil tankers represent approximately one half of the ship traffic in Route T. This study built on a combination of the grounding risk computations above with statistical accident data from Inter-tanko (1980) and oil slick movement pattern data from the study by DHI (1976). The conclusions of this study are contained in DHI Report Number MPPO 31189. The main points were that the installation of a single gas transmission line in Storebælt on Route 4, with the trenching depths indicated in the Concept Proposal, would yield the following risk increases:

(i) the existing risk of tanker disaster, due to fire and/(or) explosion involving potential loss of life or serious injury to crew, would increase by 1.2 per cent;

(ii) the existing risk of an oil pollution event would increase by between 0.065 per cent and 1.3 per cent.

The frequency of any tanker grounding event involving the pipeline was estimated at once per 7,300 years, whereas for tankers larger than 40,000 dwt, the frequency was estimated at once per 217,000 years.

Despite the expected rarity of these events however, it was subsequently agreed that some risk reduction could be achieved within reasonable economic limits by trenching to a greater depth over a limited stretch of pipeline on the slopes of Slettings Bank.

14.4 Number and Spacing of Pipelines in the Great Belt

Whereas the average repair time in the event of damage to a land pipeline is of the order of one or two days, the repair time in the event of damage to a 30-inch diameter pipeline in the Great Belt is estimated at upwards of a month, including de-watering and drying. The risk of a closure for this duration was unacceptable in the context of securing a gas supply to Zealand and Sweden. Therefore, notwithstanding the fact that the frequency of failure of a single line in Storebælt is no worse than the overall failure rate for landlines, the consequences of such failure in terms of interruption of supply make it essential to duplicate this pipeline.

To avoid damage to both pipelines from the same accident event, the spacing between them must exceed the diameter of influence of any single event. The factors affecting the choice of spacing include:

(i) anchor dragging distances;
(ii) stopping length of grounding ships;
(iii) anchor spread from lay and bury barges; and
(iv) navigational considerations.

14.4.1 Anchor Dragging Distances

Under controlled anchoring procedures a ship will drop anchor prior to losing steerage way and drifting with the current. According to DHI Report Number MPPO 30462 (1980), the mean anchor dragging distance in this situation will be less than 200 metres even for the largest vessels passing Storebælt. This distance relates to anchoring in mud (gytja). The corresponding mean dragging distance in clay is less than 50 metres. The thickness of the mud layer on the seabed in Storebælt is generally less than 4 metres with boulder clay underneath. In view of the large anchor penetration depths in mud, most anchors will reach the boulder clay and the dragging distances will be less than those for deep mud.

The safe distance between two pipelines from the point of view of anchor damage is therefore of the order of 200 metres.

14.4.2 Stopping Distance of Grounding Ships

When a vessel grounds on the seabed, it continues its forward movement and penetrates the seabed. If the course of the ship in the grounding situation is parallel with the depth contours, the ship will slide on the seabed for a considerable distance before its ultimate penetration is reached. This situation results in the largest stopping distances, but the smallest ultimate penetration or 'bite'.

A complicated mathematical model, of questionable accuracy, was developed to calculate the depth of bite for various scenarios. This model is provided in Appendix F and some of its results appear in Table 14.4.2 below. It must be emphasised, however, that informed judgement considers the accuracy of these results to be no better than 50 per cent. On this basis, the safe pipeline separation in water depths of 8 to 12 metres should therefore be regarded as not less than 75 metres, where the seabed is clay and 525 metres where the seabed is mud. For water depths between 12 and 17 metres, where mud prevails, the corresponding safe distance must be regarded as not less than 900 metres.

Table 14.4.2 The Bite of Ships Grounding in Storebælt

Ship size (dwt)	Draught (metres)	Seabed soil	Effective depth of bite (metres)	Minimum safe pipeline separation (metres)
50,000	12	clay	1.8	50
50,000	12	mud	3.8	350
150,000	17	mud	5.0	600

14.4.3 Anchor Spread from Lay and Bury Barges

Navigational procedures for lay and bury barges normally require the spread of anchor positions perpendicular to the pipeline centreline from 100 to 1,500 metres. To avoid restricting their operations therefore, the distance between two pipelines should be either less than 100 metres or more than 1,500 metres. In effect, the spacings to avoid are in the range of 100 to 1,500 metres.

14.4.4 Navigational Considerations

From the point of view of the navigator of a vessel with machine failure, about to drop anchor, it is preferable for the two pipes either to be located as close together as possible (so they can be regarded as a single crossing) or alternatively be spaced several kilometres apart.

14.5 North Sea Pipeline Protection in the Platform Area

In their initial report on the integrity of the North Sea gaspipe, Det norske Veritas (1980) gave the probability of serious damage necessitating a shutdown of supply for maintenance or repair, due to offshore activities, as 2×10^{-2} events per kilometre per year in the kilometre of pipeline closest to the platform. The corresponding risk of rupture was given as 1×10^{-2} events per kilometre per year. In effect this means that the probability of a rupture of the pipeline in the platform approach area during the course of its 30-year design life was calculated as 30 per cent while the probability of damage sufficiently serious to warrant shutdown was 60 per cent. The potential cause of this damage was fouling by the high-holding anchors of offshore construction and repair vessels. These results were checked by DHI using the model which appears in Appendix G and the DnV results were validated on the basis of no special navigational precautions being taken in the neighbourhood of the platform.

Based on evidence by Det norske Veritas, and R J Brown & Associates, it is known that high-holding anchors, weighing 5 tonnes and more, can penetrate to depths exceeding 4 metres below the seabed in sands and considerably deeper in softer soils. If the depth of penetration of such anchors increases approximately as a linear function of the distance pulled, and if all the anchors contributing to the stated risk start on the surface and come to rest at a depth of 4 metres, then it can be deduced by simple geometry that an increase of pipeline trenching depth from 1 metre to 2 metres will reduce the chance of anchor interference by two thirds. However, this level of risk is still considerably more than an order of magnitude above any acceptable level and can only be reduced to an acceptable level if the pipeline is trenched to a depth of at least 4 metres from seabed to top of pipe. For complete security the depth would have to be 5 metres or more.

There is no evidence that the risk of anchor damage can be reduced by substituting artificial backfill with selected material in place of natural backfill with sand, nor is there any evidence that the risk can be reduced by providing an excess of backfill above seabed level. This is not to say that such measures might not be marginally beneficial in reducing the risk, but their effect is thought to be of very doubtful magnitude and cannot at this time be evaluated. In these circumstances, their adoption could not be justified as effective means for reducing the pipeline damage event frequency. The use of artificial backfill in the form of gravel, stone or rock also compounds the difficulties and delays referred to below in connection with emergency repair works, and thereby reduces the security of supply. Furthermore, the provision of protective matting would not contribute any reduction to the risk of damage from high-holding anchors.

It is therefore concluded that the only means, which are theoretically possible, of protecting the pipeline physically, as opposed to operationally, are to trench the pipeline to a depth of at least 4 to 5 metres. Currently this is infeasible and would certainly be exorbitantly expensive. Moreover, there is another consideration which makes it questionable to trench so deeply, even if it were possible to do so. According to DHI Report Number MPPO 31372 (1981), in the case of a backfilled pipeline an increase of the trenching depth from 1 metre to 4 metres increases the repair time in the event of pipeline damage by 18 days as a consequence of the enormously increased quantity of excavation involved and also increases the repair cost.

In summary, the risk of pipeline damage or rupture in the platform area cannot be reduced to any substantial degree, far less to the same level as that for the rest of the transmission system, by increasing the trenching depth within technically and economically feasible limits. Increasing the trenching depth within feasible limits, to a depth of 2 to 3 metres, while yielding some improvement, would not yield a decisive improvement in risk level, and such improvement as would be obtained would be offset by the increase in repair time, and therefore of supply outage time, in the event of accidental rupture or corrosion-induced damage. Further, such protection would also compound the difficulties associated with the design of the expansion dog-leg.

It was therefore determined that the trenching depth requirement in the platform area should be the same as the rest of the pipeline west of chainage 80.00, with a trenched depth not less than 0.2 metres.

It was postulated, however, that the implementation of strict control on all vessel movement and vessel anchoring in the Tyra East platform area would obtain a reduction of the hazard rates, possibly by as much as an order of magnitude. The provision of fixed anchoring points close to the platform in the vicinity of the export pipeline and the supervision of all anchoring by divers, taken together with strict vessel control, could realise such a reduction. These operational procedures are listed in Appendix H. Although such measure greatly reduce the risk, it would still be well over an order of magnitude higher than that prevailing along the rest of the North Sea pipeline. For this reason, methods to mitigate the consequences of pipeline failure were sought and these are discussed below.

14.6 Mitigation of the Effect of Pipeline Rupture in the Platform Area

It is reckoned that the release of sour gas from the rupture of the submarine pipeline would not give rise to any significant risk of hydrogen sulphide poisoning for platform personnel or for the crew of offshore vessels, since the majority of the H_2S would be scrubbed from the gas by the seawater. Even during the initial explosive burst of the gas, during which some H_2S would reach the surface, the danger from H_2S poisoning would be extremely small by comparison with the fire, explosion, and radiation hazards which would ensue.

According to the results of the safety study contained in DHI Report Number MPPO 12554 (1980), there would also be little risk of methane suffocation, and this risk would decrease with increasing distance between the point of emanation, from the rupture, and the platform. The extent of the fire, explosion, and radiation risk is indicated in Table 14.6.1 which is based on information provided by Cowiconsultants (1981).

Table 14.6.1 Extent of Fire, Explosion and Radiation Hazard following a Submarine Rupture of the DONG Pipeline in the Platform Area Assuming the Unrestricted Blowdown of 215 Kilometres of Gas at 138 Bar

Total duration of gas release: 22 hours

HAZARD	EXTENT OF HAZARD AREA SOON AFTER RUPTURE EXPRESSED AS DISTANCE IN METRES FROM RUPTURE LOCATION	
	Typical weather conditions	Excellent weather conditions
(a) Risk of ignition of windswept gas cloud resulting in fireball affecting circle of diameter 330 metres	500–1,500	4,000
(b) Risk of death within a few minutes due to radiation from burning plume in absence of shelter	500	500
(c) Spontaneous ignition of wood due to radiation from burning plume	500	500

Excellent weather conditions, in the context of Table 14.6.1, relate to conditions of high atmospheric stability and very low wind speed. These conditions are extremely rare at Tyra East, and the probability of a pipeline rupture coinciding with such conditions are negligible. In practice, therefore, the range of the migrant gas ignition risk is seldom more than 500 to 1,500 metres, while the maximum range of radiation risk is of the order of 500 metres.

Table 14.6.2 indicates the extent of the corresponding risks when the blowdown length is restricted to 1.5 kilometres instead. In this event it can be seen that the geographic extent of the risk is only slightly improved. However, the blowdown time is only 70 seconds compared with 22 hours for the whole length of the pipeline. This means that the probability of serious consequences is much reduced, not least because the wind is likely to have a virtually constant direction during a 70-second blowdown and the chance of that direction being towards the

Table 14.6.2 Extent of Fire, Explosion and Radiation Hazard following a Submarine Rupture of the DONG Pipeline in the Platform Area Assuming Free Blowdown of 1.5 Kilometres of Gas at 138 Bar

Total duration of gas release: 70 seconds

HAZARD	EXTENT OF HAZARD AREA SOON AFTER RUPTURE EXPRESSED AS DISTANCE IN METRES FROM RUPTURE LOCATION	
	Typical weather conditions	Excellent weather conditions
(a) Risk of ignition of windswept gas cloud resulting in fireball affecting circle of diameter 260 metres	400–1,100	3,200
(b) Risk of third degree burns due to radiation from burning plume in absence of shelter	300	300
(c) Spontaneous ignition of wood due to radiation from burning plume	200	200

Tyra East flare is limited. By contrast, wind veer during a 22-hour blowdown would stand a good change of bringing the gas plume within ignition range of the flare at some stage.

If the scenario described in Table 14.6.1 could be restricted to ruptures of the pipeline further than 1,500 metres from the platform flare, (so they were beyond the range of ignition risk from flare or other platform sources during normal weather conditions, and also beyond the area of high risk of anchor interference associated with offshore activities) and ruptures closer to the platform could be restricted so they were associated only with the short-duration scenario of Table 14.6.2, then the consequences to the platform and its operational personnel of a pipeline rupture would be reduced considerably. Furthermore, the risk to the offending vessel and other vessels in the vicinity would also be reduced. Such a restriction implies the installation of a subsea check valve or other fast-closing auto-actuated valve, to prevent the egress of gas contained in the line from its point of rupture to its landfall. The installation of such a check valve realises the following additional benefits.

14.6.1 A Reduction in the Value of Lost Gas

Under typical operating conditions the pipeline will contain 8×10^6 NM^3 of gas worth DKr 15 million at 1981 prices. This is the immediate financial loss in the event of a total uncontrolled blowdown following rupture. Subject to the implementation of the recommendations of Appendix H, the probability of a rupture in the high risk platform approach area during the 30-year design life is estimated as three per cent. This corresponds with a present value of approximately DKr 0.23 million on the basis of a discount rate of five per cent per annum in real terms over the 30-year period. Effectively therefore, this value sets a limit to the cost of installing such a valve system, if its justification were based only on the reduction in gas losses.

14.6.2 A Reduction in the Pipeline Downtime

The repair of a rupture, or serious damage short of rupture, usually involves sealing off and flooding of the relevant length of pipeline. One of the more time consuming elements of such operations is the subsequent de-watering and drying process. Estimates of typical depressurising, flooding, de-watering, and drying times for the 215 kilometre pipeline are given in DHI Report Number MPPO 30877 (1980) and amount to 37 days in addition to the 15 days actually needed to repair the pipeline itself. The introduction of a check valve, however, while contributing significantly to platform safety and preventing loss of purchased gas, would not contribute to any reduction of the outage time during a pipeline repair operation, since it is impracticable to use a check valve to limit the extent of flooding. With a check valve alone, the repair of the rupture upstream of the valve would almost inevitably involve flooding the whole length of the submarine pipeline, thereby resulting in a lengthy pipeline drying process. However, the provision of a ball valve downstream of the check valve would contribute to a substantial reduction in the downtime during repair of damage or rupture in the high risk zone or during repair or replacement of the check valve. In this event, it is estimated that the corresponding time required for depressurising, flooding, de-watering, and drying would be cut from 37 days to ten days. At 1981 sales prices, the value of 27 days of lost sales amounts to DKr 435 million. Subject to implementation of the recommendations in Appendix H, the probability of serious damage, with or without rupture, in the high risk zone during the 30-year design life is six per cent. The sum to be compared with the cost of installing the ball valve is therefore DKr 13 million. In other words the main commercial, as opposed to safety, justification of providing a subsea valve assembly is to be found in outage time reduction rather than in the prevention of the escape of purchased gas, provided of course that the valves themselves do not present any adverse conditions due to problems associated with their reliability and maintainability. This point was investigated thoroughly in discussion with manufacturers, operators, and consultants and the information obtained is reported fully in the DHI Report Number MPPO 31384 (1981).

Its main conclusions are:

(i) Wheatly check valves have been used extensively subsea on a world-wide basis but particularly by US operators in the Mexican Gulf. Very little trouble has been experienced. They have been used extensively in sour gas service without problems.

(ii) Check valves should be provided with an external hinge pin so that they can be operated manually and locked open.

(iii) Cameron ball valves have been used extensively subsea on a world-wide basis. They have been used extensively in sour gas service without problems. Difficulties are normally confined to remote actuation systems provided by others.

(iv) Ball valves should be actuated through a double reduction worm gear by a diver using a hydraulic or pneumatic nut runner, or alternatively through a hydraulic actuator from a surface hydraulic power source connected up by diver, the former method being considered more reliable. Any remote actuation, if provided, should be a secondary system for grease injection, test closing, and backup.

None of the manufacturers or operators advised against the subsea use of check or ball valves in sour gas service or foresaw any serious problems with the configuration envisaged by DHI.

On this basis, it was decided that a subsea check valve should be installed at a point not less than 1,500 metres from the Tyra East flare and a ball valve should be incorporated in the pipeline immediately downstream of the check valve using diver operated power assisted tools as the primary means of actuation. It was further decided that the valve assembly should be protected against siltation, trawl board impact, and snagging by the cables of maintenance vessel anchors, though it is infeasible to protect them against anchor damage.

14.7 A Total Systems Approach to Trenching Requirements for the North Sea Gasline

So far in this chapter, we have concentrated our attention mainly on the shipping hazards affecting the integrity of the DONG pipeline, and we have explored the merits of trenching to reduce such hazards. In this final section, we discuss briefly the depths of trenching required along the route of this North Sea gasline to mitigate the effects of other hazards, some man-made, others natural. Much of this section is devoted to the propagation of sand-waves and their tendency to expose a previously buried pipeline. Also discussed, however, are the effects of stability, scour, fishing activities, the influence of an artillery range close to the pipe-route and lastly a few brief notes concerning anchors and ships' grounding.

14.7.1 The Effects of Migrating Sand-waves

Since the migration of sand-waves with any appreciable wave height can lead to pipeline exposure, spanning and problems of instability, it was considered important to investigate the likelihood of such a phenomenon and determine which zones of the pipeline could be affected. These investigations are described in DHI Reports Numbers MPPO 31082 (1980), MPPO 31083 (1980), and MPPO 31069 (1980).

A comparison of the echo soundings from the 1977 and 1980 route surveys shows a reasonably good correspondence and does not yield any evidence of detectable changes in the seabed levels. However, the migration of offshore sand-waves can be a slow process, and the variations which occur in a three-year period could well be masked by a combination of the following factors:

(i) inaccuracies in the survey vessel position logging systems used in 1977 and 1980;
(ii) inaccuracies in the depth measurements taken in 1977 and 1980;
(iii) inaccuracies of the tidal elevation reduction in 1977 and 1980;
(iv) variation in the subjective interpretations of the echograms at critical points.

Consequently no reliance has been placed on the comparison of the 1977 and 1980 surveys. Instead, the evaluation has been based on an interpretation of the 1980 survey echograms in the light of a theoretical analysis of the potential for sand-wave formation and migration.

It has been found from the echograms that in the vicinity of the 30 metre depth contour significant depth variations appear between the different survey runs at given chainages, indicating the presence of sand-waves with crests and troughs running east-west, implying a north-south sand-wave migration. The theoretical analysis confirmed that the potential for sand-wave migration is at its greatest in precisely this area.

Having established that sand-waves with amplitudes of up to 2.5 metres and wavelengths of 360 to 500 metres do occur in the 30-metre depth contour area (chainage 59.000 to 81.000), an assessment was made of the probable celerity of their migration. Calculations were made on the basis of three alternative sets of assumptions, resulting in a spread of results of considerably less than one order of magnitude. The assumptions considered most likely to be correct yielded a celerity of the order of 4 metres of horizontal movement of the wave crest per year.

These results were used to predict the seabed level variation which can be expected within the pipeline design life of 30 years. In this way the maximum likely variation has been assessed at 1.7 metres.

A similar evaluation was performed for the Tyra platform area, where the depths and hydrographic conditions could also be critical. It was found, however, that the effect had a much smaller magnitude in this area. Although wave amplitudes of 2 to 3 metres with wavelengths of the order of 2,000 metres have been observed from the echograms, the calculated migration celerity under the influence of the hydrographic conditions in this area is only 0.7 to 1.2 metre per year. Coupled with the much larger wavelength, this yields an estimated seabed level variation of only 0.1 metres in 30 years.

Sand-waves of smaller amplitude have also been observed at other locations on the pipeline route, giving rise to the maximum probable seabed level variations tabulated below in Table 14.7.1.

Table 14.7.1 Probable Maximum Seabed Level Variations during 30 Years due to Sand-wave Migrations

Chainage	Probable seabed level variation during a 30-year period
1,700 – 59,000	0.7 metres
59,000 – 81,000	1.7 metres
81,000 – platform	0.5 metres

There is no practicable way of impeding or diverting the progress of sand-waves of this type. The only feasible protection measure is to trench the pipeline to a depth equal to the maximum anticipated seabed level variation during its design life and to monitor regularly the progress of the sand-waves and the depth of the pipe. Insofar as it may prove impracticable to achieve the necessary trenching depth in the area of maximum sand-wave activity initially, the need to retrench the pipeline could arise during its design life. The initial trenching depth should, however, be such that seabed level variations, of the magnitude predicted in Table 14.7.1, will not result in a reduction of the pipeline depth relative to the instantaneous seabed position to the extent that the pipeline becomes unstable or is subjected to an increase of risk to an unacceptable level.

14.7.2 Pipeline Lateral Stability

Calculations demonstrate that with the chosen weight coating design the pipeline will be stable provided that the top of the pipe protrudes less than 0.3 metres above the surrounding seabed level during the pipeline operational lifetime, even when the pipeline trench is not initially backfilled. These calculations are provided in DHI Report Number MPPO 31150 (1981).

14.7.3 Scouring Effects

Scouring will be prevented providing the gaspipe does not protrude more than 0.3 metres above the surrounding seabed. Calculations in DHI Report Number MPPO 31123 (1980) show that backfilling is not required with the expected dimensions of the intended trench.

14.7.4 Fishing Activities
Trenching is not required for protection of the pipeline against damage from fishing gear because sufficient protection is provided by the high strength concrete weight coating required for pipeline stability.

14.7.5 High Holding Anchors
As already mentioned elsewhere in this chapter, it is infeasible to provide protection by trenching against damage from high-holding anchors associated with offshore activities, the risk of which is high in the immediate platform area. It is therefore necessary to take the other steps to mitigate the consequences of such damage, which have been discussed in detail already.

14.7.6. Ship Grounding
The ship grounding risk, associated with the North Sea gasline has been assessed using the theory developed in connection with the Storebælt crossing studies, subject to modifications which take account of the different seabed soils. It has been decided that, on average, the top of the pipeline should be flush with the seabed to give a sufficiently low probability of a grounding ship interfering with the pipeline. Furthermore, backfilling will not be required.

14.7.7 Artillery Shell Explosions
It is a fact that a proportion of the intended pipeline route will be located close to an artillery range. However, studies of the effect of artillery shell explosions have led to the conclusion that the risk of explosion damage to the pipeline will be acceptably small if the top of the pipe is 1.5 metres below the seabed level and the pipeline trench is backfilled in this particular area.

R S COLQUHOUN
J E THYGESEN

14.8 Bibliography

ASEA-ATOM, *'Sjöledning Danmark-Sverige – Haveriorsaker og Risikobedömming'*, (March 1979).

Cowiconsult, *North Sea Pipeline – Environmental Impact Study*, (April 1981).

de la Mare, R. F. and Øystein Andersen, Ø., *'Pipeline Reliability'*, Report No. 80–0572, Det norske Veritas, (August 1980).

Det norske Veritas *'Sikkerhetsanalyse for DONG's Sjøledningsprojekt*, Report No. 80–0129, Det norske Veritas, (March 1980).

DHI, *'Damage due to Anchoring'*, MPPO 30462: DHI Marine Pipelines Technical Note, (June 1980).

DHI, *'On the Possibility of Modelling the Spread and Transport of Oil at Sea'*, (September 1976).

DHI, *'Storebælt Crossing – Risk and Consequences of Tanker Grounding on Pipeline'*, MPPO 31189: DHI Marine Pipelines Technical Note.

DHI, *'Storebælt Crossing – Concept Proposal'*, MPPO 30720: DHI Marine Pipelines Report, (September 1980).

DHI, *'North Sea Pipeline – Protection Concept'*, MPPO 31372: DHI Marine Pipeline Report, (May 1981).

DHI, *'Environmental Impact Studies Phase 2, Task 2 – Outflow of Gas after Rupture of Marine Pipeline'*, MPPO 12554: DHI Marine Pipeline Report, (November 1980).

DHI, *'Subsea Valve Study*, MPPO 30877: DHI Marine Pipeline Report, (November 1980).

DHI, *'Subsea Valve Assembly'*, MPPO 31384, DHI Marine Pipelines Report, (October 1981).

DHI, *'Sand-waves in the North Sea – Discussion – Status Report'*, MPPO 31082: DHI Marine Pipelines Report, (October 1980).

DHI, *'Migration of Seabed Undulations and Level Changes (30 Metre Water Depth)*, MPPO 31083: DHI Marine Pipelines Report, (November 1980).

DHI, *'Sand-wave Migration at the Platform Area (Tyra)'*, MPPO 31069: DHI Marine Pipelines Report, (December 1980).

DHI, *'North Sea Pipeline – Lateral Stability'*, MPPO 31150: DHI Marine Pipelines Report, (January 1981).

DHI, *'Local Scour at Horizonal Pipe'*, MPPO 31123: DHI Marine Pipelines Report, (December 1980).

DHI, *'Load on the Gas Pipeline from Fishing Activities'*, MPPO 31497: DHI Marine Pipelines Report, (January 1981).

Fujii, Y., 'The Analysis of Traffic Accidents', *Journal of Navigation*, (1971).

Inter-tanko, *'Statistics of Tanker Casualties'*, International Association of Independent Tanker Owners, (Oslo, 1980).

MacDuff, T., 'The Probability of Vessel Collisions', *Ocean Industry*, (September 1974).

15

A STUDY OF PIPELINE RELIABILITY

The motivation for undertaking this study partly arose from a commercial need to assess the reliability of a particular pipeline and to determine the sensitivity of its safety to varying degrees of protective covering. It also arose because no quantitative study which could truly claim to be both rigorous and comprehensive had previously been published on the subject.

The aim of this study was to determine the reliability characteristics of North Sea pipelines. However, during the period studied, only two incidents which could qualify as pipeline failures in the context of this report occurred in the North Sea, giving rise to an average failure rate of 0.7 failures per annum per one thousand kilometres of pipeline and a 95 per cent chance that the true average failure rate was in the region of 0.067 to 2.50.

Resulting from the paucity of these failures, the scope of the study was extended to encompass other major oil and gas pipeline systems throughout the world. The rationale of this approach resulted from the hypothesis that a careful comparison of the different types of pipeline systems would reveal a sufficient similarity in their failure attributes, and that this knowledge could be used to extend the rather limited data made available for North Sea pipeline failures. As a result, the major effort of this study was to elucidate the failure characteristics of oil and gas pipelines in general.

To a large extent, the progress of this study was hindered by the inability to obtain detailed failure data in the time and with the resources available. Having completed the study, the researchers are confident that such data exists, but whether or not it could be released is probably a matter of commercial confidence. To some extent, therefore, this report is not as comprehensive and rigorous as the authors would have wished and they hold certain reservations regarding the scientific validity of the results. However, in these circumstances, and with a judicious application of statistical techniques, it is felt that these results and their conclusions are fair and reasonable.

One of the conclusions of the report was that the failure rates of pipelines were diameter-dependent and that, in certain cases, a negative exponential relationship existed. Another was that approximately 50 per cent of all the causes of pipeline failure could be attributed to factors outside and away from the pipeline. The major conclusion of the report, however, was that, for pipelines of the same diameter, there was no discernible evidence to prove that the failure rates and modes of failure for different pipeline systems involving different petroleum commodities and different environments, were different. It would therefore appear reasonable for the general results of this study to be used to extend the rather limited evidence available for North Sea pipeline failures. Put another way this means that on the basis of the evidence available, the failure attributes of North Sea pipelines were very similar to those of other pipelines throughout the world.

In two cases where the quality of the information permitted, the results of this study showed that these pipelines failed from an overall ageing (wear out) process. In other words, their probability of failing between successive and equal time periods increased with their age. If such a conclusion could be shown to be generally valid it could have considerable economic repercussions. For example, it could permit the optimisation of pipeline surveillance, inspection and testing routines.

The major recommendation of this report is that it should be considered as an exploratory study of North Sea pipeline failures and that it should be extended as and when time and resources permit, since a better understanding of their failure attributes could be of considerable benefit to designers, fabricators, operators and maintenance personnel, as well as being essential information for the proper appraisal of the insurance liability and safety of such pipelines.

15.1 Introduction

The economic viability of most capital-intensive projects is related to their ability to operate at high production/service rates for a large proportion of available time. This feature is especially true of North Sea projects, which require high rates of income to repay the massive capital and interest charges involved with their oil and gas installations. A high degree of reliability is also required of the submarine pipelines involved with such installations because of the inherent difficulty and high cost associated with their failure and subsequent repair. In addition to their economic consequences, however, the failure of such pipelines could also create environmental pollution problems and have political overtones.

Recognising the importance of the reliability of such pipelines and the fact that they are known to fail, albeit infrequently, it was decided to make a study of their reliability characteristics. Several qualitative reliability analyses of oil and gas pipeline failures have been done before, amongst them a study of deepwater pipelines in the North Sea by Sande (1974). Devine (1974), Knowless (1977) and the US Senate (1966) have published sets of failure data for a few individual pipeline systems, where the quality of the data is generally low. However, no quantitative study which can claim to be comprehensive has been published before.

The motivation for undertaking this study partly arose from the need of a commercial organisation to appraise the risks to a particular North Sea pipeline and to assess the sensitivity of its safety to varying degrees of protective covering. The fundamental objective of this research programme was therefore to study the reliability characteristics of North Sea pipelines in particular.

In any reliability appraisal, there are several alternative means for deriving an expression of the reliability of the selected components, devices or systems. These involve: fundamental engineering and materials science considerations involved with fracture mechanics, reliability synthesis using generic failure data, full-scale or model testing and the statistical analysis of the historical failure data for similar objects. Unfortunately, due to the current state of the development of reliability technology, only the last two of these methods are really amenable to the appraisal of pipeline reliability. With the exception of some limited evidence relating to the experimental testing of a short length of pipeline in the Trondheim Fjord, this study was mainly concerned with the statistical analysis of historical failure data.

Realising the intrinsic limitations of any analysis of North Sea pipeline failures, due to the paucity of their failure data, (because of their relative newness and relatively high degree of reliability), the scope of this study was extended to encompass an analysis of the failures realised by all types of major pipeline systems throughout the world. The logic of this approach was that careful comparisons of the similarities and differences in the reliability attributes of these pipelines systems could provide information which would extend an understanding of the failure characteristics of North Sea pipelines, if correctly used.

Recognising the difficulty of obtaining accurate and comprehensive information in the restricted time and with the limited resources available to this study, it was never envisaged that the final report would be a definitive treatise on this subject. However, it was thought that such a study would still be meaningful in that it would:

(i) be the first known published report of its kind and would therefore set the foundation for more detailed studies when additional data and resources became available;
(ii) expose the nature of the deficiencies in known data sources, thereby alerting providers and users to the need for better quality data;
(iii) establish the limits of the likely reliability characteristics of North Sea pipelines, which could be of direct benefit to decision makers involved with such pipelines.

15.2 Work Carried Out

Adequate and relevant failure data are obviously key requirements for a study of this kind but before a search for such data can be conducted, the scope of a reliability study first has to be decided.

Where possible the scope of this study was restricted to transmission lines conveying crude oil, natural gas or refined petroleum products. Furthermore, the failure of a transmission pipeline was defined as any event which resulted in the leakage of commodity from that pipeline. Leakages from gathering pipelines, pumping stations, valves, risers, instrumentation systems and any damage to a transmission pipeline which did not cause a leakage were precluded from this study because they did not constitute pipeline failures as defined here.

Aside from the economic considerations involved with the cost/ benefit aspects of information retrieval and dissemination, the ideal information system needed to support a thorough and comprehensive study of the kind proposed here should provide the following data:

(i) *Endogenous factors*

- (a) the identification code of the failed pipeline and its geographical location;
- (b) a detailed account of its installation, the depth of cover and the nature of its protection, along with a profile of the depth of water (if applicable);
- (c) a detailed specification of the failed pipeline including its length, diameter, thickness, method of fabrication, a pre-failure X-ray analysis, its design pressure and temperature and its test pressure;
- (d) its inspection results and commissioning dates;
- (e) the specification of the commodity conveyed including details of pH, density, percentage solids, water and salt content, degree of sourness, temperature, pressure and pipeline velocity;
- (f) the date of the failure, method of discovery, method and time to repair and test, the extent of the leakage, the costs involved and the lessons learned;
- (g) the history of the operation of the pipeline, the quantity of product handled and the number of on-off cycles.

(ii) *Exogenous factors*

- (a) the depth of water at the point of failure, the nature of landslides or soil instability, the extent and nature of third party activity (shipping, trawling, vehicles, debris – if thought to be the cause of failure);
- (b) a meteorological record;
- (c) a complete inventory of all the other pipelines of the system with each individual pipeline specified to the same degree as the failed pipeline.

At this juncture, it should be stated quite categorically that none of the data sources used by this report provided anything like this depth of information and a large proportion of the time of this study was spent seeking suitable sources of data. Unfortunately, most of them involved centralised agencies which processed the original raw data in a way which best suited their interests, so that in the majority of cases access to the original raw data was limited. All of the data sources which have been used in this report are listed in Table 15.2.1 (overleaf) along with a brief description of their capabilities. It will be noticed that the data source for North Sea pipelines listed only two failures and it was primarily for this reason that the scope of the study was enlarged to incorporate other pipeline failures throughout the world. Ideally, the scope of this research project should have encompassed the work plan depicted by Figure 15.2.1 (see page 204), but due to the severity of the deficiencies (shown as dotted blocks in Figure 15.2.1) imposed upon this study by the inability to find comprehensive and adequate data in time, the main thrust of this study was restricted to those tasks depicted by the shaded area of Figure 15.2.1.

Several different reliability criteria are used to characterise the reliability attributes of components, devices and systems. The most commonly used criterion is the Failure Rate, which is the quotient of the fraction of failures during a test and the test attribute of interest (time, number of cycles, volume or weight processed, etc). Another criterion is the Mean Time to Fail (MTTF) which, under certain restrictive statistical conditions, is the reciprocal of the failure rate. For a more fundamental understanding of the reliability attributes of objects, it is necessary to elucidate the histogram (or probability density distribution function – pdf) of their failures as a function of time (or some other variable of interest). In the majority of cases, the data available to this study constrained most of the analyses to appraisals of their failure rates. Due to the paucity of their numbers of failures, it was considered prudent to develop these average failure rate estimates in a statistical framework in order to improve the quality of the subsequent decision making. This was done using the concept of Confidence Intervals. In this study the failure rate was defined as the number of failures per year per 1,000 kilometres of pipeline at risk. In only two cases was the basic information satisfactory for developing the pdf's of the failures of these pipelines.

Besides attributes concerned with the overall numbers and rates of failure, reliability engineers also try to understand the basic causes of failure and to this end they use failure mode analysis. To perform such an analysis it is necessary to classify failures by their mode (cause) of failure and the ideal analysis would elucidate the pdf of each mode of failure. In this study it was envisaged that pipelines would exhibit four fundamental (and one unclassified) modes of failure:.

(i) *Mode A, (mechanical)*
due to constructional defects – buckling and fatigue and material

Table 15.2.1 The Various Sources of Failure Data Used by this Study

NO	GEOGRAPHICAL AREA	SOURCE	COMMODITY	NUMBER OF DATA POINTS	TIME PERIOD	COMMENTS
1.	Western Europe onshore	CONCAWE (The Netherlands)	Oil	94	1966–1977	Well specified. Possible to estimate commissioning dates. Possible to estimate length by diameter for each year using additional information.
2.	Gulf of Mexico offshore	USGS (USA)	Oil and gas	136	1967–1977	Each failure roughly specified. Possible to estimate pipeline length. Commissioning dates given for 26 of the failing pipelines.
3.	USA Transmission onshore	NTIS-report (USA)	Gas	Probably in excess of 1,000	1970–1973	No raw data. Information is result of performed analysis, by the University of Oklahoma. Failure rate by diameter given.
4.	USA Interstate onshore	US Senate	Gas	1,039	15 years before 1966	No raw data. Information is result of performed analysis. Failure rate by diameter is given.
5.	USA Transmission onshore	Dept of Transportation (USA)	Gas	1,397	1974–1976	No raw data. Distribution of causes given.
6.	USA Canada onshore	US Dept of the Interior	Gas	Unknown	1970–1974	Large diameter pipes only. Failure rates. No raw data.

Table 15.2.1 The Various Sources of Failure Data Used by this Study *continued*

NO	GEOGRAPHICAL AREA	SOURCE	COMMODITY	NUMBER OF DATA POINTS	TIME PERIOD	COMMENTS
7.	North Sea offshore	E&P Forum (UK)	Oil and Gas	2	1975–1976	Fairly well specified. Too few data for a profound analysis at this stage but of excellent potential.
8.	Gulf of Arabia offshore	E&P Forum (UK)	Oil and gas	1	1977	
9.	UK onshore	The Inst of Gas Engineers (UK)	Gas	73, only 6 significant	1970–1977	No raw data. Only incidents caused by outside force are tested.
10.	France onshore	Gaz de France	Gas	493	1967–1977	Unclear what kind of gas pipelines. Diameter of failing pipe not given. No raw data.
11.	North Sea offshore	Norwegian Deepwater Pipeline Project			1974–1977	An appraisal of North Sea pipeline integrity based on experiments.

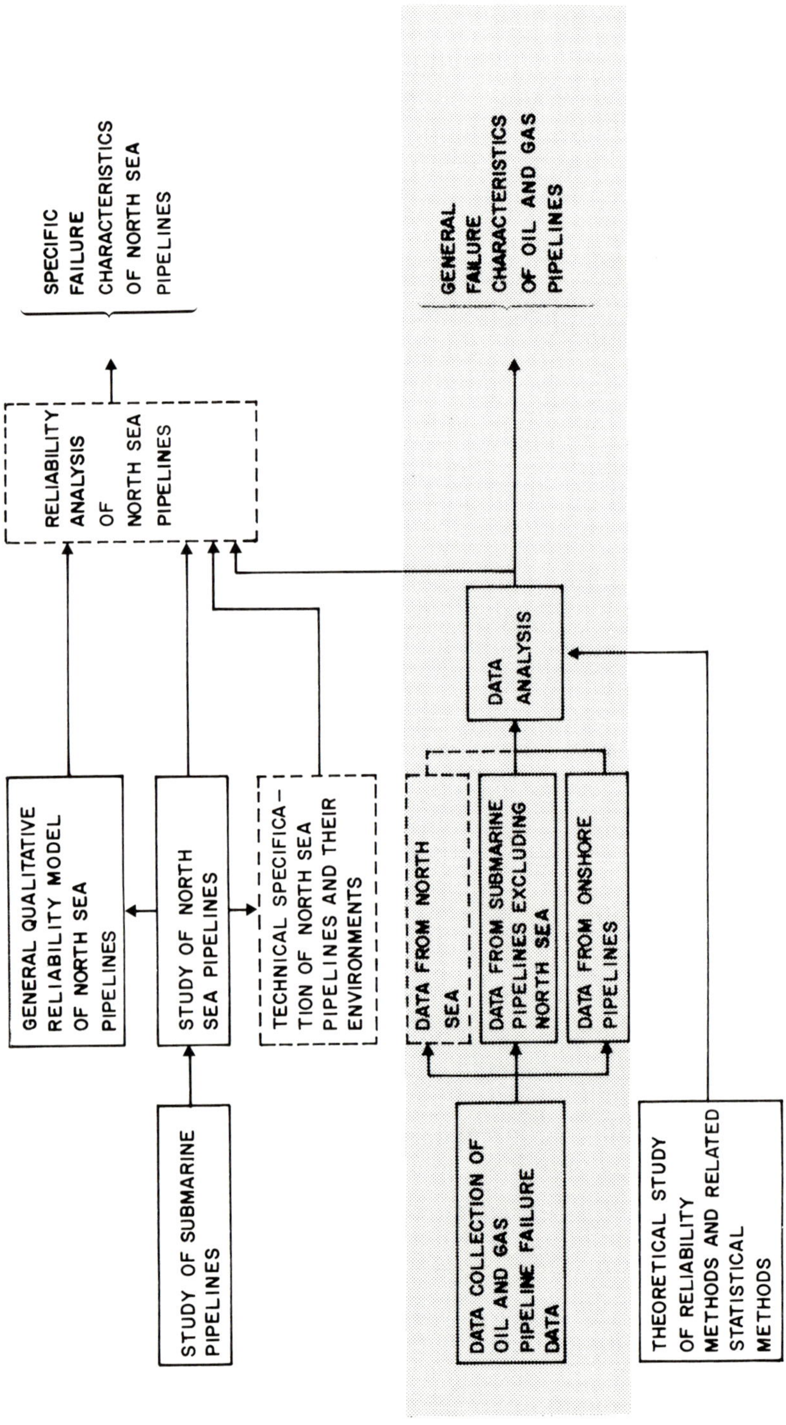

Figure 15.2.1 Work Model

defects – metallurgical and welding;

(ii) *Mode C, (corrosion)*
both internal and external;

(iii) *Mode D, (natural hazard)*
due to landslides, soil instability, storm, wind and flooding;

(iv) *Mode E, (third party)*
due to collisions by excavators, vehicles, trawls, debris and other interference;

(v) *Other*
unclassified reasons.

In this study it was quite impossible to discover the pdf's for each of these failure modes. Instead, the various percentage causes of failure were used to compare the failure attributes of the alternative pipeline systems.

As already mentioned, the objective of this study was to appraise the reliability characteristics of North Sea pipelines and to that end the scope of the study was enlarged to encompass several other pipeline systems in the hope that they would provide information which could be used to extend the limited data available for North Sea pipelines. In essence, it was hoped that a comparison of their failure attributes would indicate that they were similar, since points of difference would fracture rather than extend such a study. The basic reasons why these failure rates and modes were thought to be similar in advance of the study are laid down as hypotheses in the main report by de la Mare and Andersen (1980) and it was towards the verification or repudiation of these hypotheses that the major effort of this work programme was directed.

15.3 Discussion of Results

15.3.1 A Comparative Analysis of the Failure Rates of Various Pipeline Systems

One of the significant results of this study is the rapid diminution of pipeline failure rate with increases in pipeline diameter. The CONCAWE data showed that this relationship was approximately exponential, the US onshore gas transmission pipelines showed a reciprocal-proportional relationship, whereas the US Interstate pipelines showed a less pronounced effect. In line with intuitive reasoning, it would therefore appear proper to compare the failure rates of alternative pipeline systems as a function of their diameters. Of course, it is realised that such comparisons should be made on a *ceteris paribus* basis, realising that many other factors could influence the failure rates of such pipelines. For example, the study by Devine *et al* (1974) showed only too clearly the influence of pipeline protective coatings and cathodic protection on their failure rates. British Gas showed the effect of pipeline metal thickness and, of course, the age of the pipeline could affect other technological considerations, if nothing else. Because of the inadequate quality of the data, these other considerations have largely been overlooked in the following analysis.

The failure rates of the various pipeline systems studied by this report are shown, superimposed, in Figure 15.3.1 where the failure rates have been represented by their 95 per cent confidence intervals, as a function of pipeline diameter.

Two significant results can be derived from Figure 15.3.1. The first concerns the fact that the failure rates for pipelines with diameter of (say) less than 12 inches, were relatively well dispersed, suggesting that there might have been intrinsic differences in the failure rates of such pipelines. The second observation is the considerable degree of overlapping of the failure rates of pipelines with a diameter greater than (say) 12 inches, thereby suggesting that their failure rates might have been intrinsically similar.

A cautious appraisal of Figure 15.3.1 shows the following overall effects:

(i) There is considerable overlapping of the failure rates sustained by cross-country and submarine oil pipelines, suggesting that the failure rates of these pipelines were very similar and that there was very little difference brought about by the different external environments.

(ii) There was a large overlap in the failure rates of onshore and offshore gas pipelines with diameters exceeding 12 inches, suggesting yet again that the differences in environment made little appreciable effect to their failure rates.

(iii) There was considerable disparity between the failure rates of onshore French and US gas pipelines with diameters less than 12 inches. It must be remembered that the actual sizes of the French pipelines were unknown by the time this study was completed. However, it is known that the majority of these pipelines had diameters smaller than 12 inches; but there is no way of knowing whether they had really smaller diameters, which could possibly have contributed to the high overall failure rate of French pipelines. Resulting from this disparity, the only reasonable conclusion that can be drawn is that the failure rates of onshore gas pipelines might be some average of the two extreme cases, in which case the overall average failure rates of onshore and offshore gas pipelines would have been very similar.

A similar analysis of the other four combinations of comparisons shows that there was no discernible evidence to prove that the failure rates of the following paired comparisons differed from one another:

(i) onshore oil and onshore gas;
(ii) offshore oil and offshore gas;
(iii) offshore oil and onshore gas; and
(iv) onshore oil and offshore gas.

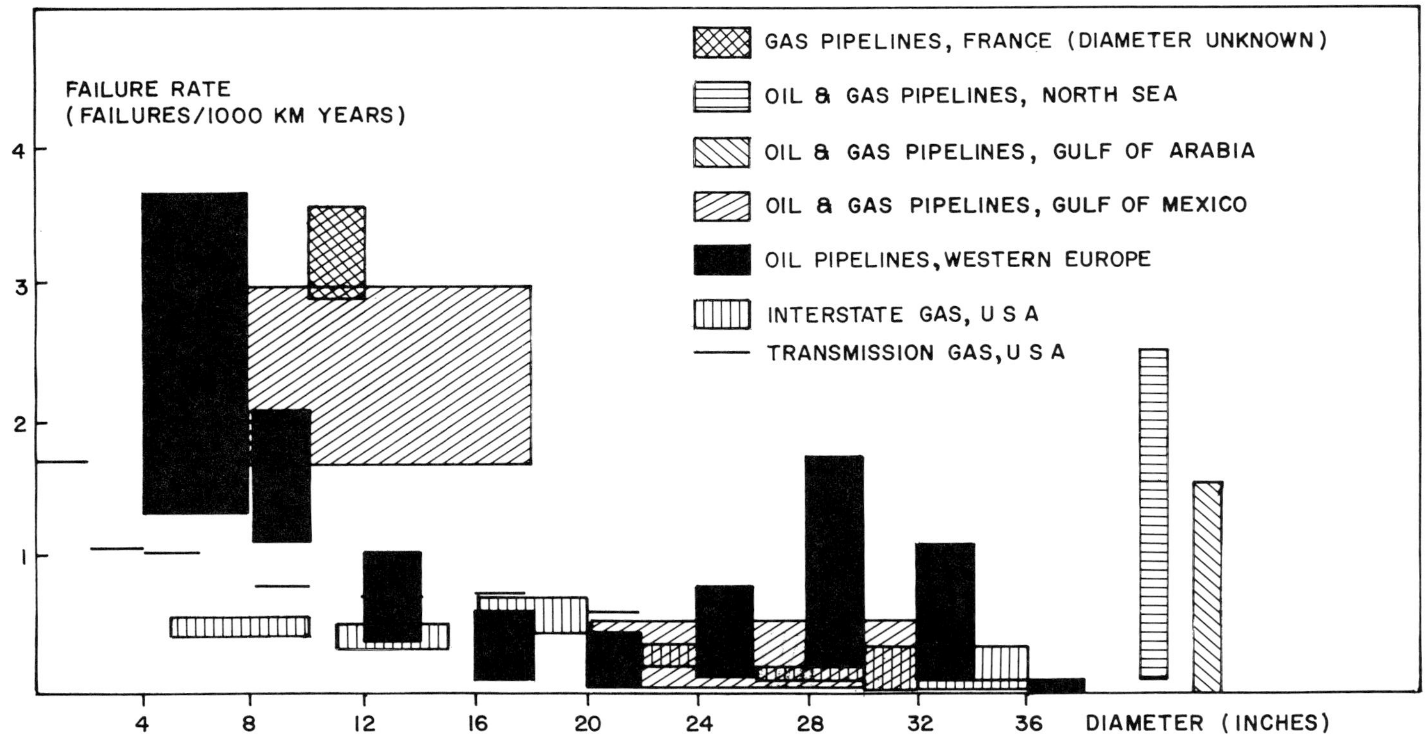

Figure 15.3.1 Failure Rates as a Function of Diameter for Different Pipeline Systems, 95 per cent Confidence Intervals

From Figure 15.3.1 it will be noticed that no attempt has been made to apportion North Sea and Arabian Gulf pipeline failures as a function of pipeline diameter.

Technically speaking, there is no reason why this should not be done, since the numbers of failures, their corresponding pipeline lengths and diameters are all known quantities. Considering the paucity of the failure data, however, it was thought that such an exercise would have been rather ridiculous at this stage until more failure data became available. However, it should be noticed that both of these geographical areas provided failure rates in line with the rest of the data of Figure 15.3.1.

To obtain a more fundamental understanding of the failure rates of these pipeline systems, it is necessary to understand how individual pipelines failed. Both the CONCAWE and USGS data showed that the times to first failure for these pipelines were log-normally distributed with increasing hazard functions over the operating time of interest as shown in Figure 15.3.2. Put another way, this evidence suggests that the probability of failure of these pipelines between successive and equal intervals of time increased with age.

If this ageing effect could be properly confirmed, then it could help management optimise the scarce resources deployed on these pipelines. In particular, it could provide a sound basis for a proper pipeline routine surveillance and monitoring schedule. This ageing process was hypothesised at the outset of this study, based on intuitive reasoning. However, it must be stated quite categorically that the evidence provided by this report does not constitute proof of this ageing process, it is only an indication. The fundamental reason for this doubt rests with the fact that the necessary statistical conditions needed to confirm or repudiate this hypothesis have not been satisfied, because of the inadequacy of the data sources used by this study.

15.3.2 A Comparative Analysis of the Failure Modes of Various Pipeline Systems

The percentage causes of failure for the pipeline systems studied by this study are shown together in Table 15.3.2 (overleaf) where their similarities and differences can be highlighted. This illustrates the relative comprehensiveness of the CONCAWE data when compared with the other data sets. It highlights the uncertainty surrounding the Gulf of Mexico statistics due to their large percentage of 'other causes', the not so large but still appreciable proportion of 'other causes' involved with North American pipelines and the incompleteness of the UK and French data. But for these reasons, the failure mode analysis could have been conducted on a similar basis as the failure rate analysis using the 95 per cent confidence interval approach. Under the circumstances, however, such an approach was hardly warranted. Despite these inadequacies, however, some useful results can be gleaned from a comparison of the statistics of Table 15.3.2.

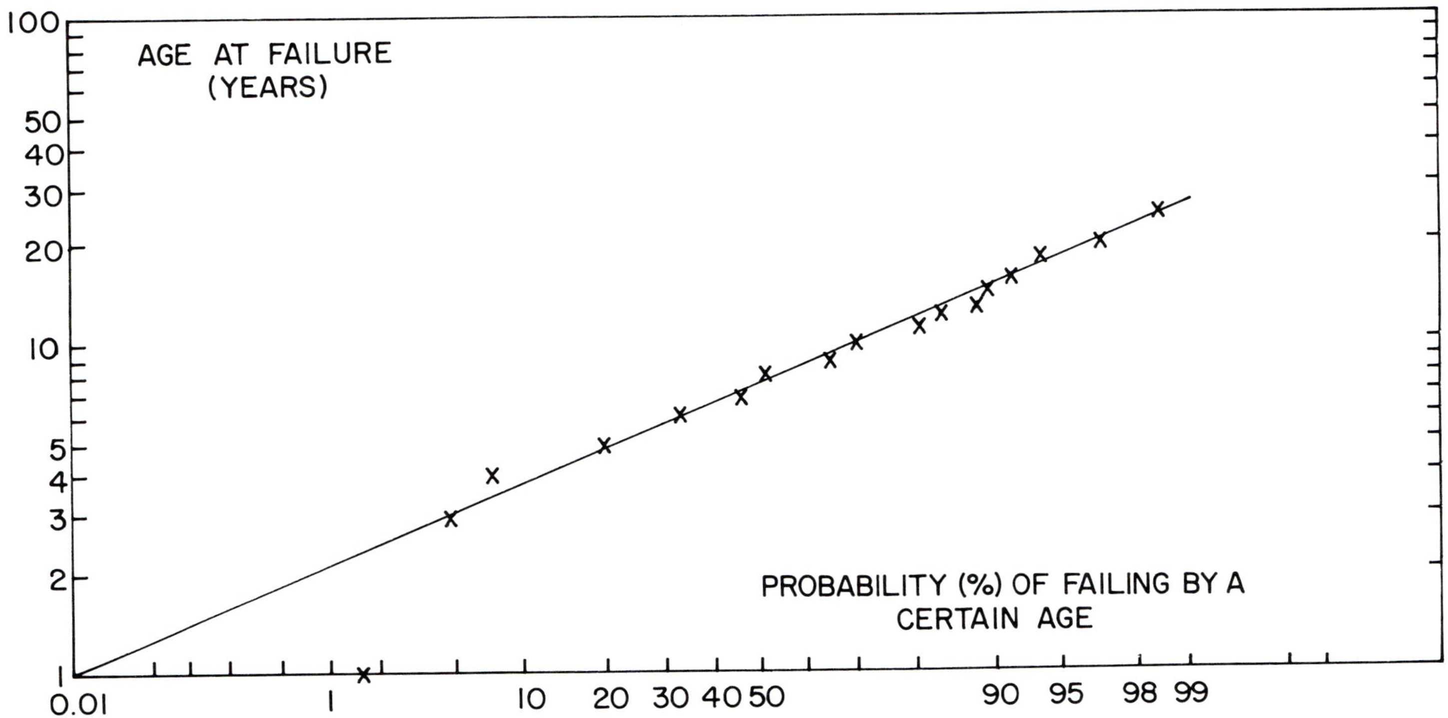

Figure 15.3.2 Log-normal Probability Plot of the Age at Failure for West European Pipelines

Table 15.3.2 Failure Mode Analysis for the Various Pipeline Systems (per cent)

PIPELINE SYSTEM	COMMODITY	MECHANICAL FAILURE	CORROSION		TOTAL	NATURAL HAZARD	OUTSIDE FORCE THIRD PARTY DAMAGE	TOTAL	OTHER CAUSES	TOTAL NUMBER OF FAILURES
			INTERNAL	EXTERNAL						
Cross-country Western European	Crude oil	24	1	23	24	12	40	52		33
	Product oil	10	1	52	53	1	36	37		61
	All oil	15	2	40	42	5	37	42		94
Gulf of Mexico	Oil	3	10	16	26	14	29	43	28	110
	Gas	8	8	27	35	11	8	19	38	26
	Oil & Gas	4	10	18	28	13	25	38	30	136
North America Transmission 1970–73	Gas	24	3	12	15			54	7	probably >1,200
Transmission before '66	Gas	25			17			52	6	1,397
Inter-state	Gas	32	3	15	20	9	30	39	9	1,039
UK	Gas							29		73
France	Gas	18	7	13	20			62		493

A search of this table for points of agreement shows that the percentage of causes of failure due to the internal corrosion of cross-country oil and gas pipelines were very similar, as one would really expect. Similarly, the proportion of failures caused by natural hazards were very similar for cross-country oil and gas pipelines and reasonable similar for cross-country oil and gas and offshore oil and gas pipelines. Where statistics are available, it would also appear that the proportions of failures caused by third party damage were similar for onshore oil, onshore gas and off-shore oil pipelines.

Perhaps the most important aspect of this failure mode analysis is the very high proportion of failures caused by outside force. Other than the discrepancy caused by the UK gas pipeline statistic, it would appear that most pipelines sustained between 40–50 per cent of their failures due to these reasons. It must of course be realised that such a statistic would undoubtedly be diameter-dependent, but nevertheless this statistic proves that despite their apparent robustness, pipelines are vulnerable to external damage which can partially destroy their integrity.

It would appear from the evidence of Table 15.3.2 that external corrosion caused a similar proportion of failures for onshore gas and offshore oil pipelines but for some unexplained reason the corresponding statistic for onshore oil and offshore gas pipelines was considerably higher. Some of the greatest differences in this study, however, are to be seen with the mechanical mode of failure, where there appear to be three regimes of statistic: low, medium and high. In the first case, the Gulf of Mexico statistic appears unbelievably low, but of course it could be amended and increased if the large proportion of 'unknown causes' could be redistributed. The CONCAWE overall statistic seems to be confirmed by the French statistic, but, by contrast, the percentage of causes of mechanical failure to US gas pipelines seems to be unbelievably high. With the quality and quantity of the data available to this study there was no means of resolving these points of difference.

Before leaving this section three other points should be mentioned. The first concerns the difficulty of discerning the initiating cause of failure for a pipeline. For example, this study provided some evidence to show that corrosion failures can often be induced by third party or natural hazard effects. To some extent, therefore, the allocation of a failure mode category to a particular failure depends on the diligence of the reporter and the importance of the failure. The second point concerns the fact that collectively all of the percentage causes of failure add up to 100 per cent so that a piece-meal analysis of the type just conducted is fraught with statistical problems, unless high confidence levels can be placed on the individual statistics. Lastly, it should be realised that none of these so-called failure mode effects has taken cognisance of differences in the dimensions of the pipelines, their methods of protection or indeed their ages, factors which could undoubtedly have a considerable bearing on such analyses.

15.4 Conclusions

This study set out to appraise the reliability characteristics of North Sea pipelines in particular but, anticipating the paucity of the failure data for this region, the aim was also to determine the reliability features of pipelines in general. Only two pipeline incidents which qualify as failures in the context of this report occured in the North Sea during the time period studied. These resulted in an overall failure rate statistic (λ) estimated by:

$\lambda =$ 0.70 failures per year per 1,000 kilometres of pipeline with 95 per cent confidence limits of 0.067 and 2.50.

Because of the small number of such failures, no attempt was made to represent this failure rate as a function of pipeline diameter or to appraise the percentage of failure causes. Instead, the bulk of the investigation was concentrated on appraising the failure rate and mode characteristics of other types of pipeline systems throughout the world, with a view to highlighting their similarities and differences, since it was thought that such information could be used to extend the study of North Sea pipeline failures, if properly used.

In line with intuitive reasoning, one of the major conclusions of this study is that pipeline failure rates are dominated by a diameter-dependent effect and some evidence of a negative exponential relationship is provided by the study results. Another major conclusion was that 40–50 per cent of the causes of pipeline failure could be attributed to forces external to the pipelines.

A comparison of the failure rates and modes of pipelines conveying different kinds of products in different external environments suggested, but by no means proved, that there was no appreciable difference in their failure characteristics, given that they were of (approximately) the same diameter, and that these results could be used to extend the scope of other pipeline reliability studies.

In two cases, the failure data were of sufficient merit to warrant investigations of their statistical distributions of times to first failures. In both cases, the log-normal distribution best modelled such failure data and both types of pipelines were shown to be experiencing wear-out (ageing) effects, over the operating times of interest.

The results also indicated that other technological factors, such as pipeline wall-thickness, the level of pipeline protection and the age of their inbuilt technology also had a direct bearing on their failure rates but, unfortunately, this evidence was too sparse to develop into a generality, even though such effects might be intuitively obvious.

Because of the incompleteness of the data sets used by this study, the author has certain reservations regarding the thoroughness of the study, the validity of the results and the conclusions derived therefrom. Despite the fact that the study has been unable to prove categorically these conclusions (in a statistical sense), nevertheless the evidence strongly suggests that these are fair and reasonable conclusions.

15.5 Recommendations

(i) It is thought that the numerical results of this report could be directly used in design and other technological studies concerned with pipelines.

(ii) It is also thought that these numerical results could be of direct use in appraising the relative economic merits of alternative pipeline design concepts, especially where pipelines of different diameters could be used.

(iii) These results could also be of use in appraising the insurance liabilities of such pipelines.

(iv) If the indication of pipeline wear-out and ageing could be verified, then such a result could be of considerable benefit to managers, since it would permit the optimal deployment of resources involved with such pipelines to be accomplished. In particular, such information would permit the optimisation of inspection, monitoring and testing routines.

(v) The major recommendation of this report is that this study should be regarded as a pilot study and that it could be in the interests of pipeline designers, fabricators, contractors, operators, insurance companies and all those other parties who might conceivably have a vested interest in the performance attributes of pipelines, if the major causes of doubt arising from this study were resolved by a much more thorough and comprehensive study. Such a study would obviously require access to good quality information, which obviously exists. Whether or not such information could be obtained would depend on the willingness of companies to co-operate in such a study and the ability of the study to maintain a required degree of confidentiality, especially on an anonymity basis.

15.6 Acknowledgement

Much of the work reported here is to be attributed to Mr Ø Andersen, who was given a one-year leave of absence by his employer, Det norske Veritas, to study these matters under the supervision of Professor R F de la Mare at the University of Bradford, England. Financial support for this project was provided by Det norske Veritas and the Royal Norwegian Council of Scientific and Industrial Research (NTNF). As author, I take full responsibility for any errors or omissions and would like it to be known that the views expressed in this chapter are my views and do not necessarily reflect the opinion of any organisations.

I would also like to acknowledge the help and encouragement given by Det norske Veritas, executives of the National Centre of Systems Reliability (NCSR) and all of those organisations whose data made this study possible.

15.7 Bibliography

de la Mare, R. F. and Andersen, Ø., *Pipeline Reliability*, Report No 80–0572, Det norske Veritas, (1980).

Devine, M. D. *et al*, *Analysis and Management of a Pipeline Safety System*, NTIS-report No PB-238 828, (1974).

Knowless, A. E. *et al*, 'The Background and Implication of IGE/TD/1', *43rd Autumn Meeting, The Institution of Gas Eng,* (London, 1977).

Sande, T. and Takle, J., *Reliability of Pipelines*, Appendix 24, NOU 1974: 40, Oslo, Norway, (Governmental report, in Norwegian), (1974).

US Senate (Anonymous), 'Safety of Interstate Natural Gas Pipelines', *Proceedings, 89th Congress, Second Session– US Senate Bill 1553,* (1966).

R F DE LA MARE

16

OPERATING GAS PIPELINES AT HIGH PRESSURES

The operation of gas pipelines at high pressures, in what is known as the 'dense phase', is one way to reduce the overall life cycle cost of offshore natural gas/crude oil systems. This chapter describes the rudiments of 'dense phase' operation and the benefits which it offers.

16.1 The Thermodynamic Implications of Gasline Design

A basic appreciation of the thermodynamic properties of gases can be gleaned from Figure 16.1.1. This is a typical phase diagram for any natural gas or gas mixture, but its exact shape depends entirely on the composition of the gas(es).

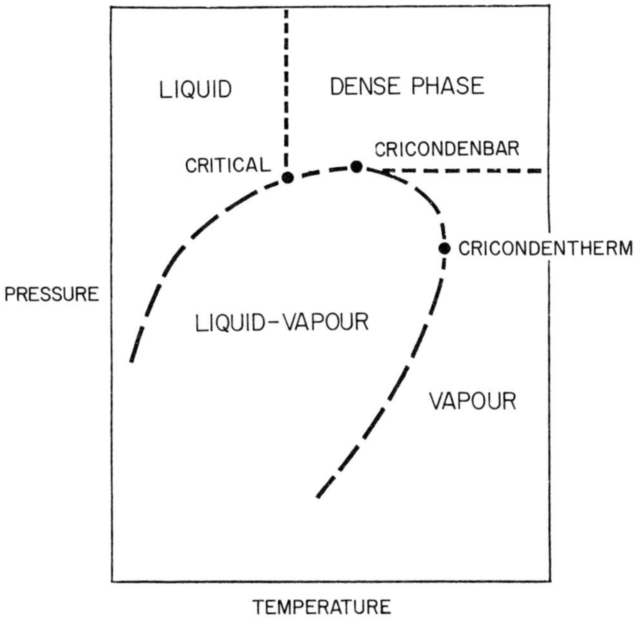

Figure 16.1.1 Typical Phase Diagram for Natural Gases

H

It will be noted that the 'cricondenbar' is defined on this phase diagram as the highest pressure at which separate liquid and vapour phases can co-exist and the 'cricondentherm' is the highest temperature at which liquid can form regardless of pressure.

The 'dense phase', as defined by the critical point and the cricondenbar, is neither liquid nor vapour, but it comprises a dense phase-less fluid.

Generally, two-phase flow in long distance pipelines is avoided because it necessitates substantial costs in pigging and liquid separation and it reduces the capacity of the pipeline. One of the major design problems, associated with two-phase flow, is the difficulty of accurately designing a pipeline's capacity, especially with changes in pipeline elevation, and this leads to over-designed equipment and high costs. The problem of 'slugging', which is associated with two-phase flow, is currently being researched by the Norwegian Institute of Technology.

For conventional gaslines operating in the vapour phase only, the gas(es) must have a cricondentherm below the lowest operating temperature, otherwise condensation will occur. In the North Sea, this could entail temperatures below zero centigrade, due to the expansion of the gas(es) caused by pipeline pressure drop. The alternative to two-phase flow for gases which would otherwise condense partially, is to recover Natural Gas Liquid (NGL) on the offshore processing units, but this can be extremely expensive, both in capital and disposal costs.

The phase characteristics of the gas produced from a platform are controlled by the composition of the reservoir fluid and the design of the platform separation facilities. Generally, platform processes are the simplest that can be devised to meet oil and gas product specifications. Every effort is made to restrict the number of pieces of equipment because of space and weight constraints. Cryogenic processing, distillation, and fired process heaters are also avoided because of complexity and safety considerations. Typically, an offshore separation process includes two to four stages of oil-gas separation, corresponding stages of gas recompression to pipeline pressure and some form of gas conditioning. Figure 16.1.2 is a simplified diagram of a generalised separation process found on offshore platforms. There can be many variations, particularly in the recompression arrangement.

Crude oil vapour pressure and gas hydrocarbon dewpoint are the principal product specifications which control platform separator design. Directionally, the process is increasingly complicated by lowering either vapour pressure or dewpoint. The most rigorous separation requirements would be placed on a platform which had to meet a 14 psia crude Total Vapour Pressure (TVP) for offshore oil loading and a 30°–40°F (at 1,000 psig) gas dewpoint specification for transportation in a conventional pipeline. For some wellstream compositions this would require NGL to be produced. This could be used as a fuel, or it might have to be reinjected

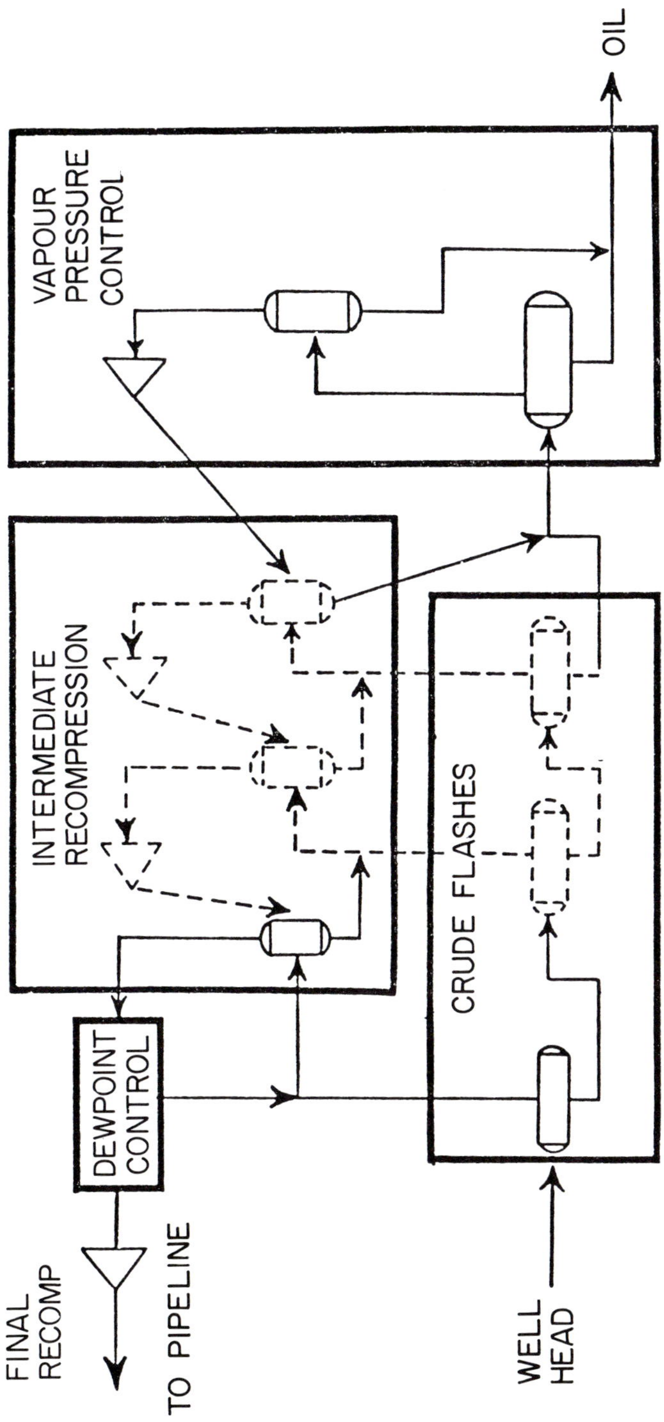

Figure 16.1.2 Block Flow Diagram, Typical Cascade Separator Process

or flared. The costs imposed by low dewpoint specifications can make gas sales unprofitable for some small operations.

16.2 The Cost Implications of Gasline Design

To establish precisely the optimum gas quality to produce on a platform requires a trade-off between the costs associated with the platform, the pipeline and any onshore processing facilities. This requires a systems approach to ensure that the life cycle costs incurred by the total system are minimised. Such an approach was used in the study of the 'North Sea Gas Gathering System', which was conducted jointly by the British Gas Corporation and Mobil on behalf of the UK Department of Energy. That study concluded that the most economic means of landing the mixtures of gases from the many fields was to operate the gas gathering and distribution pipeline in the 'dense phase'. It was recognised that a pipeline designed for dense phase operations would be more expensive to build and operate than one designed for conventional vapour phase operations, but that the former would obviate the need for very expensive offshore gas conditioning equipment. For the same gas volumes, it was estimated that the capital costs of the former would be only £13 millions more than the latter, (according to the details shown below), but the added capital cost for gas conditioning equipment designed for ten fields or more was estimated conservatively to be at least £100 millions.

Case	Maximum pressure of platforms, psig	Minimum pipeline pressure, psig	Pipeline capital cost (£ millions)
Dense Phase	2,500	1,600	666
Vapour Phase	2,500	1,200	653

A further advantage of the dense phase option was that it could accept gases from a wide variety of fields for which some of the gas compositions were not properly known.

Having established that dense phase operation was the more economic mode of operation, a further study was made of the optimum design pressure for the system. This is illustrated in Figure 16.2.1, which shows that the optimal inlet design pressure is about 2,500 psig and that there would be little cost advantage in designing for higher pressures, whereas the technical and safety problems would increase considerably.

16.3 Establishing the Phase Diagram for Gas Mixtures

The calculations involved with establishing the boundaries of the phase diagrams for various gas compositions are subject to some uncertainties in the areas of the critical pressure and the cricondenbar, because correlations of gas-liquid equilibrium ratios are especially sensitive in these regions.

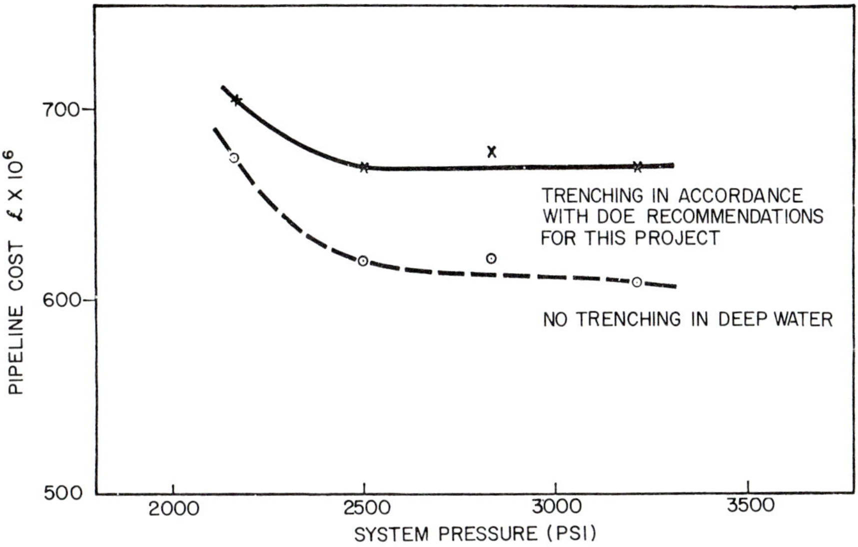

Figure 16.2.1 Pipeline Cost *versus* Maximum System Pressure

The case history involving the Beryl Field in the UK sector of the North Sea is a useful example to demonstrate this matter. This field was operated with gas reinjection, but Mobil was interested in the prospect of transporting the gas for sale. An experimental programme was therefore established in 1979 to determine the precise boundaries of its phase diagram. A comprehensive sampling programme was conducted of the various flow-streams on the Beryl platform. The gas samples were analysed using gas chromotography, whereas the liquid samples were fractionated using low temperature distillation and the 'heavier' components were determined using temperature-programmed gas chromotography. The heavy components were of particular interest because of their significant effect on the phase envelope. Table 16.3.1 (overleaf) shows the composition of the 2,500 psig gas-stream that would be delivered to a gas pipeline. This gas is known to have a hydrocarbon dewpoint of 93°F at 1,040 psig because liquids were last separated from it in an interstage separator operating at those conditions.

A portion of the 2,500 psig gas sample was introduced at 3,000 psig into a visual PVT cell mounted within a constant temperature bath. After the cell had come to the desired temperature, the pressure was slowly reduced to determine the dewpoint pressure of the gas at a constant temperature. These experiments were made at 5°F increments, between 30°F and 50°F, to cover the range of pipeline temperatures of interest, and the experiments were repeated several times at each temperature. The results are shown in Table 16.3.2 (overleaf).

Mobil was also interested in knowing the volume of the liquid dropout *versus* pressure below the dewpoint, at constant temperature, since calculated results in this region of the phase envelope are often unreliable. Cathetometer observations of levels in the calibrated visual cell were made at temperatures of 30°F and 45°F. These results are shown in Table 16.3.3.

The calculated phase envelope for the gas composition of Table 16.3.1 is shown in Figure 16.3.1, with the experimental dewpoint pressure points from Table 16.3.2 superimposed on the calculated curve. It will be

Table 16.3.1 Extended Gas Analysis 2,500 psig Separator Gas

COMPONENTS	COMPOSITION MOL %
Nitrogen	1.043
Carbon Dioxide	3.793
Methane	70.236
Ethane	11.536
Propane	8.834
Isobutane	0.923
Normal Butane	2.592
Isopentane	0.412
Normal Pentane	0.468
Neo-Hexane	0.001
2-Methyl Pentane + 2, 3-Dimethyl Butane	0.053
3-Methyl Pentane + Cyclopentane	0.013
Normal Hexane	0.035
Methyl Cyclopentane	0.010
Benzene	Tr
Cyclohexane	0.015
Heptanes	0.015
Methyl Cyclohexane	0.007
Octanes	0.007
Nonanes	0.007
Decanes	Tr
Undecanes +	Tr
	100.00

Table 16.3.2 Experimental Dewpoint Pressure 2,500 psia Separator Gas

TEMPERATURE °F	DEWPOINT PRESSURE (psia)
30	1,455
35	1,475
40	1,465
45	1,490
50	1,455

noticed that the variation was only about 50 psi. It is also interesting to note that the calculated dewpoint temperature at 1,040 psig, from this curve, is only 4°F from the known dewpoint temperature.

Table 16.3.3 Experimental Liquid Dropout *versus* Pressure

TEMPERATURE °F	PRESSURE (psia)	VOLUME/RATIO LIQUID/VAPOUR
30	1,455 (DP)	.0--
	1,448	.010
	1,438	.040
	1,425	.062
	1,397	.096
	1,376	.109
	1,350	.128
45	1,490 (DP)	.0--
	1,483	.001
	1,477	.008
	1,472	.012
	1,430	.048

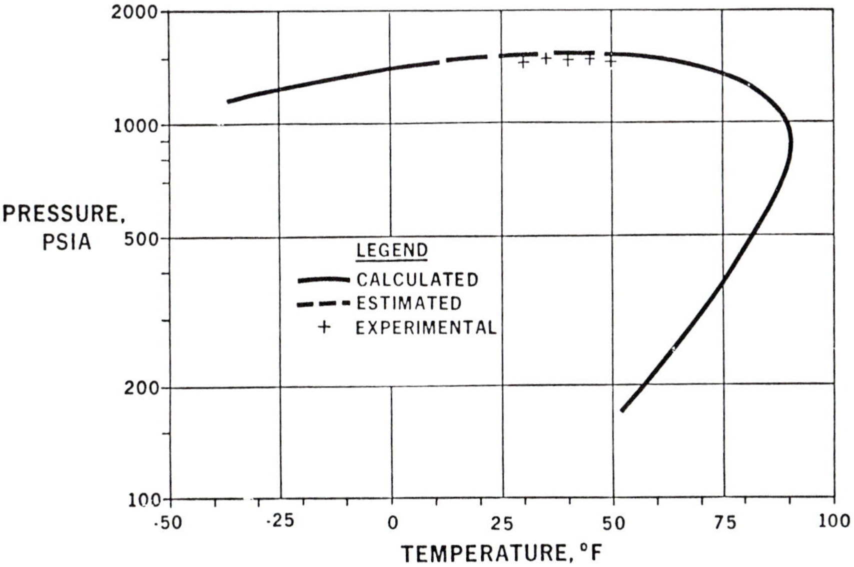

Figure 16.3.1 Beryl Gas Envelope, Calculated and Experimental

16.4 The Practical Significance of Phase Diagram Determination

These results show that Beryl gas could be transported in a dense phase pipeline with a minimum pressure of about 1,500 psig. Since data were not available from all other fields from which gas might be transported, a study was undertaken to develop a generalised correlation of phase envelope characteristics of gas from various fields and separation processes. The study examined the wellstream composition of eleven North Sea reservoirs with Gas-Oil Ratios (GORs) ranging from 200 to 14,000 scf/bbl. On average, the reservoir composition followed the trends shown in Figure 16.4.1.

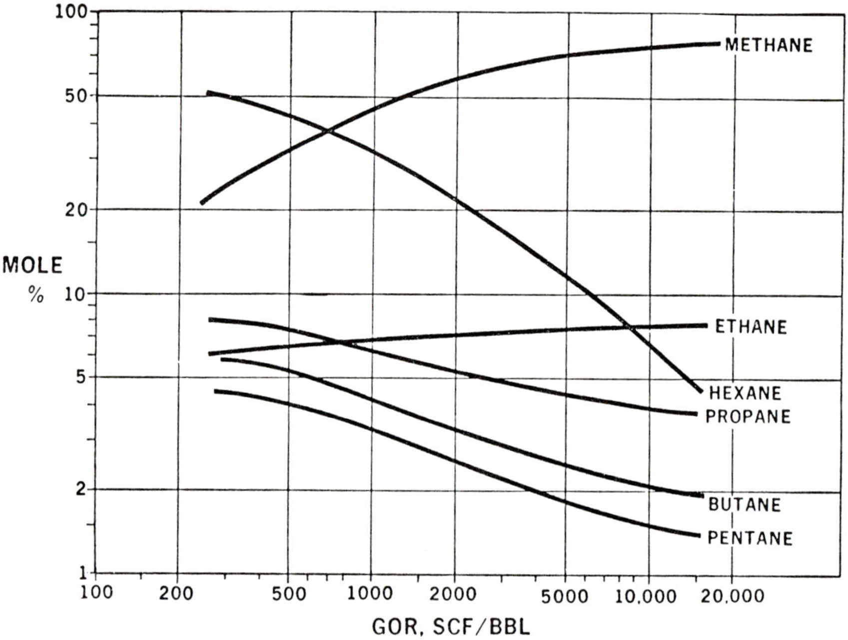

Figure 16.4.1 Typical Reservoir Composition

A process design computer model was developed to simulate the separation processes which are common to many North Sea platforms; an example of this was shown in Figure 16.1.2. The principal process variables (number of stages, wellhead pressure and temperature and oil vapour pressure) were screened to determine which processes gave the leanest and richest gas. Four processes were chosen, as tabulated below, which covered the range of initial separation pressure and crude oil vapour pressure, which were found to be the most significant variables. The oil vapour pressures are typical for tanker-loadable crude (14 psia) and high vapour pressure pipeline crude (100 psia).

Process	Number of crude flashes	Initial separation pressure, psia	Crude oil vapour pressure @ 100°F, psia
A	4	1,000	14
B	2	200	14
C	3	1,000	100
D	2	250	100

The computer model then calculated the gas composition, phase envelope, and platform process horsepower for the range of wellstream compositions, with various gas dewpoint requirements, for each process. Figure 16.4.2 illustrates a typical family of phase envelopes for one wellstream composition and process, at various gas dewpoints. Correlations were then made from all the numerous phase envelope data generated by the computer model.

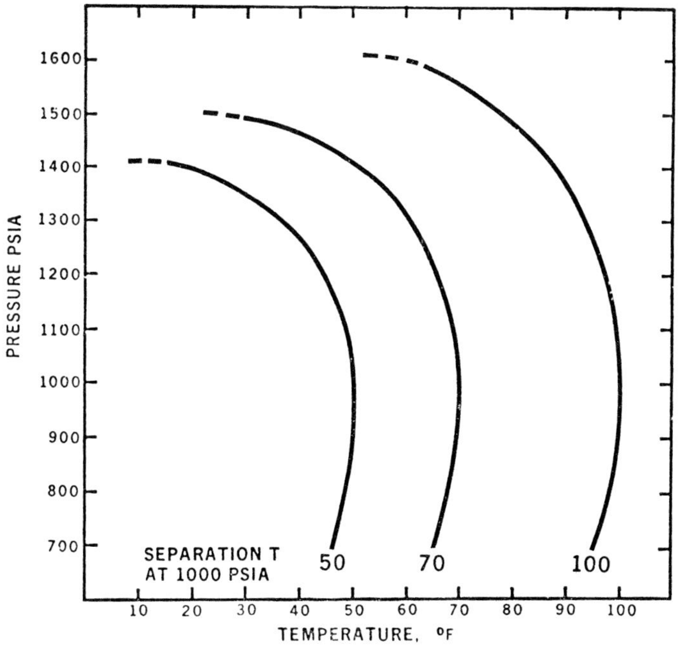

Figure 16.4.2 Family of Gas Envelopes. (200 psi wellstream)

Figure 16.4.3 (overleaf) summarises the gas envelope correlations derived from the study. The cricondenbar is chiefly a function of gas dewpoint control on the platform, and, to a smaller extent, a function of wellhead flowing pressure. The hydrocarbon dewpoint of the gas at 1,000 psia is a convenient correlating parameter, which also can be physically measured. The cricondentherm was found always to be within ± 4°F of the gas dewpoint at 1,000 psig. For conventional pipeline operation, specification of hydrocarbon dewpoint at 1,000 psia of 4°F less than the minimum expected gas temperature will prevent condensation of liquids at all pressures. For dense phase pipeline operation, hydrocarbon dewpoint at 1,000 psia can also be used as a gas specification, with the system pressure kept above the cricondenbar pressure shown by the correlation. Such a system should be operated above 1,550–1,600 psia if a gas dewpoint specification of 100°F at 1,000 psia is established.

The cricondenbar and cricondentherm correlations are also independent of the vapour pressure of crude oil being produced, for all compositions for which process separation is feasible. Cricondentherms approaching the seabed temperature are not feasible for low vapour pressure crude at a GOR less than about 500. A more complex process than the one used in the model is required to improve platform separation for low GOR reservoirs to make low vapour pressure crude, because there is insufficient methane to carry away the high vapour pressure NGL components.

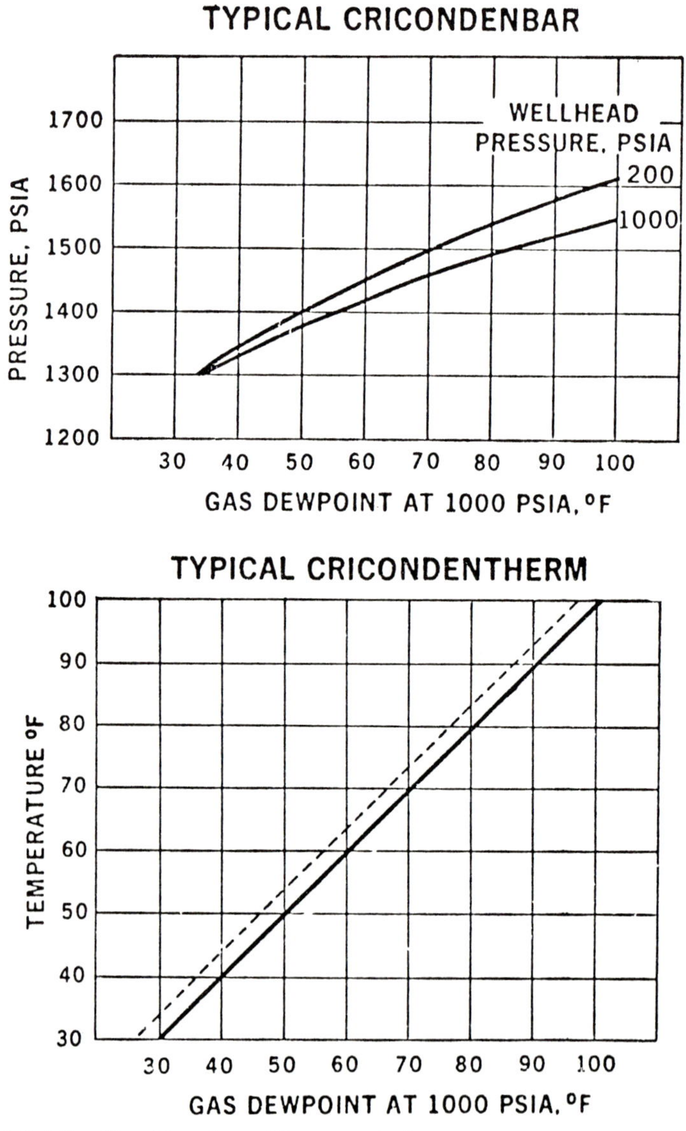

Figure 16.4.3 Gas Envelope Characteristics

16.5 Conclusions

The study showed that using the fairly simple processes, which are incorporated in the model, gases from the variety of wellstreams could be produced which would be satisfactory for transportation in a pipeline operated in the dense phase at pressures above 1,600 psi. The only gas specification which would have to be placed on the platforms producing such gas is that it would need a hydrocarbon dewpoint less than 100°F at 1,000 psia. This means that during the platform process the gas must be cooled to 80–100°F at 700–1,200 psi and the condensed liquids must be separated. Such temperatures as these can easily be achieved using seawater as a coolant, or possibly with air in the North Sea.

A gasline delivery pressure approximating 2,500–3,000 psig for dense phase operations requires little more horsepower than the delivery pressure required for conventional vapour phase pipeline operation. Moreover, the lower production platform costs associated with dense phase operation may also allow development of more marginal fields for delivery of gas into the pipeline system. For gas condensate fields there are less liquids produced to help support the high cost of offshore gas conditioning facilities, and for very low GOR oil fields without a high vapour pressure crude outlet, separation to a low dewpoint gas may not be feasible, in which case dense phase gasline operations could become economically viable.

It is also significant that platform processing to the 35°F dewpoint range for conventional pipeline operation, while sufficient to avoid liquid dropout in the gasline at any pressure, is nevertheless insufficient to avoid further gas processing onshore to meet British or European gas system specifications. In effect, the gas must be processed twice for NGL recovery, both offshore and onshore. Since we do not really want to recover NGL offshore, the offshore processing results only in a trade of pentane and heavier components in the gas for propane and butane components, resulting in a lower dewpoint. Dense phase pipeline operation allows more NGL components to be transported to shore in the gas pipeline, where they can be recovered and more economically utilised. Also, the high pressure pipeline system is less susceptible to upsets due to temporary delivery of off-specification gas because of problems in platform gas conditioning operations.

In effect, therefore, the operation of gaslines in the dense phase is an efficient way to solve some of our gas transportation problems. Whilst this case describes its application to the cold waters of the North Sea, the technique may also be applicable to onshore pipelining from remote cold areas and in less severe offshore areas.

D P PENICK
W B THRASHER
R GIBBONS

17

A CASE IN PIPELINE DESIGN

At this juncture, it is appropriate to reconsider the implications of material from preceding chapters in the context of an actual pipeline design. This will enable us to explore the interactive effects of many variables taken collectively in a total systems approach to pipeline design. Furthermore, the pipeline description contained within this chapter forms a basis for the more quantitative material which was presented in previous chapters, especially those concerned with the risk management implications of pipeline design.

17.1 Introduction

Plans to further develop the hydrocarbon resources in the Danish sector of the North Sea were realised recently with the construction of two pipelines of 214 and 219 kilometres apiece from offshore fields to the west coast of Jutland.

The 30-inch gas pipeline will be owned and operated by Dansk Olie & Naturgas A/S, henceforth referred to as DONG A/S, and the 20-inch condensate and crude oil pipeline will be owned and operated by Dansk Olierør A/S, otherwise known as DORAS, which is a wholly owned subsidiary of DONG A/S.

Survey and engineering work for the 30-inch gas pipeline started in 1978 and the project received its formal authorisation by the Natural Gas Act passed by the Danish parliament in 1979. Engineering activities for the 20-inch crude oil pipeline commenced in early 1981 and this project received its formal authorisation by the Crude Oil Pipeline Act later in that year.

The crude oil pipeline is intended to serve as a trunk line to transport crude oil and condensate from existing fields, which are administered by the Danish Undergrunds Consortium (DUC), as well as other fields which might become economically viable in the future. The overall routing of these pipelines is shown in Figure 17.1.1 (overleaf). The crude oil pipeline will connect the existing Gorm Field and crude oil processing centre

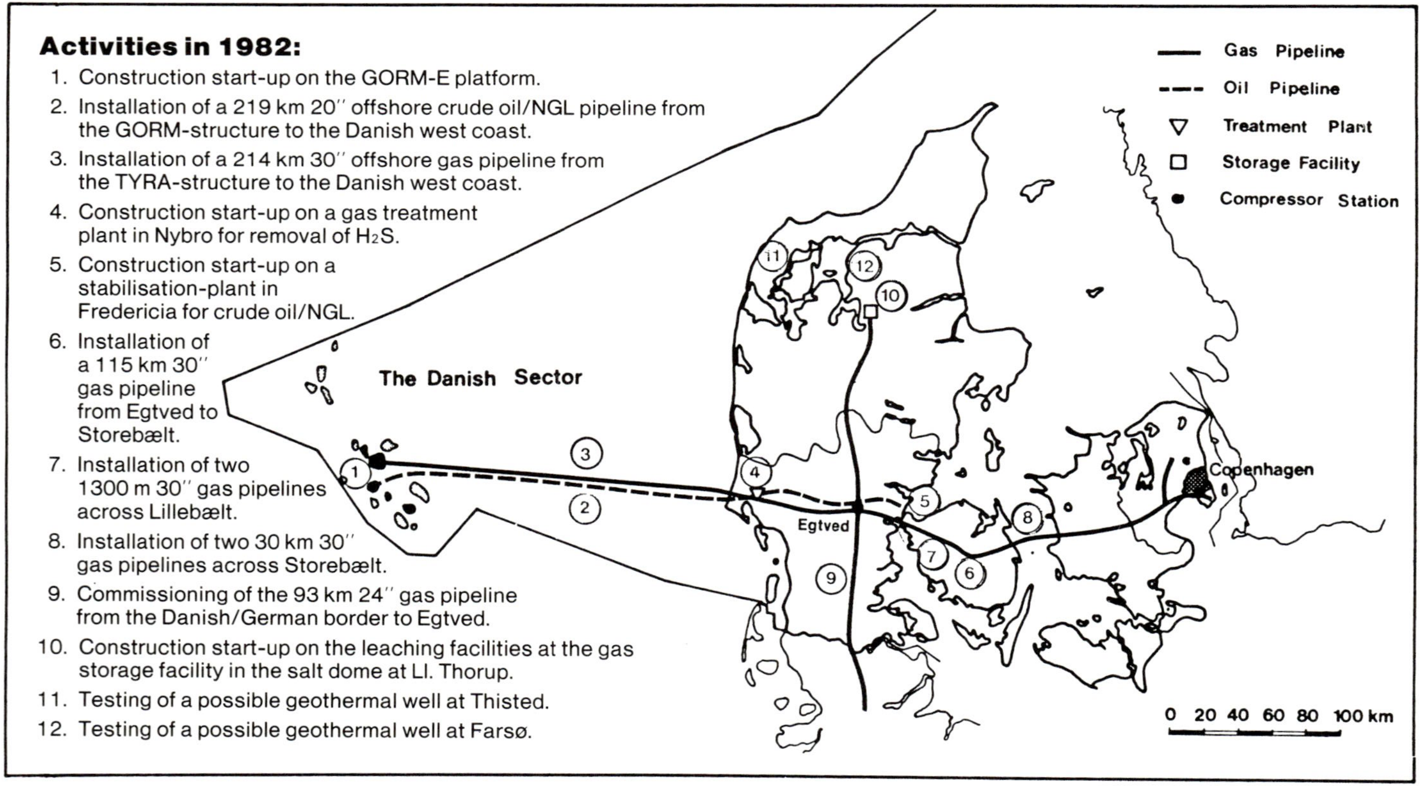

Figure 17.1.1 **Overall Routing and Details of Danish Oil and Gas Pipelines.**

with the Shell Refinery at Fredericia on the east coast of Jutland. The gas pipeline will connect the Tyra Field main gas processing centre with the DONG A/S Nybro onshore gas treatment plant, where further dewpoint control will be exercised. The pipelines will share a common shore approach on the west coast of Jutland.

The main hydrocarbon processing scheme is shown in Figure 17.1.2 (overleaf). The hydrocarbon dewpoint of the gas will be controlled mainly at the Tyra gas processing centre and the extracted condensate will be pumped ashore through the crude oil pipeline for further processing at the Fredericia stabilisation plant and subsequently to the Shell Refinery. Furthermore, condensates extracted at the Nybro gas treatment plant will be injected into the crude oil pipeline at Nybro.

The close operational integration of the system has required both pipelines to be ready for commissioning early during 1984 to meet the deadlines laid down in the DONG A/S–DUC Gas Agreement of 1979.

The Danish North Sea gas transmission pipeline and North Sea oil pipeline together with the Great Belt gas pipeline dual crossing were installed under a single contract by the Italian contractor Saipem SpA during 1982 and 1983 using their third generation laybarge *Castoro 6* supported by laybarge *Castoro 5* and trenching barge *Castoro 8*. The lines were constructed according to regulations specified by the Danish Energy Agency, whose rules generally comply with ANSI B31.8 and B31.4 for the gas and oil pipeline respectively, and current DnV pipeline rules.

Since know-how on design and construction of large diameter marine pipelines was limited in Denmark when the design of the 30-inch gas pipeline was commenced in 1978, extensive survey and engineering activities were initiated to establish environmental data for a safe and reliable design.

For example, assays of the gas showed the presence of hydrogen sulphide (H_2S) as well as carbon dioxide (CO_2). Because of the potential corrosion problems associated with these gases, detailed studies of pipe materials were initiated. When design of the condensate and crude oil pipeline was initiated early in 1981 it was found that the established seabed conditions, wave and current data, as well as the material specifications, proved extremely valuable, and resulted in considerable time and cost savings.

The North Sea pipelines have been routed to minimise crossing of hard seabed areas. The majority of the seabed will be easily trenched sand with maximum water depths not exceeding 55 metres. The shore approach has required special engineering attention. Wave and current actions generate huge material transportation along this coast and result in a constantly varying bottom profile in the surf zone. Consequently, the pipelines are trenched below the lowest recorded bottom envelope in the surf zone to protect them. They were towed to a laydown area approximately 2 kilometres from shore and then pulled to shore and installed in a pre-excavated/dredged pipeline trench.

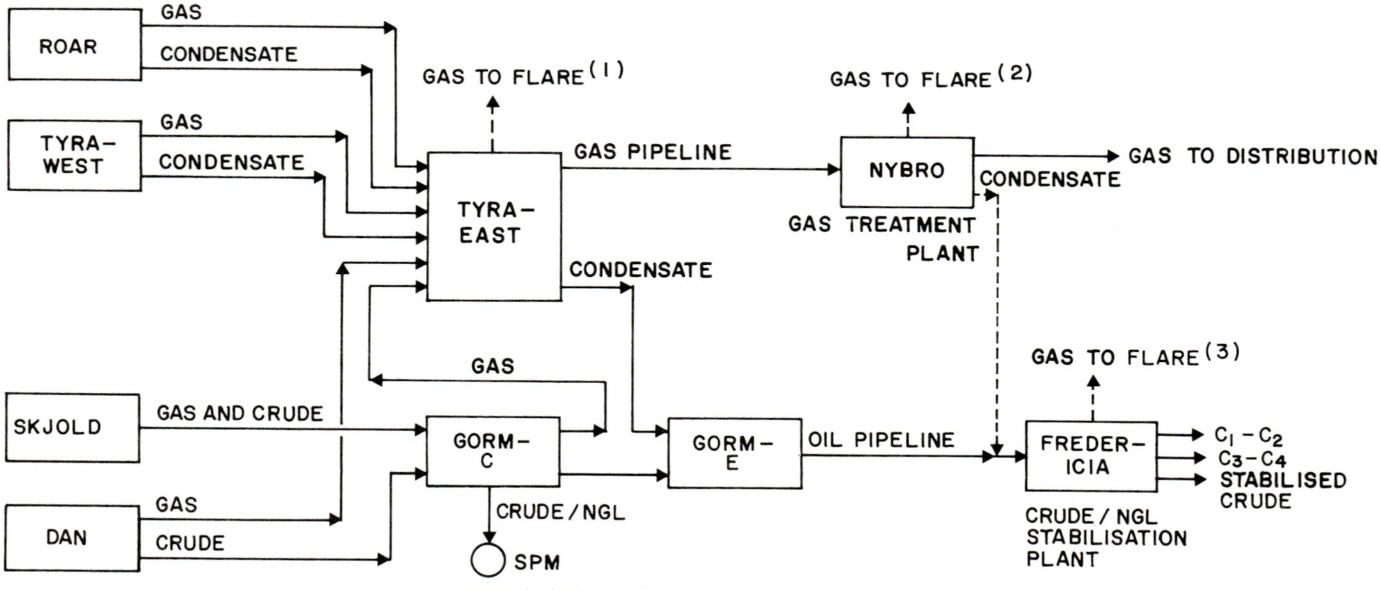

LEGEND

—— NORMAL PROCESS FLOW

- - - - ABNORMAL CONDITIONS

NOTES: (1) IF a 12 RVP condensate is produced at TYRA – E CPA, it will be necessary to flare intermediate Hydrocarbons or send a rich gas through the gas pipeline to NYBRO

(2) IF for any reason Hydrocarbon liquids are received or formed at NYBRO, the liquid stabilisation process will most likely require the flaring of some gas.

(3) IF for any reason stabiliser offgas exceeds gas handling equipment capacity and fuel requirements, some flaring may occur.

Figure 17.1.2 Schema for the Danish Hydrocarbon Processing System

The 30-inch gas pipeline is designed for a maximum operating pressure of 139 bars and it will use grade X 60 material (762 millimetre OD × 18.7 millimetre wall-thickness) supplied by Mannesmann and Vallourec. The 20-inch crude oil and condensate pipeline is designed for a maximum operating pressure of 105 bars and it will use grade X 60 material (508 millimetre OD × 12.7 millimetre wall-thickness) supplied by the Italsider pipemill. Both pipelines have expansion loops at their platform tie-in points and the gasline is also equipped with a subsea valve arrangement.

Both pipelines are corrosion and weight coated and they are trenched to provide stability and protection against accidental damage. Trenching provides a depth from seabed to top of pipe of approximately 1 metre for the 30-inch gas pipeline from the shore approach to approximately 80 kilometres offshore and at least 0.2 metres for the remaining part, whereas the 20-inch crude oil pipeline is trenched to provide approximately 1 metre of cover for the entire route.

The 1.5 kilometre-long Lillebælt gasline crossing between Jutland and Fyn and the 30 kilometres long Storebælt gasline crossing between Fyn and Zealand, have given rise to special engineering problems relating to the provision of a sufficient degree of protection against traffic-induced damage. A dual, active redundancy pipeline design has been employed to ensure that gas supply can be maintained under conditions when one pipeline may have to be taken out of service for mechanical repair, even though the probability of such an occurrence is small.

The Lillebælt crossings shown in Figure 17.1.3 (overleaf) were installed by bottom-tow in a pre-dredged trench cut into the seabed Lillebælt clay, which is a heavily overconsolidated clay of extreme plasticity. Some engineered backfill may be required to satisfy the authorities that proper protection is provided. By contrast the Storebælt crossings shown in Figure 17.1.4 (overleaf and page 234) were installed by conventional lay-barge techniques.

Extensive safety studies were made to establish the optimal routing of the gasline. Some of these studies are described in Chapter 14. Because ships negotiating the Storebælt regularly exceed 200,000 dwt, and due to the muddy seabed conditions, it has been impossible to devise a completely reliable pipeline protection concept. It is reckoned that the most important risk factors at these locations will be damage caused by ships grounding or anchor damage caused by ships executing emergency stopping procedures, as might occur, for example, after collisions.

Given the particular environmental conditions, it was discovered that the single most effective way to enhance the safety of the pipelines was to locate the crossings in an area with the least complex and intensive traffic pattern. In the main, both of these pipelines have been trenched 1 metre below the natural seabed level, but 2 metres of trenching and engineered backfill were required in exposed areas with increased risk of ships grounding.

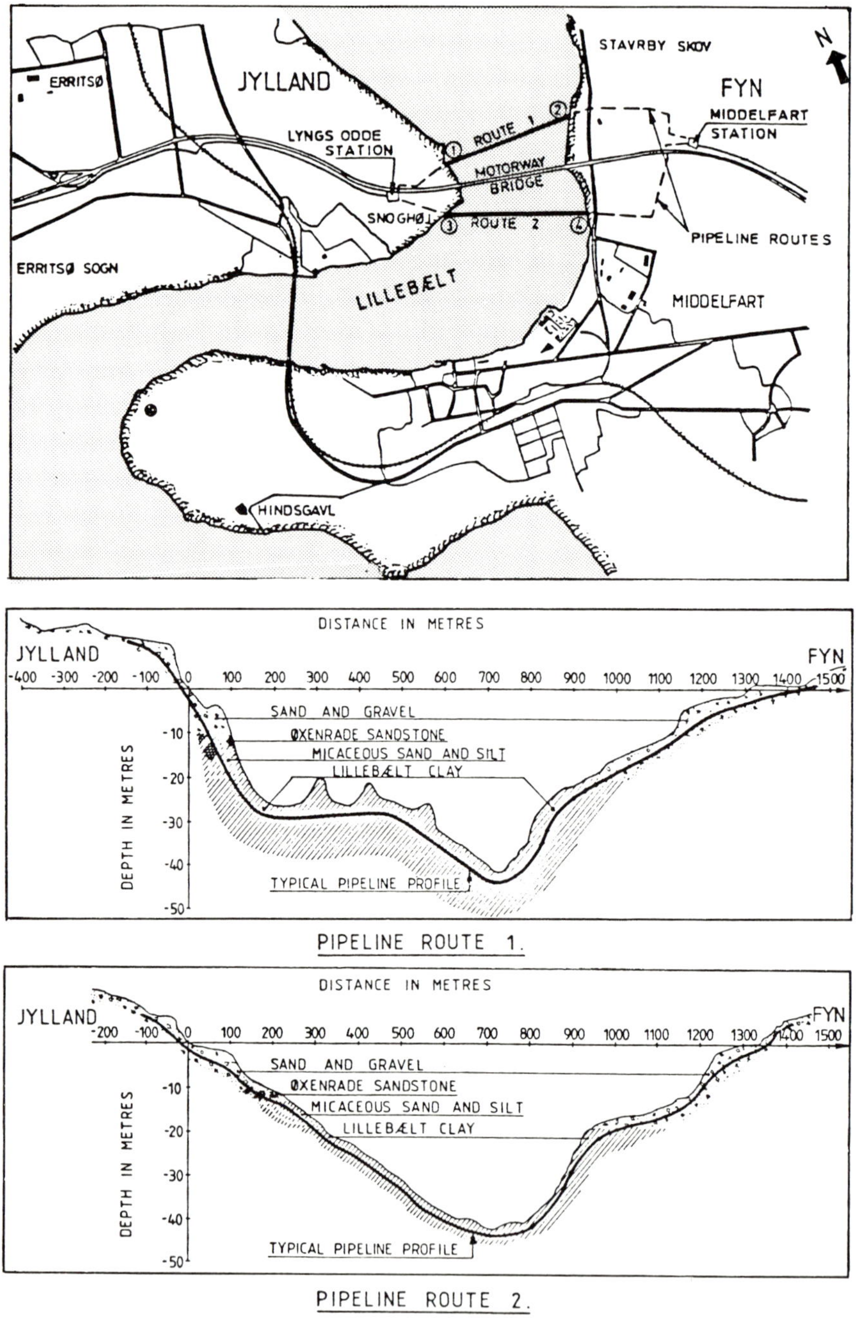

Figure 17.1.3 Bathymetric and Geological Profiles for the Danish Natural Gas Transmission System, Lillebælt Crossing

17.2 The Topography of the Optimum Pipeline Route in the North Sea

Bottom topography and soil conditions in the North Sea were determined by bathymetry and soil surveys using conventional echo sounding, side-scan sonar, sub-bottom seismic profiling, vibrocore and drop sampling techniques in addition to cone penetrometer and plate load tests *in situ*. The friction factors deduced from these surveys and tests are shown in Figure 17.2.1 (see pages 236 and 237).

The deepest part of the selected route is in the Helgoland Channel, which comprises a 35 kilometres broad depression running north–south and with a maximum water depth of 53 metres. To the west lies the relatively flat North Sea bank, known as the Dogger Bank, with water depths

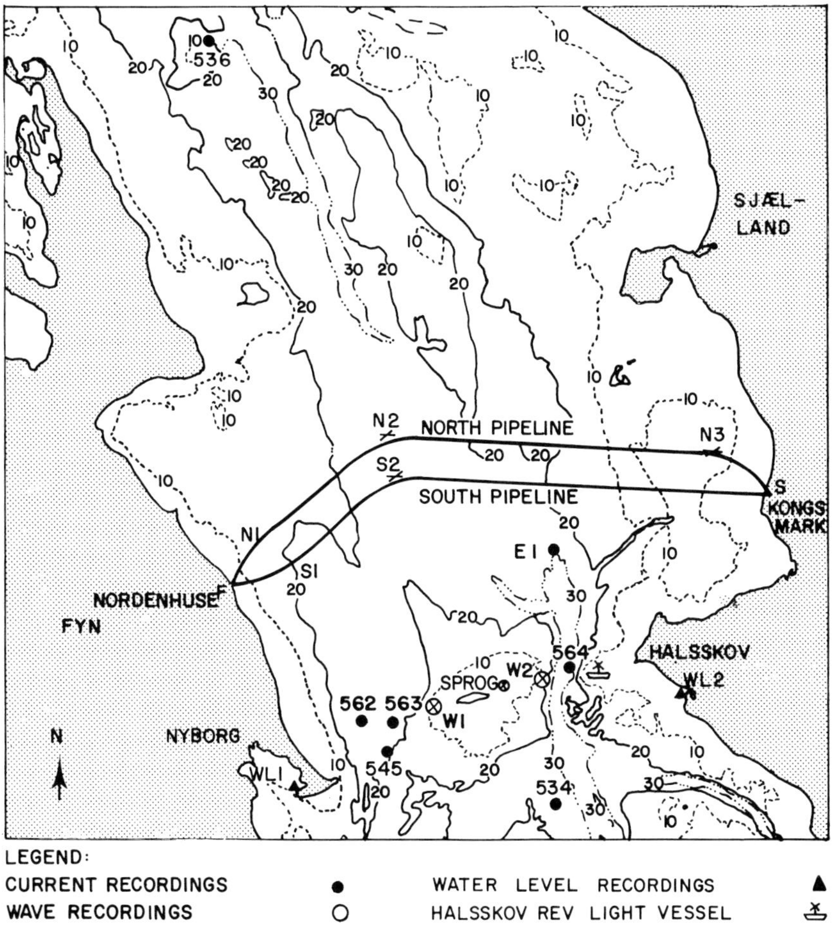

Figure 17.1.4 The Route, Chainage, Residual and Tidal Currents of the Danish Natural Gas Transmission System, Storebælt Crossing. (Part (a); Route)

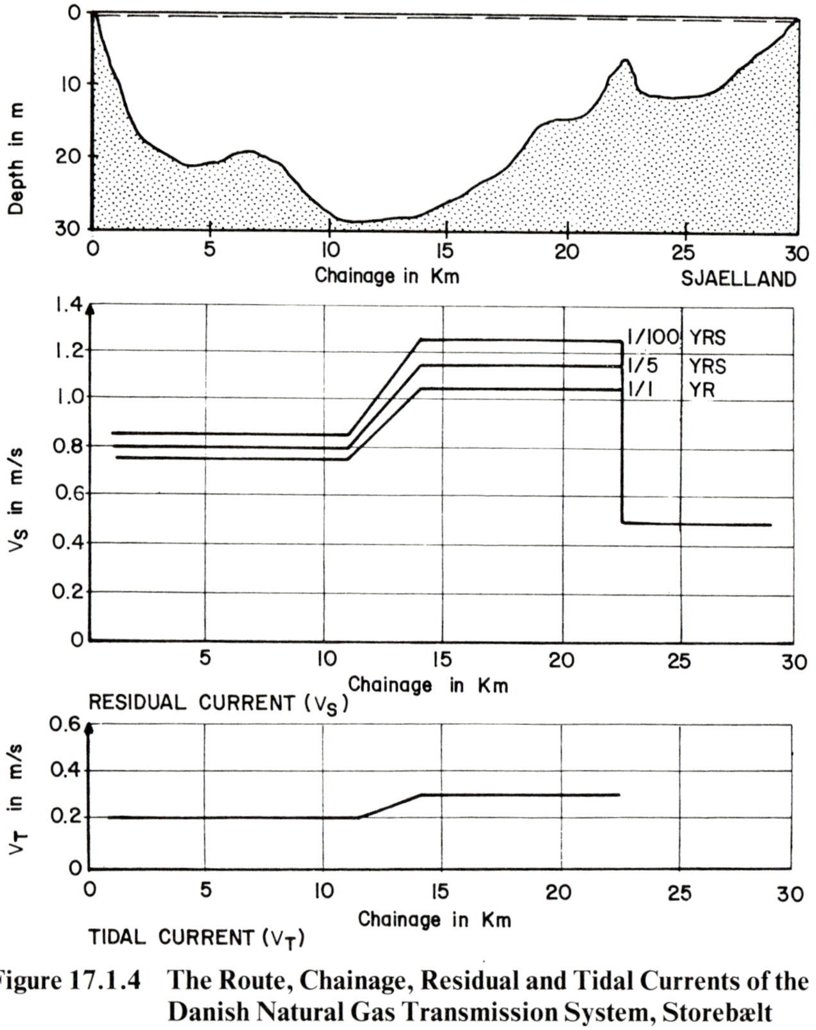

Figure 17.1.4 The Route, Chainage, Residual and Tidal Currents of the Danish Natural Gas Transmission System, Storebælt Crossing. (Part (b); Chainage, Residual and Tidal Currents)

of 37–40 metres. From the Helgoland Channel the seabed slopes gently upwards to a depth of 32 metres at Ch.80. Between Ch.80 and Ch.40, the seabed rises in a more irregular manner to a depth of 18 metres. Over this length, undulations with 6 to 7 metre variations in bottom elevations over distances of approximately 2 kilometres are met. A plateau with depths between 15 and 20 metres exists between Ch.35 and Ch.4, and the route terminates with a 4 kilometres section to shore.

Two migrating sand-bars are found within 1 kilometre from the coastline, with heights between trough and bar top approximating 4 metres for the outer bar and 1.5–2 metres for the inner bar. The predominant bottom feature along the pipeline route is the smooth seabed of recent marine sand deposits with silts and soft clays in places. Older sandy moraine clay outcrops are also found locally in some places. In the Helgoland Channel (Ch.142–Ch.191.5), however, the dominant soil type on the seabed is either 'silt, sandy' or 'sand, silty' containing 10–50 per cent silt. For these reasons the soil friction factors given in Table. 17.2.1 were used in the pipeline design studies.

Table 17.2.1 Soil Friction Factors Used in the DONG North Sea Pipelines Study

LOCATION	SOIL FRICTION FACTOR (f)
Shore to Ch 109	0.8
Ch 109 to Ch 142	0.6
Ch 142 to Ch 191.5	0.4
Ch 191.5 to platform	0.75

17.3 The Analysis of Wave and Current Conditions

The study programme of the wave and current conditions pertaining to design of the North Sea pipelines included the following main tasks:

(i) collection and analysis of existing wave and current data;
(ii) hindcast of extreme wave and current conditions during historical storm events;
(iii) field measurements and analysis of waves, currents and wave-induced orbital velocities at the seabed during the period November 1979 through April 1980; and
(iv) a combined evaluation of the results of tasks (i) through (iii) to arrive at design values of the various parameters characterising waves and current.

Many of the detailed considerations of these studies are included in Chapter 10.

The most important qualitative results of these studies are listed below:

(i) The numerical model of waves and currents gave an accurate description of the angle of incidence and the peak distribution of currents. Furthermore, it showed that the simultaneous occurrence of high waves and high current is relatively non-frequent.
(ii) A water depth of 30–40 metres is shallow for the deep water extreme wave generated in the North Sea. This results in a wave height reduced by as much as 20 per cent, a reduced wave period, and a different wave spectrum compared with that which would otherwise prevail in deep water.
(iii) The significant wave-induced velocities in the shallow water of the North Sea exceed those calculated by the first order theory and the

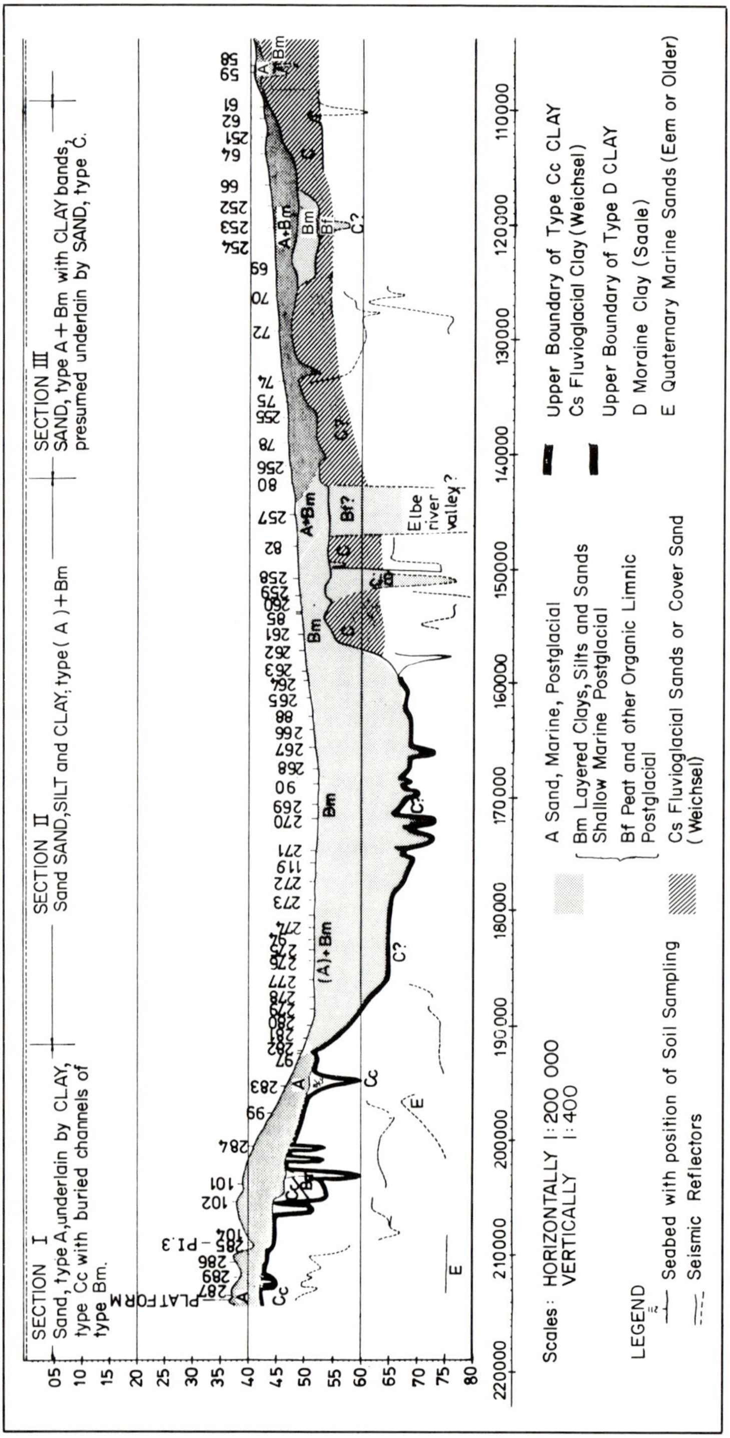

Figure 17.2.1 The Chainage, Topography and Soil Conditions along the North Sea Pipeline Route

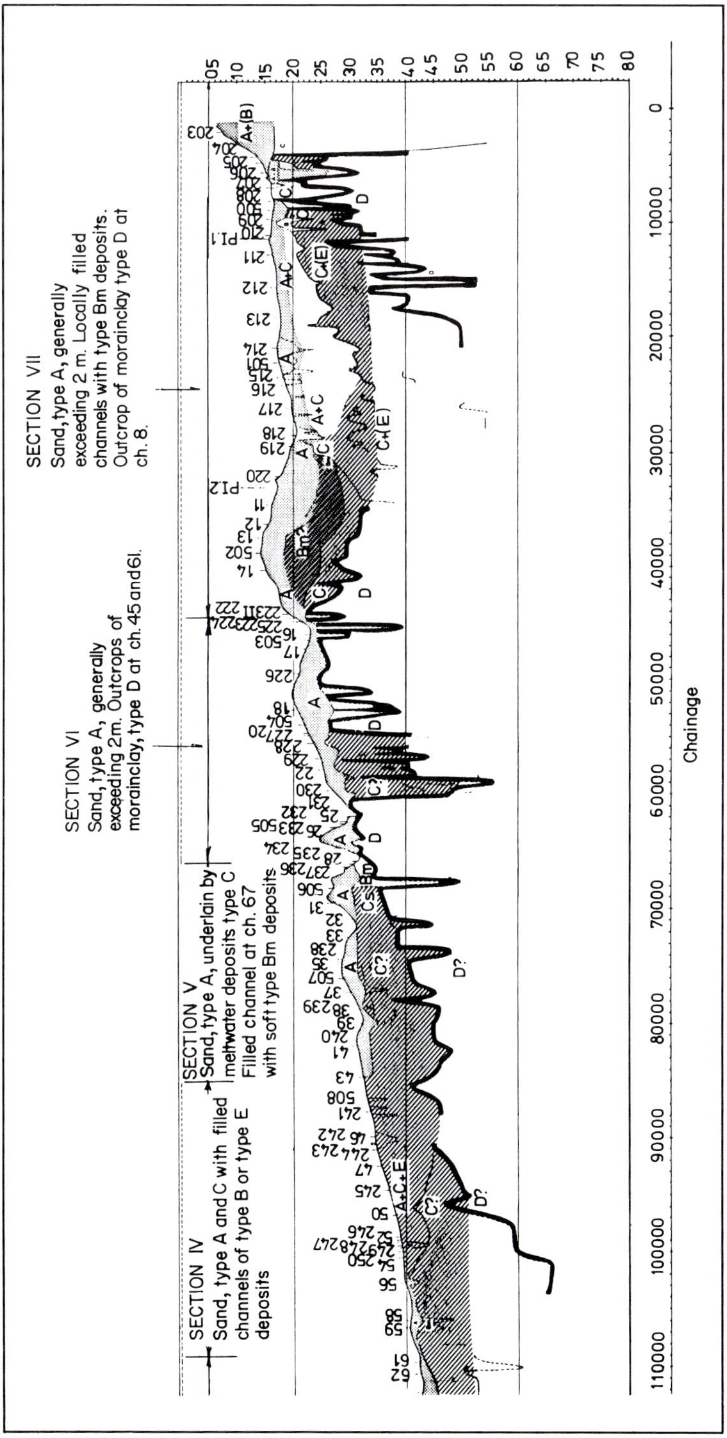

Figure 17.2.1 The Chainage, Topography and Soil Conditions along the North Sea Pipeline Route *continued*

Pierson-Moskowitz spectrum by up to 15 per cent. (See Pierson *et al*, 1966).

Collectively, these results affect profoundly the stability of the pipeline. This matter receives proper attention in Chapter 10 but some of the more interesting results are shown below in Figure 17.3.1.

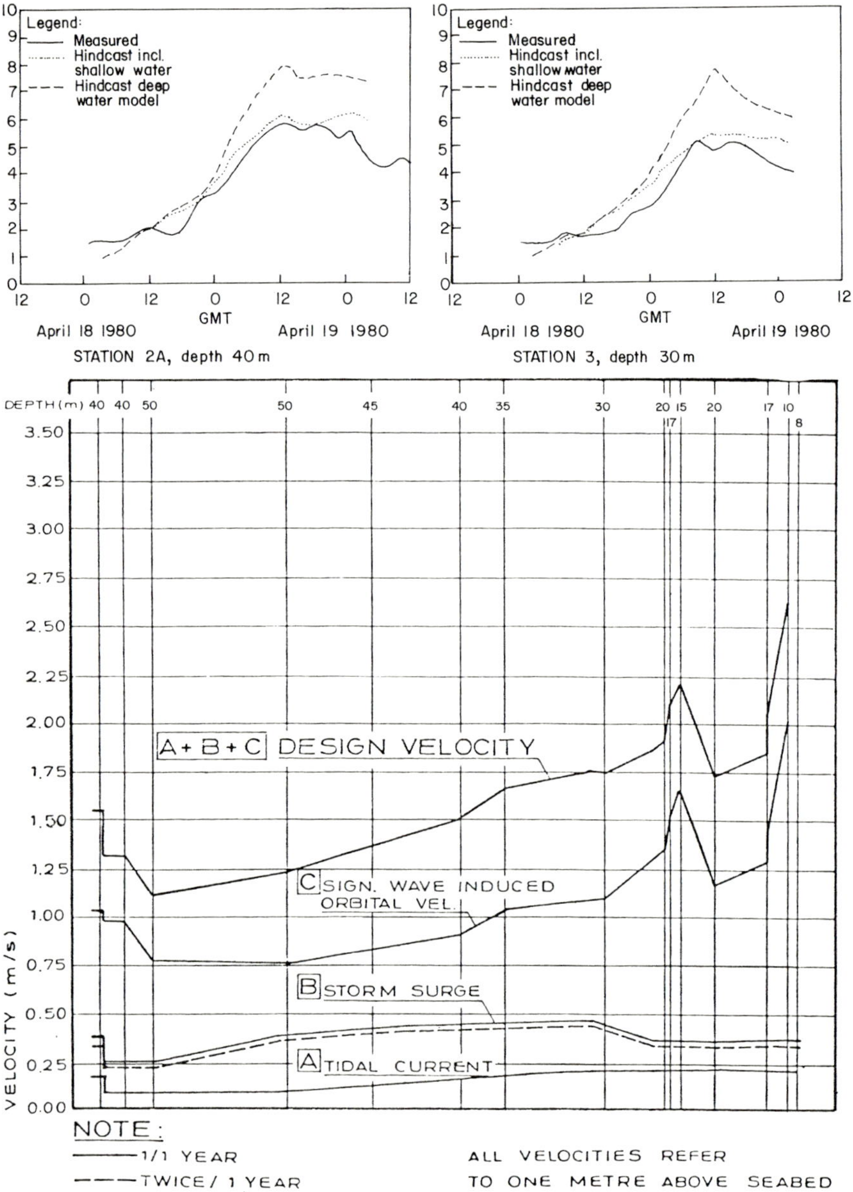

Figure 17.3.1 Comparison between Hindcasted and Measured Significant Wave Heights

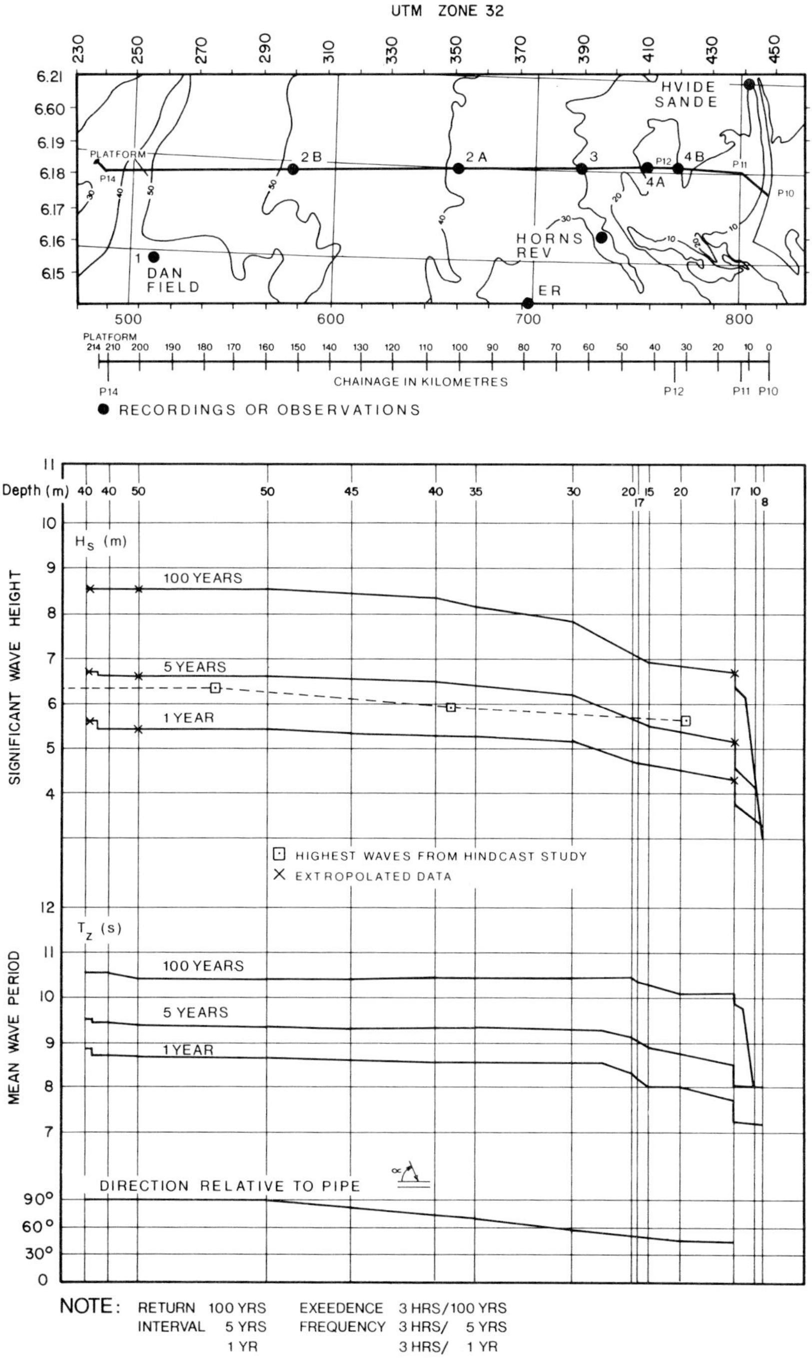

Figure 17.3.1 Comparison between Hindcasted and Measured Significant Wave Heights *continued*

17.4 Gasline Design and Construction

The North Sea gasline originates at the Tyra Field TCP-E platform and runs through three turn points to a landfall point near Kærgaard and continues to the Treatment Plant at Nybro. It circumvents the route to be taken by the future Dan-Tyra gas pipeline and avoids the Oksbøl artillery range, bringing the total length of the offshore pipeline section to 214 kilometres. In addition to the offshore part of the pipeline, there is a 15 kilometre onshore section from the landfall near Kærgaard to the treatment plant at Nybro. From the foot of the platform riser, the pipeline runs through an expansion offset and through a subsea valve assembly comprising a check valve and a block valve, located approximately 1.5 kilometres from the platform complex, and then continues to the onshore tie-in point connection into the overland pipeline leading to the treatment plant.

Isolating couplings have been installed at the riser, at the tie-in points between the offshore and onshore pipeline section, and again at the interface line to the treatment plant. A block valve complex has been installed in the landline section at a short distance from shoreline.

During normal operation the gases flowing ashore through the North Sea pipeline may be a mixture taken from the Dan, Gorm, Roar and Tyra fields located in the Danish sector of the North Sea. The principal contributor to the total gas flow-rate however, will be the Tyra gas/condensate field, the other gas-streams being associated with crude oil production. The gas is treated before it enters the pipeline to ensure that water vapour and hydrocarbons will not condense under normal operating pipeline conditions. The estimate compositions of each gas-stream at inlet to the pipeline are as shown in Table 17.4.1.

Table 17.4.1 Compositions of the Various Gas Feedstocks to the 30-inch DONG Gasline

	NORMAL OPERATION			UPSET CONDITION
GAS COMPONENT	Dan Field	Gorm Field	Tyra Field	Tyra gas before CTE-A chilling
N_2	0.297	0.60	0.25	0.25
H_2S	0.000	0.00	0.08	0.08
CO_2	0.000	0.74	1.13	1.13
C_1	92.887	82.50	89.62	89.01
C_2	4.113	9.14	5.18	5.22
C_3	1.230	5.67	2.25	2.36
iC_4	0.466	0.42	0.43	0.48
nC_4	0.507	0.75	0.69	0.82
iC_5	0.298	0.09	0.16	0.22
nC_5	0.112	0.08	0.13	0.20
C_6	0.074	0.01	0.08	0.21
C_7	0.014	Trace	—	0.03 (C_7+)
C_8	0.001	—	—	—

The pipeline system has been designed to transport potentially sour gas at a maximum operating pressure of 139 bar with a capacity between 25 and 30 million Nm³/day. Initially the maximum contracted flow-rate will be 10.88 million Nm³/day. The principal design features of this pipeline are provided in Table 17.4.2.

Table 17.4.2 Principal Design Features of the 30-inch DONG North Sea Gasline

Operations, design lifetime	30 years
Pipe material	X 60
Line pipe type	API 5LX longitudinal welded (SAW)
Pipe outer diameter	762 mm
Pipe wall-thickness	18.7 mm
Maximum operating pressure	139 bar (gauge)
Maximum operating temperature	50°C
Minimum yield strength	413.7 MPa
Corrosion protection	Zinc anodes; coal tar wrap
Concrete thickness	73–107 mm

Under normal operation, the gas will be dry and non-corrosive. However, during conditions when the gas composition does not comply with the sales gas specification, free water may form in the pipe and create a corrosive environment in conjunction with H_2S and CO_2. Corrosion inhibitors will be injected continually into the gas under these conditions to reduce the risk of corrosion. Furthermore, pipe metallurgy and weld specification have been selected carefully to contain the sour gas corrosion.

External corrosion protection is provided by a double wrap coal tar enamel comprising one layer of fibreglass reinforced wrap and one of asbestos felt wrap with applications of coal tar enamel between them. Additional corrosion protection is provided by a cathodic protection system using zinc anodes placed every ten pipe joints. Closer anode spacing will be provided near the platform, the subsea valve assembly, and the shore approach.

The pipeline is weight-coated by heavy aggregate concrete to provide lateral stability during the most severe loading conditions which, for an untrenched pipeline, occur during the construction period under the environmental loading of the one year storm. The required specific gravity of the pipeline varies from 1.4 to 1.6, corresponding to weight-coating thicknesses in the range of 73 millimetres to 107 millimetres respectively.

The majority of the pipeline was laid by a third generation laybarge. The shore approach was constructed by pulling the pipe to shore from the laybarge through a pre-dredged trench. From here, the laybarge was layed towards the platform to a laydown target area within 300 metres of the platform location. A spool piece including the expansion offset was used to connect the pipeline to the riser. The pipeline was trenched into the seabed to prevent excess environmental loading. Sufficient trench was generally accomplished by one pass of a high pressure jet sledge. In limited areas, two passes were required. The pipeline was trenched empty to avoid overstressing the pipe in the area with SG greater than 1.4.

In the platform area, the pipeline is left on the seabed in order to ensure the proper function of the expansion offset. Protection is provided by controlling all marine activities in this area.

Backfilling of the pipeline trench by natural processes is expected to occur throughout most of the pipeline route, but as backfilling is not required to ensure pipeline stability, no artificial backfilling is anticipated.

The method of constructing the shore approach is the one where the pipeline string is fabricated offshore, on board a laybarge, and then pulled ashore along a partly or fully pre-dredged trench by an arrangement of land-based winches. An open cut was established through the dunes and down to the high-water mark. A sheet piled cofferdam, internally braced and connected to an access bridge, was established and excavated to a point beyond the second row of bars, a location that could be reached by marine dredgers. In this way longshore sediment transport was diverted seaward to a point where it could be intercepted by the dredgers. The fully pre-dredged method was chosen in preference to the method of establishing a pulling profile and then lowering the pipe into its final position after the pull by means of post-trenching. The schema shown in Figure 17.4.1 illustrates most of these features.

17.5 Pipeline Design and Construction for the Storebælt Crossing

The topography of the Storebælt crossing is shown in Figure 17.5.1 (overleaf). From the landfall on Fyn the water depth increases to 20–30 metres over a distance of 3–4 kilometres. Local depressions or accumulations are found in the nearshore areas, whereas in the deeper parts with mud deposits the seabed is generally smooth, but with some minor irregular portions including moraine clay outcrops. From the centre of the Belt the seabed rises towards the east to a minimum depth of 6 metres at Musholm Ridge, approximately 7 kilometres from the Sjælland landfall. In the eastern and the nearshore western parts of the Belt there is a very stony moraine clay surface, which is covered by a thin layer of sand.

Based on this topography, the friction coefficients shown in Table 17.5.1 (see over) have been used to design this pipe crossing.

Immediately to the south of the pipeline corridor, the flow in Storebælt concentrates in two channels, separated by an island. In particular the eastern channel, reaching a depth of about 70 metres, causes the flow to concentrate and become almost jet-like. The Belt is part of a system of straits where North Sea water of high salinity is exchanged with less saline water from the Baltic. Under normal conditions this creates a situation of

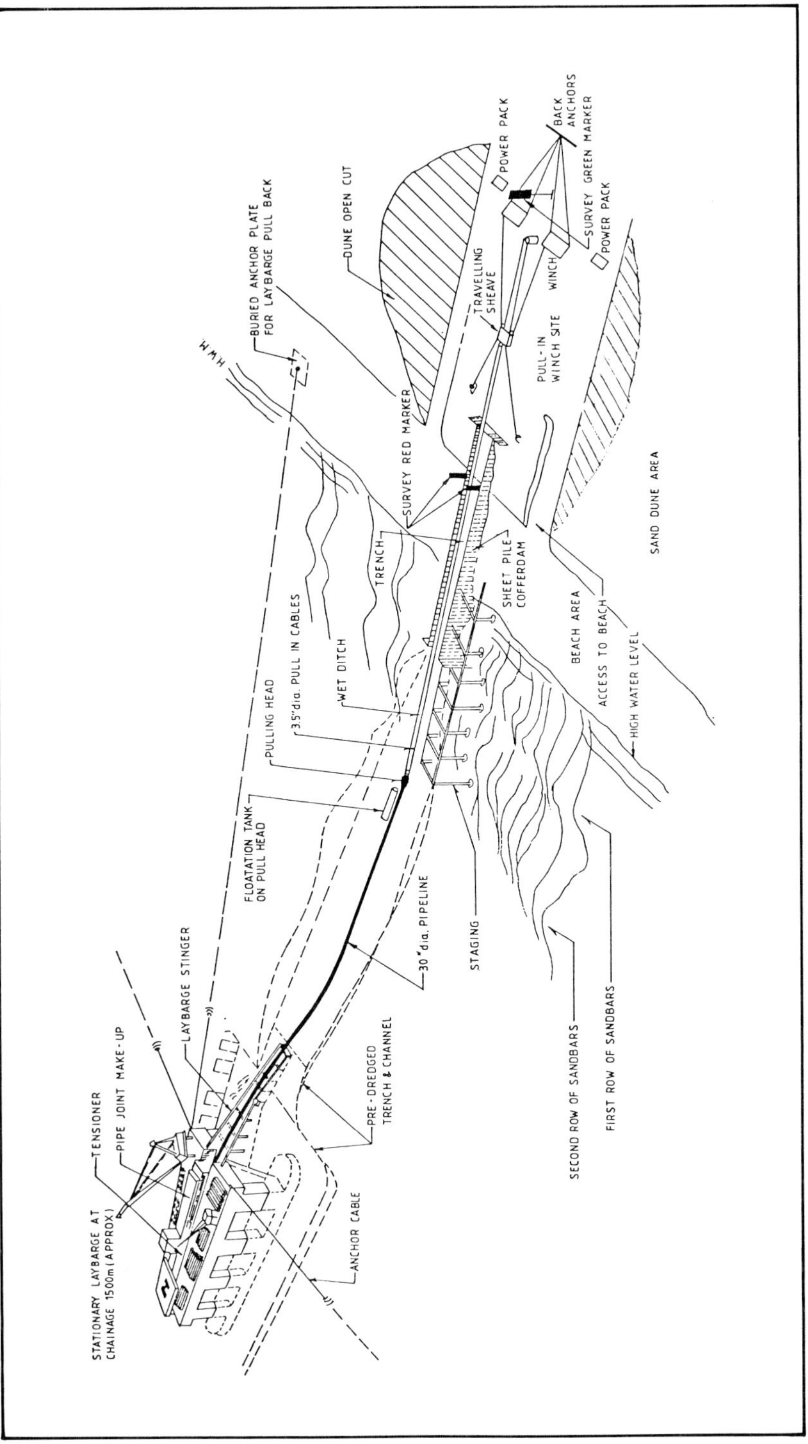

Figure 17.4.1 A Schema of the Method of Installation Using Pre-dredging

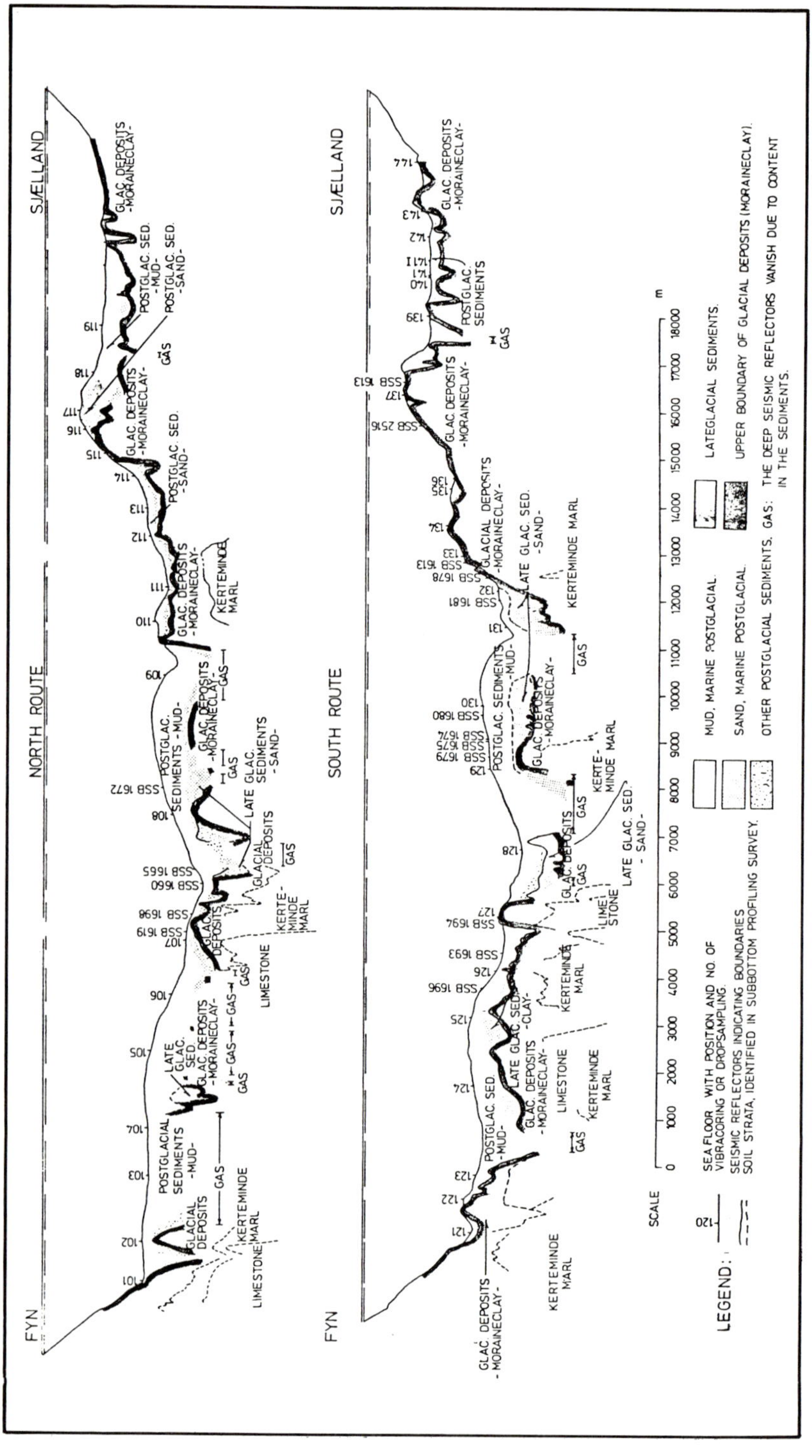

Figure 17.5.1 Topography and Soil Conditions for the Storebælt Crossing

Table 17.5.1 Soil Friction Factors Used in the Storebælt Crossing Study

AREA	SOIL FRICTION FACTOR (f)
Sand, lateral stability	0.6
Moraine clay, lateral stability	0.6
Mud, lateral stability	0.4
Sand and clay pulling	1.0

layered flows, but under extreme conditions water from either North Sea or Baltic dominates.

The current predictions, which are shown in Figure 17.5.2 (overleaf) combined with waves, showed the stability of the pipeline to be critical and, at the shallow eastern ridge, trenching and backfilling might be required. It was therefore decided to improve sea-current predictions using DHI's System 21 – a very general and versatile mathematical model for current and water level predictions. This has resulted in better predictions of peak values and an improved distribution along the pipeline routes. As a result, it has been possible to reduce the concrete weight coating in several regions of the pipeline.

The principal design features of the Storebælt pipeline are provided in Table 17.5.2.

Table 17.5.2 Principal Design Features of the Storebælt Pipeline

Operation design lifetime	30 years
Pipe material	X 65
Linepipe type	API 5LS Spiral Welded (SAW)
Pipe outer diameter	762 mm
Pipe wall-thickness	12.7 mm
Pipe wall-thickness (Sjælland end)	14.3 mm
Maximum operating pressure	80 bar (gauge)
Maximum operating temperature	50°C
Minimum yield strength	449 N/mm^2
Number of pipelines	2
Corrosion protection	Zinc anodes; coal tar wrap
Concrete thickness	80–115 mm

The corrosion coating for this crossing is coal tar enamel with two layers of fibreglass reinforcement. The cathodic protection is achieved by means of anode bracelets located on the pipeline at every fourth pipe joint (every 48 metres). Pipeline welds (field joints) are corrosion coated by a cold tape wrap of bituminous rubber on a PVC carrier tape. The mechanical protection of the field joints is improved further by a hot bitumastic which is set around each joint.

The pipeline is trenched in Storebælt to a minimum depth of 1 metre. However, it should be noted that for a considerable length of the deep-water part of the crossing, trenching is not really necessary for stability against hydrodynamic forces. In certain areas, where the risk of ship groundings is considerable, the pipeline is trenched to approximately 2 metres. Laboratory simulation tests have been done to study the forces induced on a pipeline located in an open trench resulting from the passage of a 300,000 dwt tanker with a clearance of only 1 metre from its keel to the seabed. These tests have shown that the pipeline will remain stable. This aspect is covered in more detail later in Section 17.8.

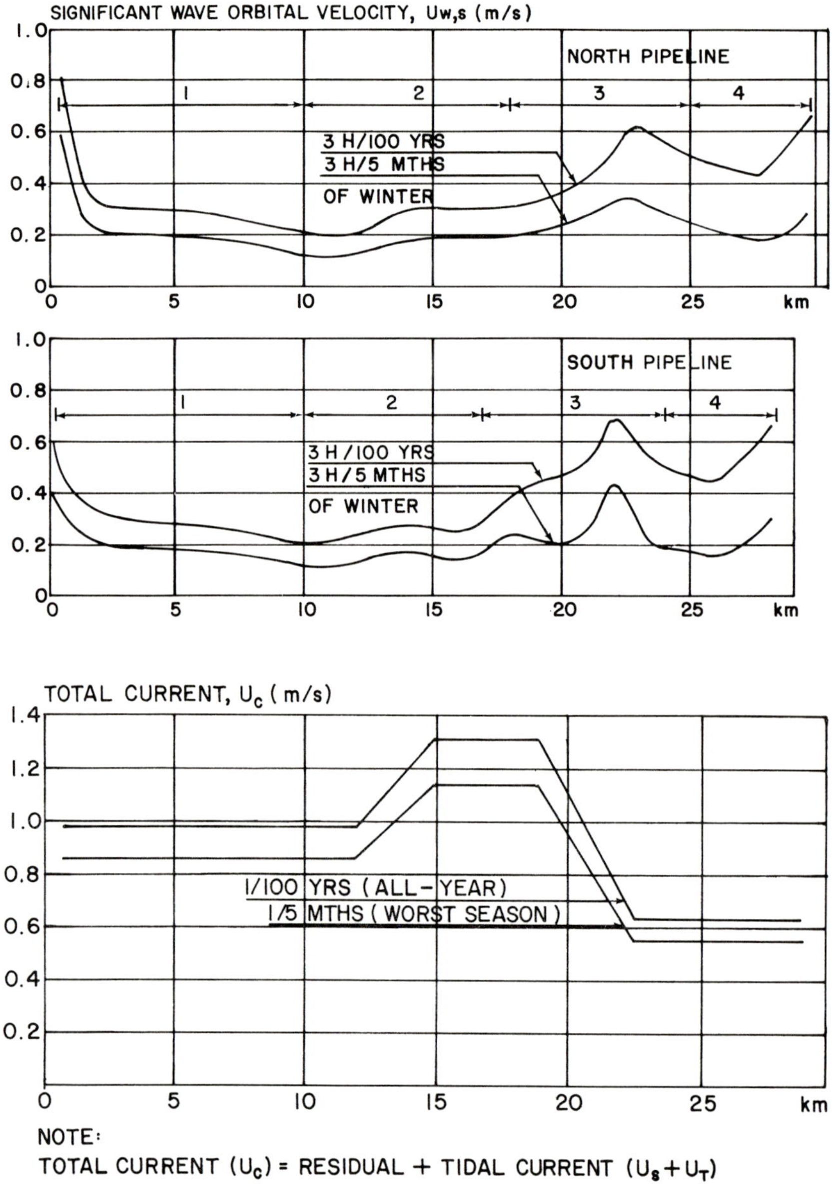

Figure 17.5.2 Current and Wave Predictions for the Storebælt Crossing

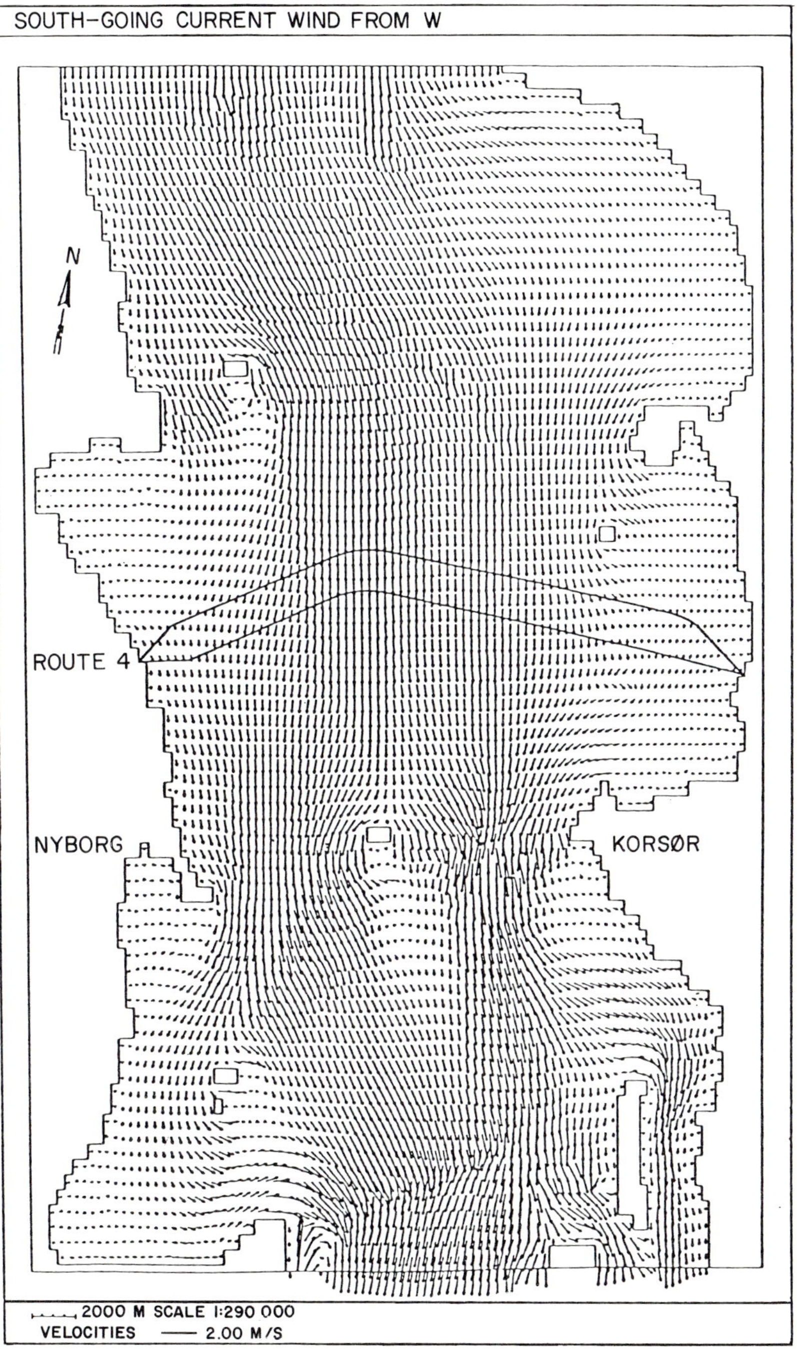

Figure 17.5.2 Current and Wave Predictions for the Storebælt Crossing *continued*

J

The Storebælt pipelines were installed by the usual laybarge method. Landfalls at Fyn and Sjælland were made by towing the pipe-string ashore, through a pre-cut trench. Following its installation, the rest of the pipe was trenched to an average depth of 1 metre and inspected for unacceptable free spanning. Hereafter the pipelines were pressure-tested to 134.4 bar, equivalent to 90 per cent SMYS. At the two shore approaches out to the 8 and 10 metre water depth contours respectively, the trenched pipeline is covered with backfill material.

17.6 Pipeline Design and Construction for the Lillebælt Crossing

The topography of the northern route of this crossing is shown in the top section of Figure 17.6.1. In the western part an uneven plateau is found at a water depth approximating 25 metres and hereafter the seabed slopes gradually into the V-shaped central part of the Belt, where a maximum water depth of 41 metres is found. The eastern slope is fairly regular.

The topography of the southern route of this crossing is shown in the bottom section of Figure 17.6.1. From the western bank the seabed slopes evenly to a depth of 42 metres of water at the centre of the Belt. The eastern slope contains a 250-metre wide plateau with water depths between 10 and 18 metres.

The predominant soil type is the Eocene Lillebælt clay, above which is found Oligocene/Miocene mica deposits consisting of clay, silt, and sand with possible concretions of Øxenrade sandstone. The mica deposits are covered with unsorted glacial deposits and melt water sediments.

The friction coefficients used in the design of this crossing are shown in Table 17.6.1. The maximum tidal currents are constant, having a typical magnitude of 0.8 m/s and the two-year residual current has a velocity of approximately 2 m/s which gives Lillebælt the character of a river, except that the currents can come from both directions.

Wave-induced currents for the two-year condition are negligible for water depths greater than 5 metres and about 0.8 m/s for water depths of 2 metres. Both wave-induced and residual currents have a pronounced seasonal variation.

Table 17.6.1 Soil Friction Factors Used in the Lillebælt Crossing Study

AREA	FRICTION FACTOR (f)
Sand, lateral stability	0.7
Mud, lateral stability	0.4
Sand and clay, pulling	1.0
Moraine clay, Lillebælt clay, lateral stability (for a pulled pipeline)	0.6

The principal design features of the Lillebælt pipeline are shown in Table 17.6.2 (overleaf).

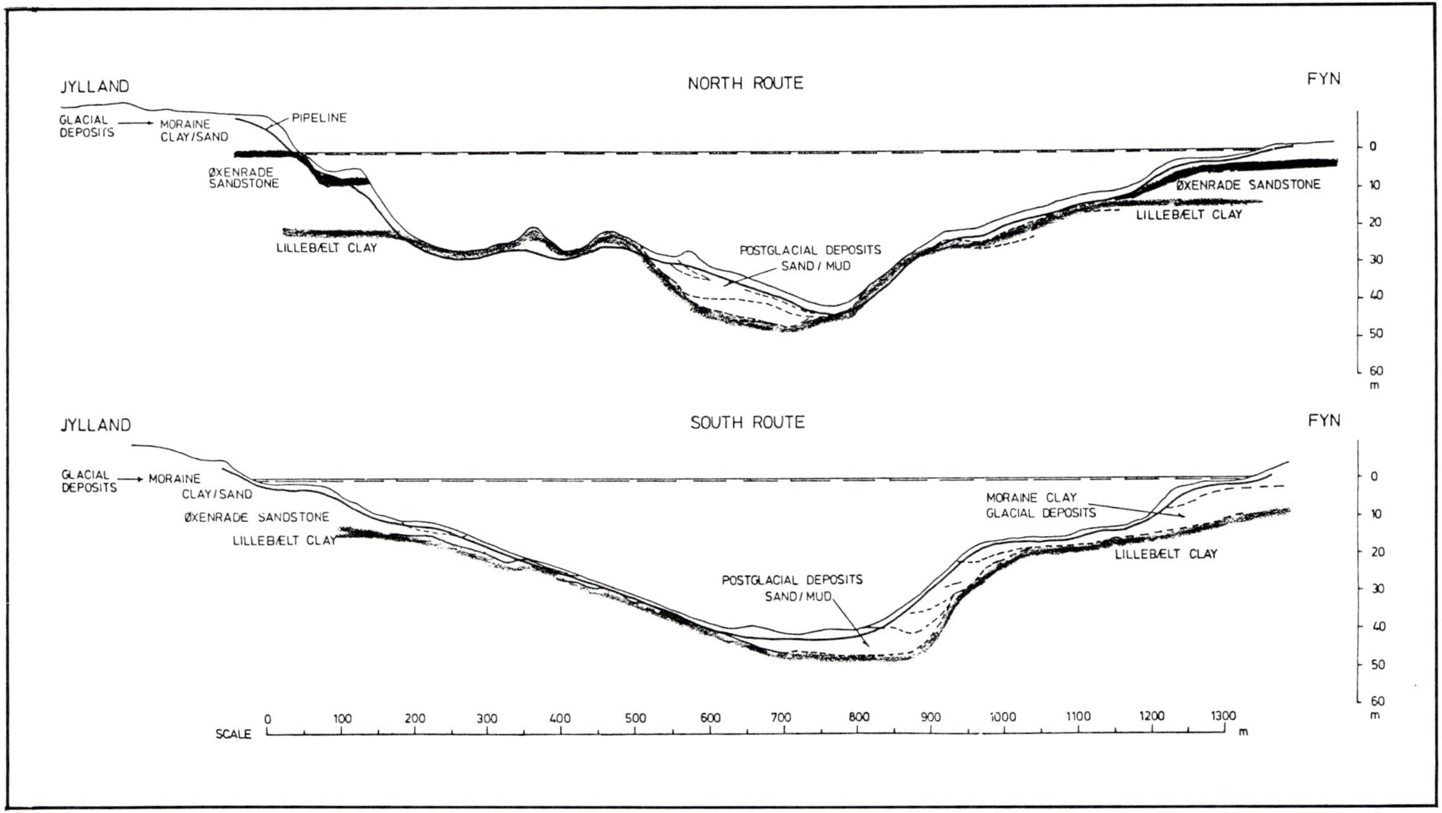

Figure 17.6.1 Pipeline Topography and Soil Conditions for the Lillebælt Crossing

Table 17.6.2 Principal Design Features of the Lillebælt Pipeline

Operation design lifetime	30 years
Pipe material	X 65
Linepipe type	API 5LS Spiral Welded (SAW)
Pipe outer diameter	762 mm
Pipe wall thickness	14.3 mm
Maximum operating pressure	80 bar (gauge)
Maximum operating temperature	50°C
Minimum yield strength	449 MPa^2
Number of pipelines	2
Corrosion protection	impressed current, coal tar wrap
Concrete thickness	80 mm

The corrosion coating is coal tar enamel with two layers of fibreglass reinforcement. Cathodic protection is achieved by means of impressed current. Field joints are corrosion coated by a cold tape wrap of bituminous rubber on a PVC carrier tape. The mechanical protection of the field joints is made by casting reinforced concrete around the joint.

The pipeline is pre-trenched and backfilled. Generally the ship traffic is not heavy in Lillebælt. However, the Ammonia Harbour close to Route 1 receives two to three 30,000 dwt ships and another 20 in the 7,000 dwt category each year. The potential risk for anchor damage to the pipeline is therefore a main point in the requirements for its protection.

In general, the backfill requirement of both routes is for 1 metre of cover, and this protection is sufficient for Route 2. For Route 1 however, additional trenching to provide a cover thickness of 2.5 metres in water depths greater than 5 metres is necessary.

The route between the shore crossing points was pre-formed by trenching to provide a bottom profile which complies with a minimum bending radius of 400 metres. For both lines, pipe assembly took place at a construction site located on the western bank because of the larger area available. Pulling was carried out by the conventional winch method. Upon completion of the pipeline installation, a survey was performed to ensure that the pipeline is within the trench. Hereafter backfilling was executed.

17.7 Pipeline Design Requirements for Horizontal and Vertical Stability

17.7.1 Designing for Horizontal Stablity

The most important design refinement during the detailed engineering stage has been associated with the concrete weight-coating requirements for on-bottom horizontal stability, where important savings have been achieved. This has been made possible by improved wave and current data, improved soil information obtained from *in situ* tests, and from using a three-dimensional wave spectrum model to analyse maximum wave stresses.

The design methodology incorporated the traditional approach using the 'significant wave height' criterion to establish the specific gravity requirements of the pipeline. This method was then checked using the 'maximum wave model' described by Gravesen and Nielsen (1981) to ensure that when the pipeline moves laterally such movements will not result in pipe stresses exceeding 60 per cent SMYS.

The rationale of the horizontal stability design of the pipeline is shown schematically in Figure 17.7.1.1 and its specific gravity requirements are shown in Figure 17.7.1.2 (overleaf).

As an aside, it is worth mentioning that if the Danish Energy Agency

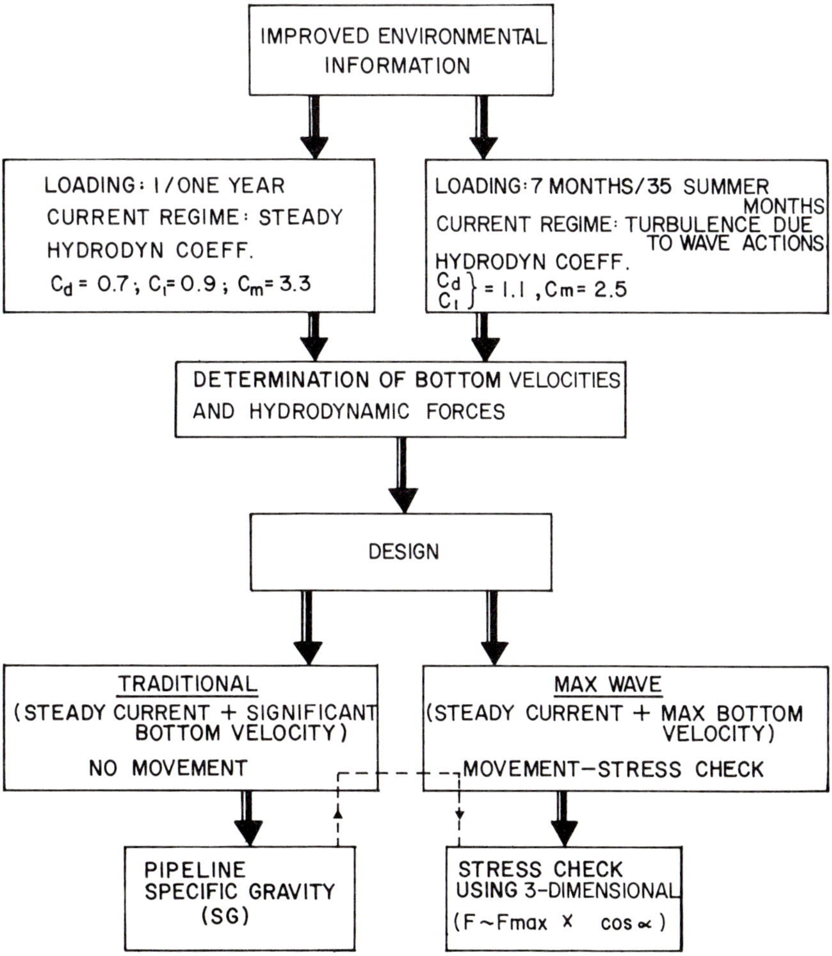

Figure 17.7.1.1 Typical Design Sequence for Horizontal Stability

had insisted on the adoption of coefficients and procedures as prescribed in the new DnV 1981 Rules, not only would this have necessitated considerable extra expenditure on additional concrete weight-coating, but the pipeline would have required a laybarge with tensioning capabilities beyond those currently available.

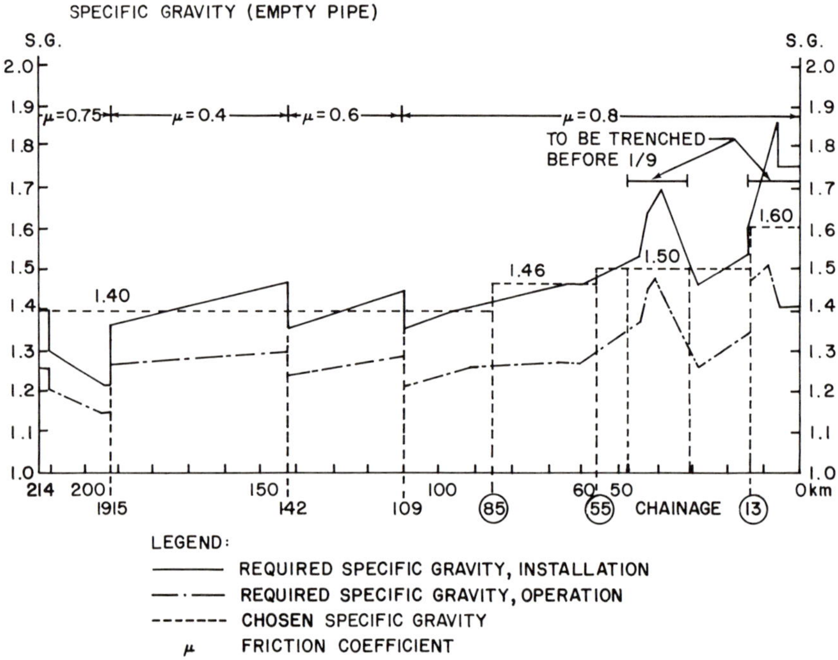

Figure 17.7.1.2 Specific Gravity Requirements for Horizontal Stability

17.7.2 Designing for Vertical Stability and Liquefaction

A preliminary analysis of the vertical stability of the pipeline showed that the possibility of flotation from Ch.38.5 to Ch.58.5 could not be dismissed. However, it was realised that the preliminary results were conservative insofar as the effects of dissipation of the excess pore pressures had not been taken into account. When these effects were considered properly it was shown that no significant risk of flotation existed during the installation season. The zone where there existed the greatest risk of liquefaction was at the shore approach, which could have caused problems if the sand had been finer. During the first winter period the pipeline was flooded however, so that it attained a specific gravity approximating 2, which gave adequate safety against flotation even in the event of liquefaction.

17.8 The Influence of Ship-induced Forces on the Stability of the Pipeline

A study was made of the loads exerted on a pipeline due to the passage of large tankers with a very small clearance between their keels and the pipeline. The investigation included the predicted worst case of a 300,000 dwt supertanker making 12 knots over a pipeline with a clearance between keel and seabed of 1 to 3.5 metres. Trench depths of 1 and 2 metres from natural seabed to top of pipeline were included in the model tests performed at the Danish Hydraulic Institute in association with the Danish Ship Research Laboratory. Using scale model tests and normal pipeline stability criteria it was shown that the velocity field created by the passing vessel could possibly cause lateral pipe movements in the trench. However, small movements of the pipe are harmless, providing the induced stresses are safely below the specified minimum yield strength. To appraise this problem, forces in the pipe induced by the passing vessel were measured directly so that the stresses and pipe movements could be examined by traditional static structural analysis.

As shown in Figure 17.8.1, (overleaf) three main types of forces are exerted on a pipeline during the passage of a vessel, namely: a force in the forward ship direction before the vessel reaches the pipe, a combined lift and backward force just after the vessel bow has passed the pipe and a combined lift and forward thrust when the stern of the vessel passes the pipe. All three forces are of appreciable magnitude, as illustrated by Figure 17.8.2 (overleaf).

From these force combinations it was possible to analyse the induced stresses in the pipe as well as the maximum lateral pipe movement for the pipe under operational conditions. This evaluation showed that:

(i) increasing the trench depth will not reduce the effect of the passing vehicle significantly;

(ii) the ship clearance has a significant influence on the magnitude of the predominant forces. These forces will be more than halved if the ship clearance increases from 1 to 2 metres;

(iii) the total combined stress in the pipe due to internal pressure and forces induced by the passing vessel does not exceed 50 per cent of the specified minimum yield strength, so there will be no danger of overstressing the pipeline owing to the passing of such vessels; and

(iv) when the clearance is only 1 metre the maximum pipe lateral movement will be in the order of 1 metre.

17.9 Stresses Incurred in the Pipeline during Installation

In most pipeline projects, only the static pipe stresses incurred during the laying operation are considered. This complies with the DnV Rules, Section 4.3.2.6 A and B, under the assumption that the contribution to the bending moment M due to environmental loads does not exceed

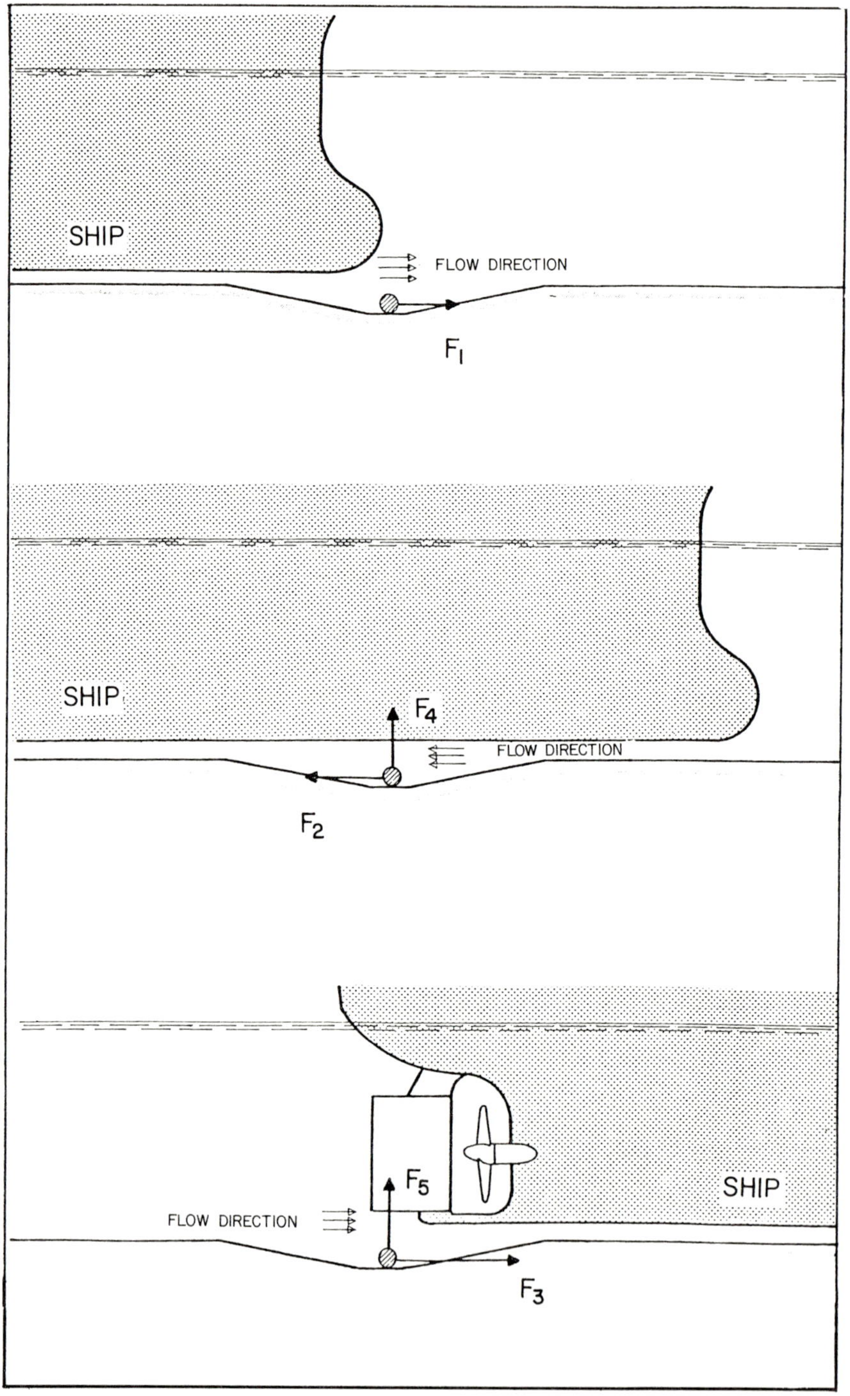

Figure 17.8.1 Forces Induced on a Pipeline by a Ship's Passage

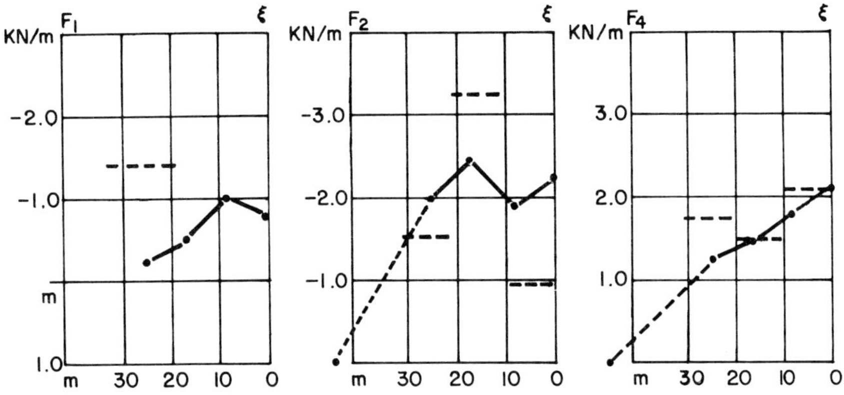

Force Combination a, Trench 1.

Force Combination b, Trench 1.

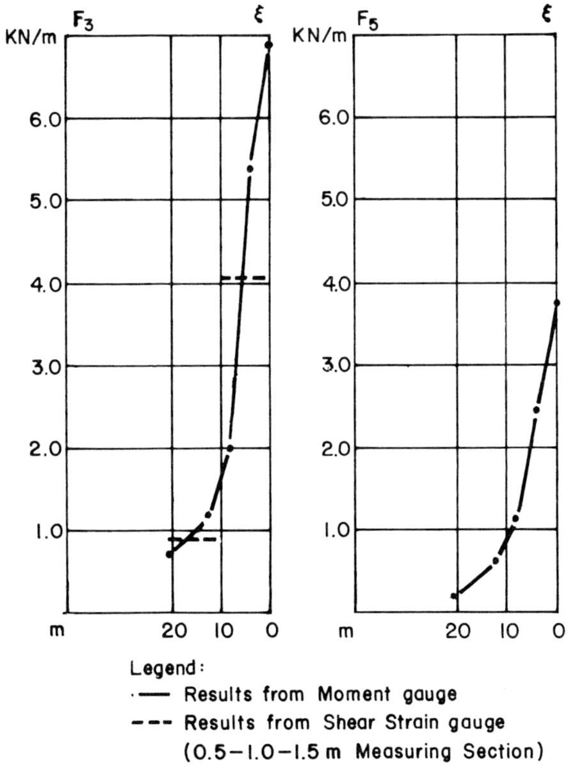

Force Combination – Trench 1.

Figure 17.8.2 The Magnitude of the Forces Induced on the Pipeline by a Ship's Passage

33 per cent of the equivalent moment due to functional loads. However, in view of adverse weather conditions in the North Sea, it was considered prudent to examine this assumption further. A computer model was used to examine the variation in pipe sagbend stresses when laying with typical installation equipment exposed to normal cyclic motions. This matter was described in Chapter 2.

The model barges were of two types, a third generation type equipped with a fixed and rigid stinger and a conventional type fitted with an articulated stinger. Realistic limits to normal operational conditions were reckoned to correspond with a maximum cyclic amplitude of 1.5 metres. Using these cyclic movements and typical combinations of water depths and pipeline specific gravities, the model simulations showed an increase of bending moment M in the sagbend of up to 60 per cent relative to the bending moment in the static situation. Some of the more important results of this study are shown in Figure 17.9.1 and Table 17.9.1.

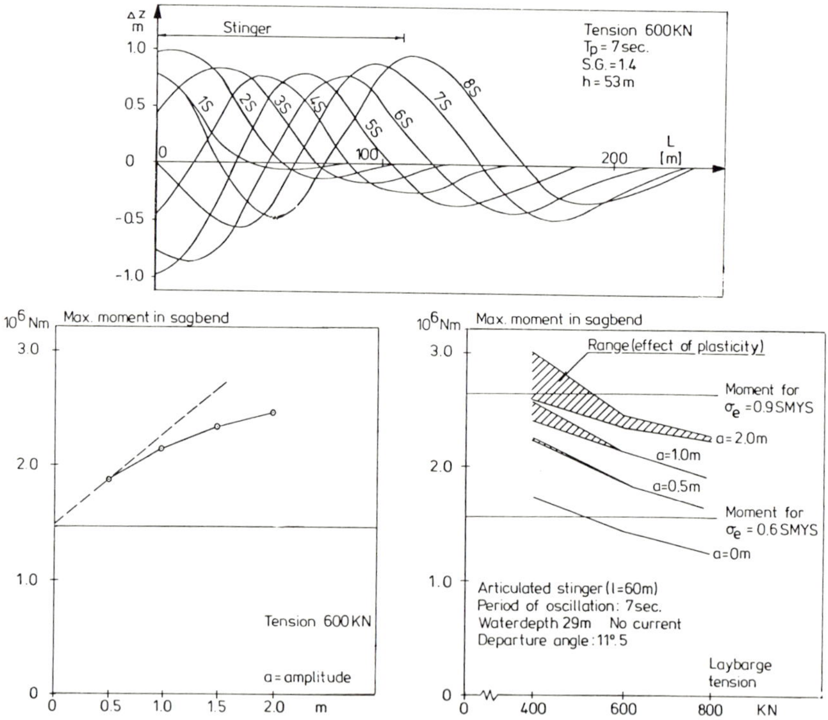

Figure 17.9.1 Stresses Incurred in the Pipeline during Installation

According to Table 17.9.1, these results suggest the adoption of a usage factor of 0.60 when computing pipe lay-stresses during functional or static loads. The combined stress, as defined in DnV Clause 4.3.2.5, allows a reduction factor of 0.85 on the actual bending moment for the purpose of calculating the bending stress. This means that if the above usage factor $\eta = 0.60$ is inserted and:

(i) axial stresses are neglected, since they normally fall below five per cent of SMYS; and

(ii) hoop stresses are neglected,

then the resulting maximum permissible static bending stresses, according to DnV, are approximately 0.7 × SMYS, by contrast with a value of 0.6 × SMYS deduced from the DHI modelling.

It should be noted that due to the sour gas service anticipated for this pipeline this requirement on pipe bending stresses during laying was tightened in the construction tender documents, so that the static longitudinal fibre stress in the pipe under normal conditions was not to exceed 60 per cent SMYS in the sagbend and, during the most unfavourable conditions of wind, wave, and current, the sagbend stress was not to exceed 90 per cent SMYS as shown by dynamic analysis and monitoring. Monitoring of lay-stresses by means of accelerometers and tension measurement was performed during pipelaying.

Table 17.9.1 Bending Moments, Usage Factors and Permissible Static Bending Stresses

Incremental bending moment M expressed as a percentage of the bending moment due to functional (ie 'static') loads	Usage factor according to the DnV Rules (ie 'static' loading conditions)	Resulting maximum permissible static bending stress, σ in % of SMYS $\frac{\sigma}{M} \approx \frac{\eta}{0.85}$
33 or less	0.72	85
50	0.64	75
60	0.60	70
70	0.56	65
80	0.53	62
100	0.48	56

17.10 The Pipeline Shore Approach on the West Coast of Denmark

The Danish west coast is exposed to the full force of the North Sea, and the average sediment transport in the approach area was estimated as 150,000 cubic metres per month during the construction phase. Moreover, during the worst day in the same period the full extent of such transportation might exceed 10,000 cubic metres.

Sedimentation in the nearshore area occurs mainly in the location of the constantly drifting sand-bars and as much as 65–90 per cent of the overall amount is deposited here. Further, the greater portion is deposited on the sand-bars, and during calmer weather conditions, along the inner rows.

Figure 17.10.1 illustrates the amount of precipitated sediments in suspension *versus* the steady current at the location of: (a) the dredged channel and (b) in the sand-bar area. However, it is anticipated that for the gas pipeline at the shore approach, the outgoing current arising from the dredged channel would reduce this problem a little.

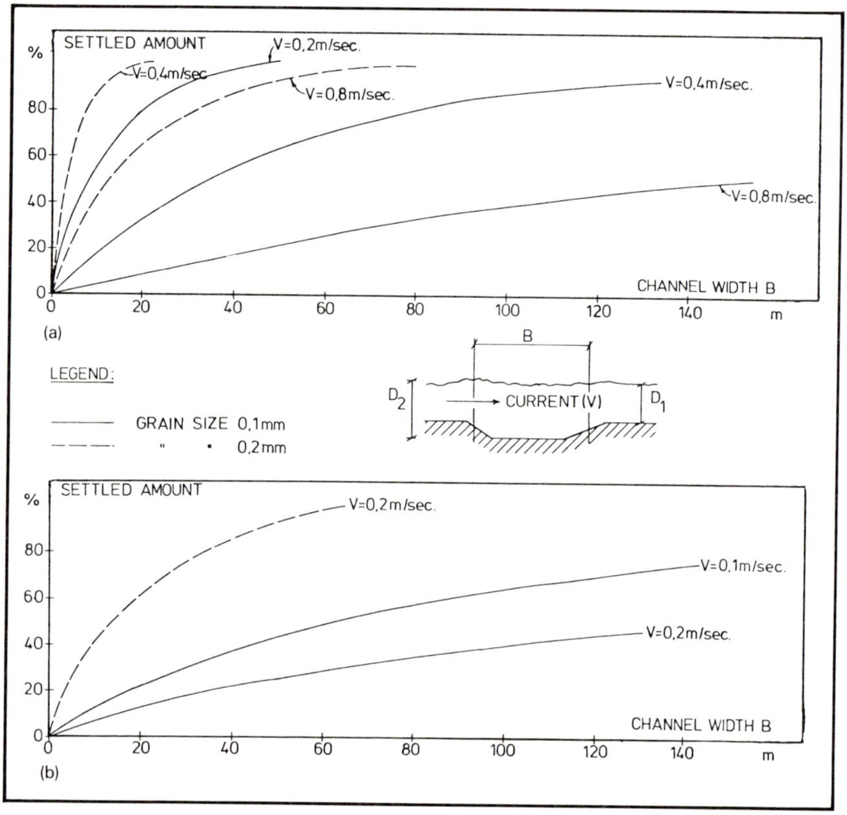

Figure 17.10.1 Sediment Precipitation at the Location of: (a) the Dredged Channel and: (b) the Sand-bar

The erosion of the beach during storms has been investigated by the Coast Authority (Kystinspektoratet) at Aargab, 30 kilometres north of the site. Unfortunately their results are unpublished, but the following information is available. According to their experiments, coloured sand-columns were arranged in lines across the beach, each line consisting of four columns. The tops of the sand-columns were levelled and restored approximately every other month during a two-year period. According to their results, which are shown overleaf in Figure 17.10.2, the maximum erosion was found to be 1.2 metres at the foreshore and 1 metre at the foot of the dunes. The corresponding average of registered erosion depths was 0.3 and 0.1 metres respectively.

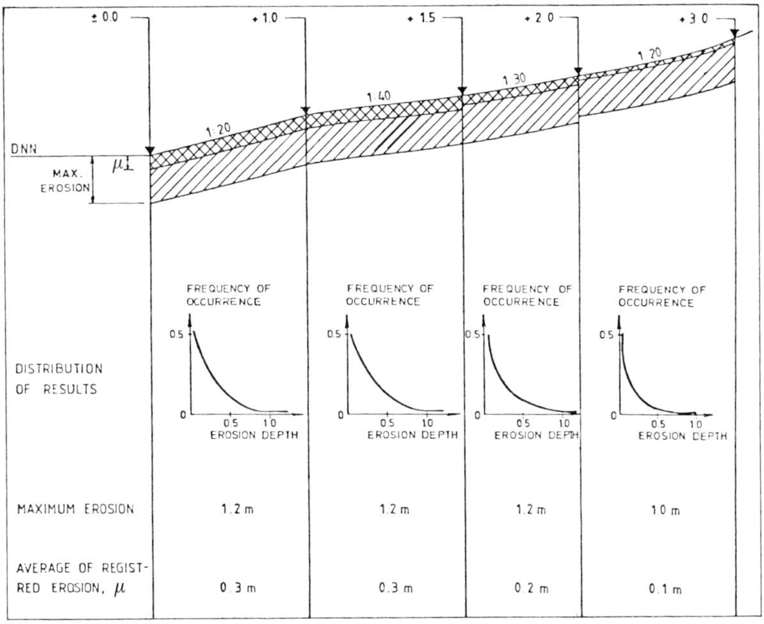

Figure 17.10.2 The Extent of Beach Erosion.

To sustain these sediment transport phenomena, the pipeline profile has been chosen so as to obtain at least 1.5 metres of cover under the lowest registered profile (1 metre on the beach) which yields a minimum of 2 metres below the present lowest seabed level (3.5 metres on the beach), as shown by Figure 17.10.3 (overleaf).

17.11 Other Design Features of the Pipeline

A pipeline expansion offset will be installed to prevent overstressing of the pipeline, riser, and riser supports when the pipeline expands due to an increase of internal pressure and temperature. The analysis was performed in accordance with the DnV Rules, 1976 edition, but with due regard for the stricter requirements prescribed in the 1981 edition. Thus, a conservative longitudinal friction factor between concrete and sand of 0.4 has been used to allow for creep effects, while the permissible von Mises stress has been reduced to 0.5 for loading condition A (functional) and 0.67 for loading condition B (functional + environmental). The spool pieces will be installed on the seafloor, and trenching will start after the last bend.

Considerable effort has been directed to environmental impact studies. These have covered the impact of sour gas release in the event of pipeline rupture, the effect of the release on the marine environment of chemically

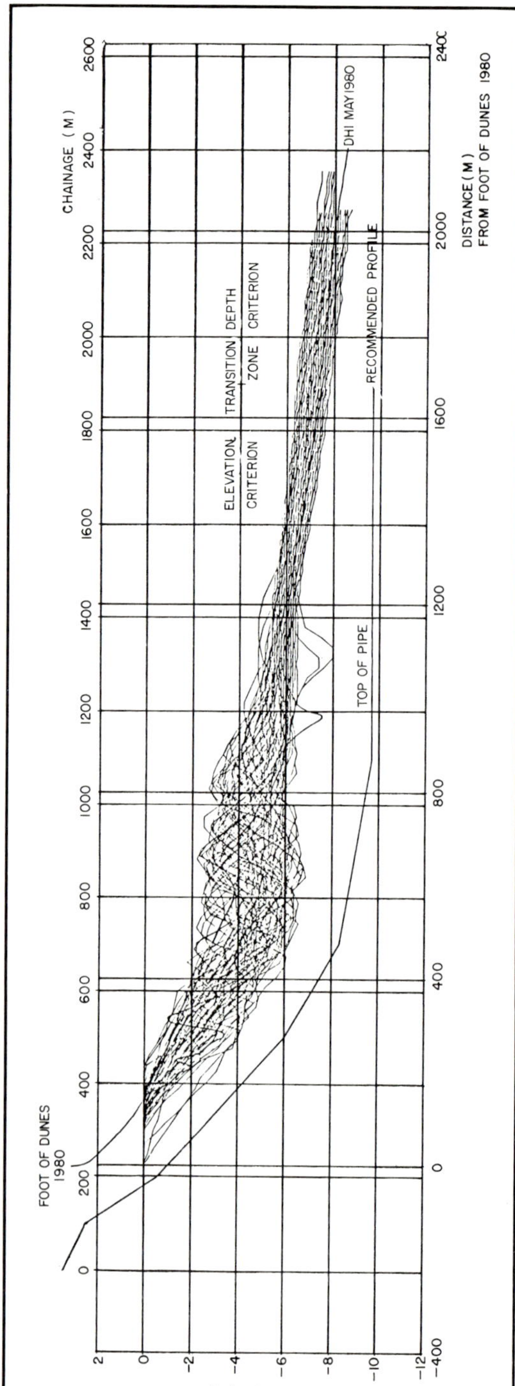

Figure 17.10.3 Selected Pipeline Profile and Envelope of Likely Pipeline Erosion.

treated pipeline test-water, the impact on fisheries during the construction and operational stages, and the effect of the shore approach construction works on the coastal sediment transport regime and the Kærgaard dune area. The main conclusions of these studies are:

(i) The H_2S content of the gas does not result in any significant increase in the personal hazard level or environmental impact, and therefore compliance with the ASME Guide and Danish Supplement thereto may be regarded as satisfactory for the achievement of an acceptable safety level, subject only to such special precautions as may be necessary in connection with the zone of higher rupture risk in the platform area.

(ii) The impact of a rupture on the marine environment will be very limited in time and geographical scope (harm to stationary organisms is unlikely to occur outside a circle of diameter 100 metres).

(iii) The impact on the marine environment of the release of chemically inhibited pressure test water from the pipeline, which is a 'once for all' event prior to commissioning, will be very limited in time and geographical scope.

(iv) The pipeline is not expected to have any serious adverse effect on fishery, either during the construction or the operation phase.

(v) The shore approach construction activities will not have any long-term impact on the coastal sediment transport regime. The dunes will be fully reinstated after construction.

17.12 Bibliography

Gravesen, H. and Nielsen, R., 'Some Environmental Design Aspects of Marine Pipelines for Danish North Sea', *Proceedings, 1981 European Seminar Offshore Oil & Gas Pipeline Technology*, (London, 1981).

Pierson, W. J., Tick, L. J. and Baer, L., 'Computer-based Procedure for Preparing Global Wave Forecasts and Wind Field Analysis Capable of Using Wave Data Obtained by a Spacecraft', *Proceedings, Sixth Naval Hydrodynamics Symposium, 499–532,* (1966).

17.13 Acknowledgement

This chapter is an edited version of the paper given by A Damsgaard (DHI), H J Rasmussen (DONG), J E Thygesen (DONG) and H Gravesen, which was presented at the Oyez Offshore Oil & Gas Pipeline Seminar, held in Basel during 1982.

H GRAVESEN

18

RISER DESIGN

The purpose of this chapter is to broadly describe the types of riser which are used in the North Sea and to outline the criteria which are employed in determining the specific form of riser system best suited to a particular application.

Because the stress levels in a riser must conform to the requirements of the design codes stipulated by certifying authorities, a discussion of these requirements and the methods of stress analysis is contained herein.

A description of commonly used riser supports, with suggestions on their efficient design and correct operation, is also given along with a brief account of the methods of riser protection against corrosion.

18.1 A Description of the Types of Riser and their Tie-ins

The type of riser which is best suited to a particular application will depend largely on the pipeline diameter, the platform type, the direction of approach of the pipeline and whether the riser is installed with the platform or at some later date.

Two types of riser have proved effective in North Sea applications. The first consists of the conventional type involving a continuous length of pipe which is clamped to the platform structure and then joined to the pipeline by flanges, mechanical connectors or hyperbaric welds. The second involves pulling the riser pipe through a pre-installed guide system, such as a 'J' tube, bending shoe, flexible pipe-guide or a one-atmosphere pull-in welding chamber. Figures 18.1.1 and 18.1.2 (overleaf) illustrate both categories of riser and the alternative tie-in methods.

A major advantage of the conventional riser system is that it offers the widest choice of tie-in method. Depending on the movement of the structure and pipeline, the tie-in section can be either rigid, as in the case of a straight spool piece, or flexible as in the case of a dog-leg type or articulated spool piece, which uses swivel joints. The spool piece can be secured using flanges, mechanical connectors or hyperbaric welding.

The second type of riser system is limited in its application to small

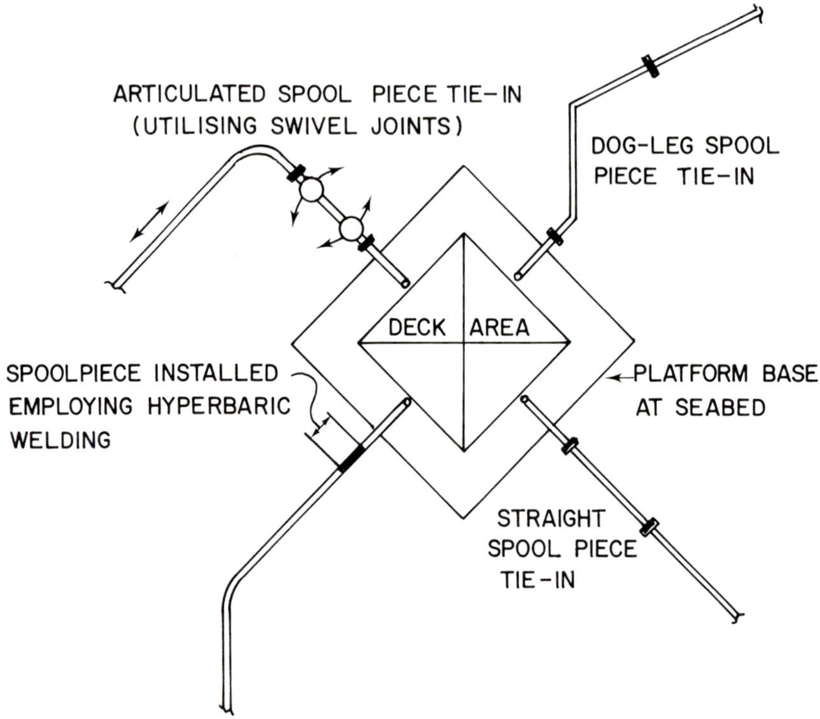

Figure 18.1.1 Conventional Type Riser Systems
Note: Spool piece/pipeline & spool piece/riser connections can be accomplished by utilising flanges, mechanical connectors, or welding techniques

diameter pipes and its use is dependent upon the loads which can be placed on the platform structure. Generally speaking, pipes with diameters less than 8 inches can be pulled without a great deal of structural modification. The tension needed to pull the pipe varies considerably, from elastic deformation, as required by the 'J' tube application, to only a few tons, as involved with the flexible pipe method.

The installation of a riser on a concrete structure requires a hybrid of these two riser systems. The most common variation is when the riser assembly is pulled into a guide tunnel independent of the actual pipeline with the subsequent tie-in using flanged spool pieces as shown in Figure 18.1.3 (overleaf). Once the riser has been drawn into the concrete structure, the guide tunnel is sealed and the water expelled.

18.2 Riser Tie-in Design Concepts

The first consideration in riser design is the nature of the platform involved. Where the riser is installed with a steel structure, a conven-

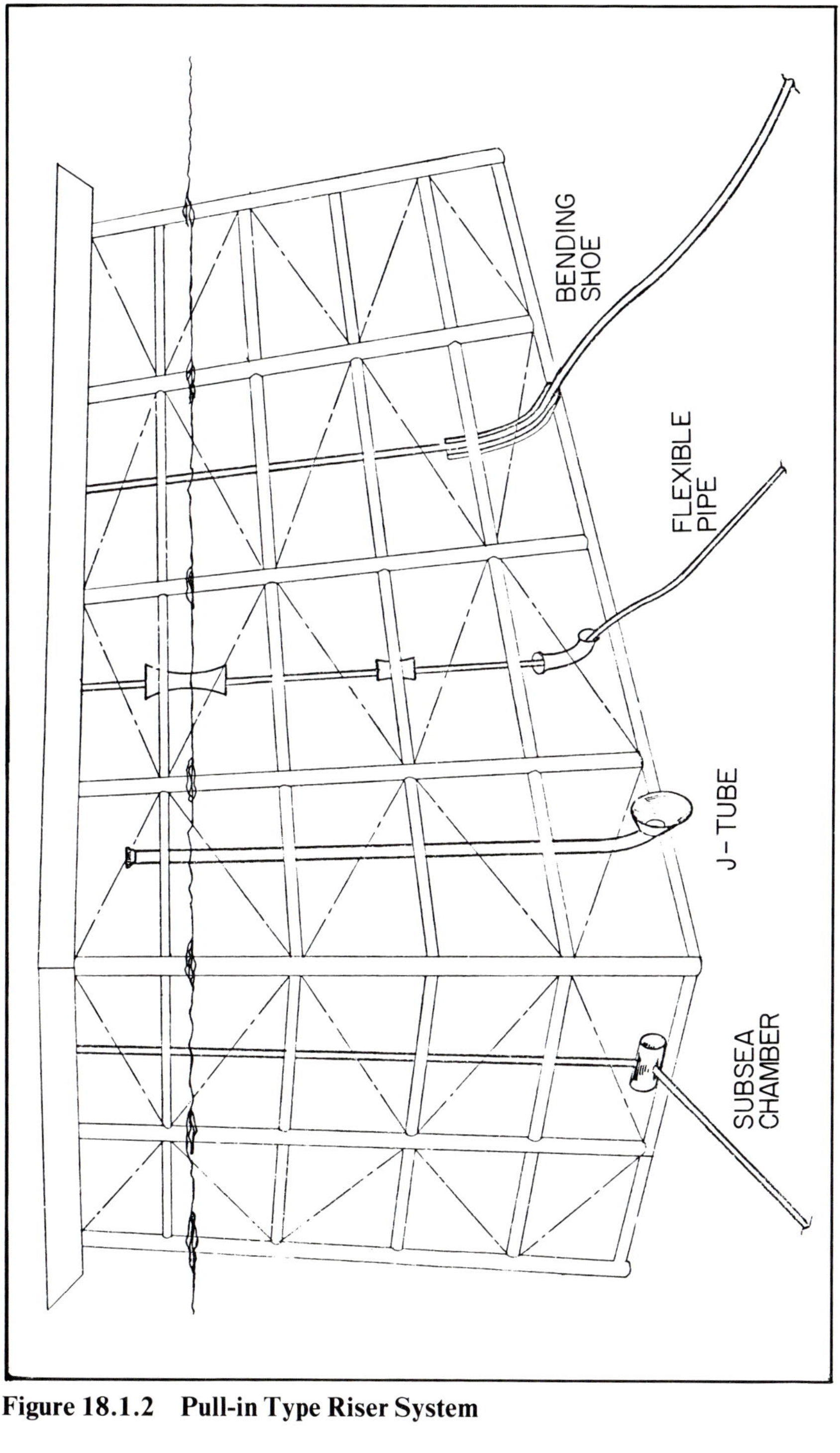

Figure 18.1.2 Pull-in Type Riser System

tional riser system, supported on main steel braces, is the most popular design concept. For concrete gravity structures, however, dry risers passing through sealed tunnels and up through dry access shafts are more common. The placement of an external riser in a concrete gravity structure may at first appear quite feasible but complex geometry, the need for cathodic protection and post installation inspection and maintenance generally preclude this alternative. Duplicate tunnel penetrators, or even complete spare risers, are usually incorporated into the design of concrete platforms to obviate the need for difficult offshore riser installations at some later date.

Many factors influence a riser's design when it needs installing on a pre-existing platform. On steel jackets it is unlikely that prior provision will have been made for the new riser, which will invariably be bolted onto the structure. For small diameter pipes, a decision must be made between a 'J' tube and a conventional riser. In the case of bundled pipes of small diameter, 'J' tubes may be grouped together.

The method of riser/pipeline tie-in will depend principally on the extent of the pipeline expansion, the likely degree of structure settlement and the freedom of the bottom section of the riser to deflect. Connections to concrete structures should allow sufficient flexibility to absorb settlement and wave-induced rocking motions without significant stress variation.

Pipeline expansion due to temperature and pressure is an effect which has often been underestimated. In the case of pull-in techniques, using 'J' tubes, subsea chambers and tunnel-pulls, the effect of axial pipeline expansion may cause buckling if the expansion significantly surpasses the pipeline pretension. Where the pipeline is expected to undergo significant axial expansion, this should be accommodated by providing an expansion loop, a dog-leg or an articulated connection to minimise the load transfer to the riser, otherwise the riser must be designed to withstand the load by inhibiting the expansion without its overstressing. In some cases, the spool piece geometry may be designed to take advantage of greater flexibility in the riser about another axis. For example, if a conventional straight riser were connected directly to a straight pipeline which operated at high pressure and elevated temperatures, the expansion might approximate 3 to 4 metres. In this case, the riser would be subjected to a very high bending stress and considerable displacement. However, if a dog-leg connection were used, advantage could be taken of the torsional flexibility of the riser as well as its bending flexibility in both planes, noting that the former is generally much higher than the latter, since the riser is only restrained partially by its supports. Settlement effects can also be accommodated using a dog-leg or expansion-loop spool piece, because the displacement is taken up by torsion in the pipeline rather than bending in the spool piece.

The initial cost of fabricating an expansion loop is greater than a

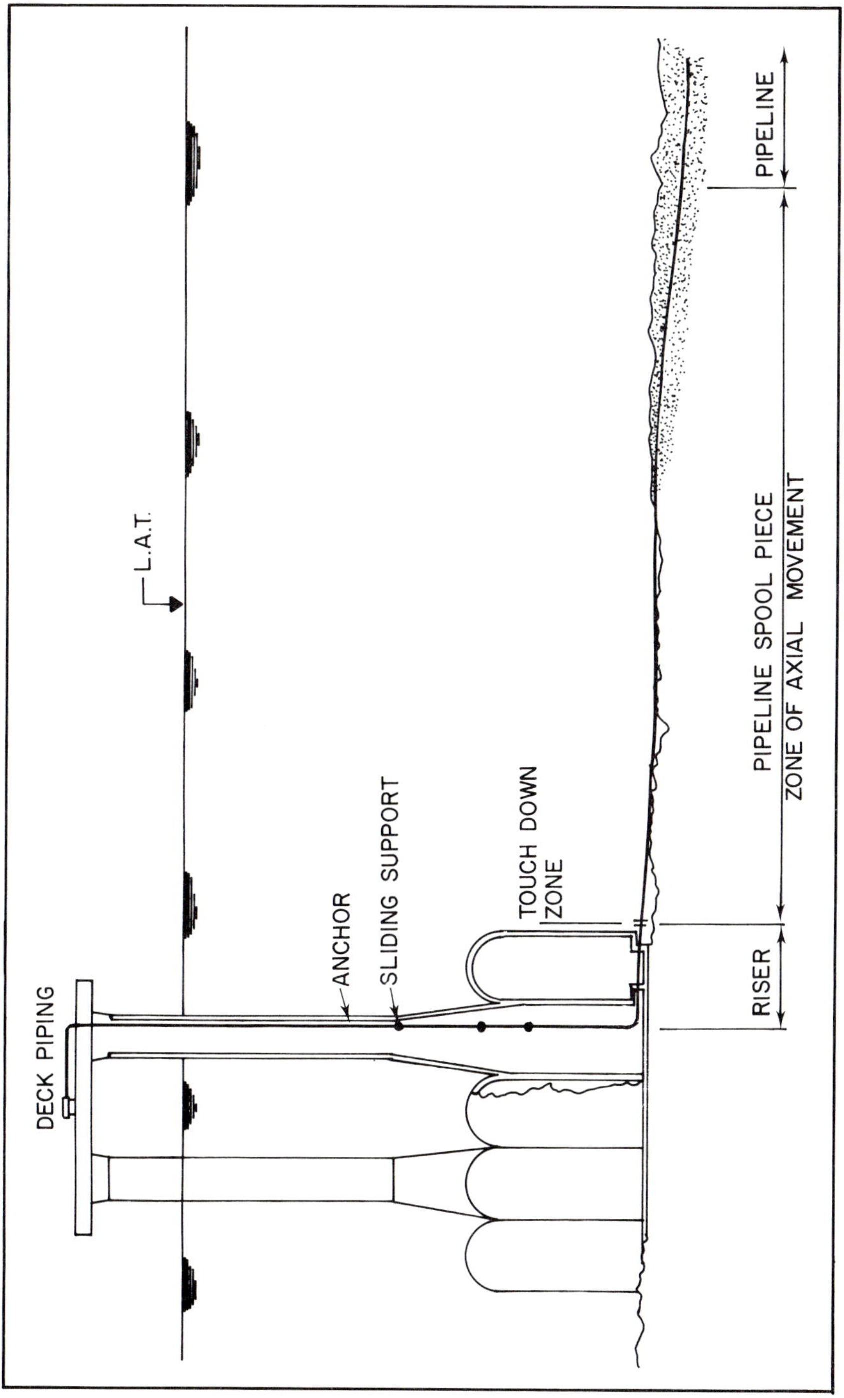

Figure 18.1.3 Typical Configuration for Gravity Platform

straight spool piece but if properly engineered and planned it should cost no more to install. Furthermore, its potential savings could be enormous if it prevented the need for subsequent diving to rectify a riser having come into contact with a platform member, as a result of expansion.

Pipeline expansion due to temperature and pressure effects comprises four separate components:

(i) thermal growth;
(ii) shrinkage due to the Poisson effect from hoop stress;
(iii) expansion due to the endcap pressure effect; and
(iv) frictional resistance to pipeline motion.

According to Nielsen and Stephens (1978a, 1978b) the internal compressive force in a fully restrained pipeline is given by the following relationship:

$$F_T = \pi(D-t)\,t\left[E\alpha T + \frac{-P}{2t}\mu(D-t)\right]$$

where: D = inside pipe diameter (16 inches in the example below)
t = wall-thickness (0.625 inches)
T = operating temperature (123°F)
P = operating pressure (2,010 psi)
E = Young's modulus (29×10^6 psi)
μ = Poisson's ratio (0.27)
α = expansion coefficient (0.65×10^{-5}/°F)

An example which illustrates the effects and the forces involved with pipeline expansion is given below, using typical operating conditions which are displayed with the previous notation.

Using these values it follows that:

$$F_T = 422{,}171 \text{ lb}$$

and the endcap effect due to tensile forces at the end of the pipeline is given by:

$$F_p\ 343{,}455 \text{ lb}$$

The equivalent length of trenching required to sustain these forces by friction is given by:

$$L = \frac{W}{Wsco} \times Cf = 18{,}404 \text{ ft}$$

where: $W = F_T + F_p = 765{,}626$ lb
Wsco= weight per foot run of the submerged concrete coated pipeline (104 lbs/ft).
Cf = coefficient of sliding friction between pipe and the trench (0.4).

The resulting expansion of pipe is given by the expression:

$$L = \frac{L \times W}{2 \times A \times E} = 96.56 \text{ inches}$$

18.3 Load Conditions which Affect Riser Design

The loading conditions which must be incorporated in the stress analysis of a riser/tie-in design concept are clearly defined in the codes of practice, as listed in Table 18.3.1 (overleaf). These load conditions are generally divided into two categories, 'sustained' and 'sustained plus occasional' stresses.

Sustained loads are those which act without significant or rapid variation over an extended period of time. Included in this category are dead-weight, design pressure differential, design temperature differential, and any applied displacements or sustained external forces. Some codes treat the temperature effect separately but this can lead to erroneous results.

Occasional or environmental loads are those which fluctuate with time or are not applied for extended periods. Included in this category are wave and wind effects and any occasional loads which may be applied from external sources.

A hydrotest is not normally considered a load condition because the hydrotest is normally determined by the riser wall-thickness. In addition to the normally accepted loading conditions, it is recommended that two further types should be considered. The first is the aggregate load comprising all the individual loads working in concert with a factor applied to each based on the statistical chance that the particular load condition may be exceeded. For example, self weight is fairly well defined and may only warrant a safety factor approximating 1.05–1.10, whereas temperature differential may not be so well known and may warrant a factor of up to 1.50. The results of such an analysis should then be used to determine the maximum possible stress condition and, more importantly, the maximum possible displacement condition. Under this condition no stress failure or structural interference should be permitted. The second type is a form of failure analysis where once again all likely load conditions are superimposed but without applying multiplying factors. In this case, however, the analysis should investigate the effect of losing one riser support, thereby allowing the riser greater freedom to move. This condition should not cause any stress failure or structural interference, however.

The ANSI and IP6 codes follow a similar format in that they place limits on the allowable hoop and longitudinal stresses and they also define a stress range for temperature effects. In some cases, restrictions are also imposed on shear stress and equivalent tensile stress, as defined by the relationships which appear in Table 18.3.2 (overleaf). The DnV code takes an entirely different approach in that it requires a limitation only on the von Mises distortion energy stress.

All codes use a similar approach in their calculations of hydrotest pressure for certification purposes. A full discussion of these code requirements is provided by Stephens (1977), and a summary of the limits imposed on hoop and other stress may be found in Tables 18.3.3 and 18.3.4 (see pages 272 & 273).

The hoop stress is used to determine the minimum allowable wall-

Table 18.3.1 Code Requirements – Load Conditions and Related Stresses

LOAD CONDITION			LOAD						CHECK STRESS	
CODE	No	DESCRIPTION	DEAD-WEIGHT INCLUDING CONTENTS	DESIGN PRESSURE DIFF.	DESIGN TEMPERATURE DIFF.	APPLIED DISPLACEMENTS SUSTAINED EXTERNAL FORCES	LIVE/OCCASIONAL LOADS	HYDROTEST PRESSURE	REQUIRED	RECOMMENDED
ANSI B31.4	1	Operational (Clause 402.3.3a)	X	X		X	X		SH, SL SV	
	2	Additive Longitudinal (Clause 402.3.2d)	X	X		X			SL, SV	
	3	Equivalent Tensile (Clauses 402.3.2c & 419.6.4b)	X	X	X	X	X		STE	SL, SV
	4	Expansion (Clause 402.3.2.c & 419.6.4c)			X	X			SE	
	5	Hydrotest (Clauses 402.3.3b & 437.4.1a)	X					X	SH	SL
ANSI B31.8	6	Sustained & Occasional (Clause 833.4bc)	X	X		X	X		SH, SL	
	7	Sustained & Occasional (Clause 833.4abc) + Temperature	X	X	X	X	X		SL	STE, SV
	8	Expansion (Clause 833.3)			X	X			SE	
	9	Hydrotest (Clause 841.322b)	X					X	SH	SL
IP6	10	Environmental (Clause 11.4.3)	X	X	X	X	X		SH, STE	SL, SV
	11	Longitudinal (Clause 2.6.4)	X	X		X			SH, SL	
	12	Expansion (Clause 2.6.4)			X	X			SE	
	13	Hydrotest (Clause 6.4)	X					X	SH	SL
DnV	14	Functional (Clause 3.1.2.1a)	X	X	X				SEQ, SH	
	15	Functional & Environmental (Clause 3.1.2.1b)	X	X	X	X	X		SEQ	SL, SV
	16	Hydrotest (Clause 8.8.4.2)	X					X		

Table 18.3.2 Elementary Pipe Stress Formulæ

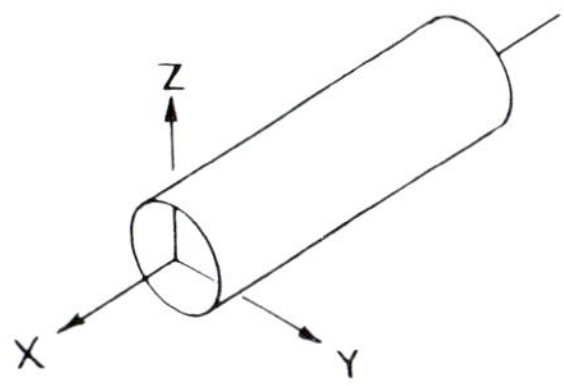

A = Cross-sectional area of pipe
Z = Section modular of pipe
D = Nominal OD of pipe
t = Pipe wall-thickness
i = Stress intensification factor
Fx Fy Fz = Forces along XYZ axis
Mx My Mz = Moments about XYZ axis
p = Internal pressure differential

Hoop stress

$$SH = \frac{pD}{Zt}$$

This calculation is common to all codes

Axial stress

$$SAX = \frac{Fx}{A}$$

This calculation includes flexure and endcap forces

Bending stress

$$SB = \frac{i\sqrt{My^2 + Mz^2}}{Z}$$

Longitudinal stress $\quad SL = SAX + SB$

This stress may vary from max tensile through zero to max compression

---

Shear stress

$$SF = \frac{i\sqrt{Fy^2 + Fz^2}}{2A}$$ Force component

$$ST = \frac{Mx}{2Z}$$ Torsion component

$SV = SF + ST$

Expansion stress

$$SE = \sqrt{SB^2 + 4ST^2}$$

Equivalent Tensile Stress

$$STE = SH + SL$$

This stress is the algebraic sum of hoop stress and maximum longitudinal compressive stress

---

von Mises equivalent stress $\quad SEQ = \sqrt{SL^2 + SH^2 - SL\,SH + 3ST^2}$

Table 18.3.3 Code Limitations on Hoop Stress

HOOP STRESS			
Code	Condition	Pipelines	Risers
ANSI B31.4	Operating Test	0.72 0.90	0.72 0.90
ANSI B31.8	Operating Test	0.72 0.90	0.72 0.90
IP6	Operating Test	0.72 0.90	0.60/0.72(1) 0.90
DnV	Operating Test	0.72 0.75/0.90(3)	0.60(2) 0.75/0.90(3)

Notes:
1. Limit of 0.60 applies to gas risers on manned platforms (Clause 11.6).
2. Limit of 0.60 is implied by limit on von Mises equivalent stress (Clause 4.2.2.1)
3. Limit depends on pipe diameter (Clause 7.2.5)

thickness for the riser pipe and spool piece for any given grade of steel. In addition, a corrosion allowance should be added. This corrosion factor should not be used in the stress analysis but only in consideration for the additional self weight that would result. Once the design wall-thickness is determined, the limitations on the other stress values will govern the geometry of the riser and the position as well as the nature of its supports.

It should be noted that the codes are not all in agreement on the limit to be applied to hoop stresses. The ANSI codes require that it should not exceed 0.72 Fy, whereas IP6 allows 0.72 Fy for oil risers, but reduces the limit to 0.6 Fy for gas risers on manned platforms. DnV does not restrict hoop stress in risers directly but places a limit of 0.6 Fy on the von Mises equivalent stress for both oil and gas risers. This can be the most conservative code of all since it implies a limit of 0.6 Fy or even less on the hoop stress component if the longitudinal stress component is very small or compressive in nature.

It is generally agreed by all codes that the calculated hydrotest pressure should not exceed 0.9 Fy, although lower values may be required in some cases.

One area of the ANSI and IP6 codes which requires further clarification and possible revision concerns the imposition of separate limits on the longitudinal stress caused by sustained loads and the expansion stress resulting from temperature variations. When the law of linear superposition of loads applies, the current approach may be acceptable, but with pipeline tie-ins it is thought that nonlinear boundary conditions may exist which make the current approach invalid.

Table 18.3.4 General Code Stress Limitations

GENERAL CODE STRESS LIMITATIONS			YIELD STRESS FACTORS				
LOAD CONDITION		STRESS TYPES					
CODE	No	DESCRIPTION	LONGITUDINAL STRESS (SL)	EXPANSION STRESS (SE)	EQUIVALENT TENSILE STRESS (STE)	VON MISES' EQUIVALENT STRESS (SEQ)	SHEAR STRESS (SV)
B31.4	1	Operational	0.80	—	—	—	0.45
	2	Additive Longitudinal	0.54	—	—	—	0.45
	3	Equivalent Tensile	(1.00)	—	0.90	—	(0.45)
	4	Expansion	—	0.72	—	—	—
	5	Hydrotest	(0.90)	—	—	—	—
B31.8	6	Sustained + Occasional	0.75	—	—	—	—
	7	Sustained + Occasional + Temperature	1.00	—	(0.90)	—	(0.45)
	8	Expansion	—	0.72	—	—	—
	9	Hydrotest	(0.90)	—	—	—	—
IP6	10	Environmental	(1.00)	—	0.90	—	(0.45)
	11	Longitudinal	0.54	—	—	—	—
	12	Expansion	—	0.72	—	—	—
	13	Hydrotest	(0.90)	—	—	—	—
DnV	14	Functional	—	—	—	0.72 (1) 0.60	—
	15	Functional + Environmental	(1.00)	—	—	0.96 (2) 0.80	(0.45)
	16	Hydrotest	(0.90)	—	—	—	—

Notes: (1) First value applies to pipelines, the second to risers. (2) Values in brackets are recommended only.

Shear stress is ignored by the codes, excepting ANSI B31.4 which imposes a limit of 0.45 Fy for sustained plus occasional loads. This value is in accordance with comparable structural codes for shear stress and should be generally acceptable for the other piping codes. It is recommended that in areas of large movement or high support reactions the shear stress should be checked.

For the two composite load cases, which were recommended previously, a von Mises equivalent stress check should be made and the maximum stress should not exceed 1.0 Fy.

18.4 Riser Support Designs

The most common form of riser support is the bolted support clamp as shown in Figure 18.4.1. This is fabricated as an integral part of the platform, in the shape of a saddle and hinged door with the saddle connected to the platform *via* welded shear plates, either directly to a jacket brace or to an anchored steel plate in the case of concrete structures. Neoprene approximating 6 millimetres in thickness is normally used for electrical insulation at the support and to reduce local bending effects in the pipewall. For large diameter pipelines, 10 millimetres-thick neoprene is used for its good structural damping properties; when circumferential grooving of the neoprene is available, thicknesses up to 25 millimetres may be

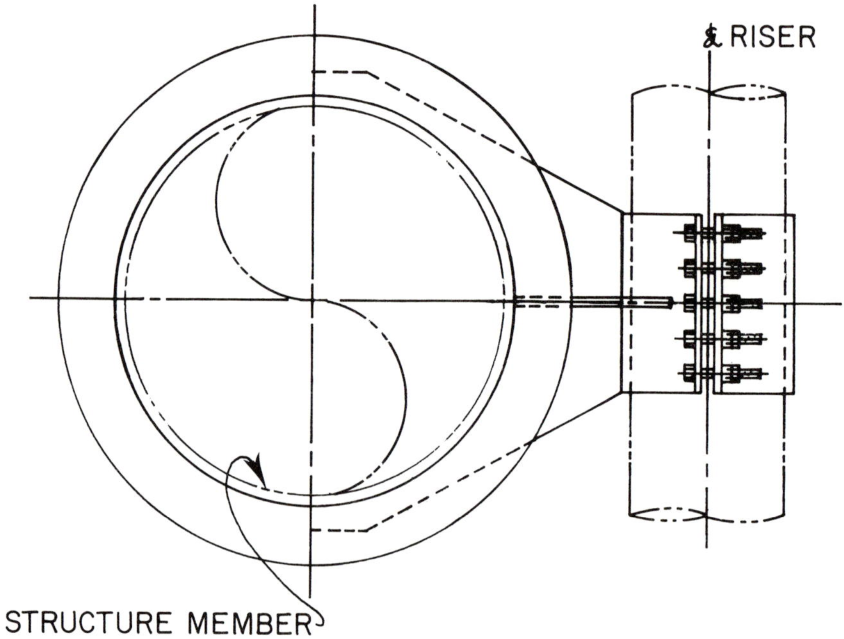

Figure 18.4.1 Welded Type Riser Clamp

preferred. By these means it is possible to minimise the transmission of mechanical vibration and reduce the possibility of the clamp bolts shaking loose.

The same form of clamp may be installed after a jacket has been in service for some time. In this case, provision must be made for aligning the clamps. This is achieved by connecting the riser clamp saddle to an extension arm, which passes through a stand-off clamp which is also clamped to the structure brace. In this case the stand-off clamp and the brace clamp should not be lined with neoprene. This form of clamp should not be subjected to high lateral loads since it is likely to move. Where the clamp support is designed to act as a sleeve with minimal longitudinal or torsional restraint, a Teflon liner, approximating 3 millimetres in thickness, is used between the pipe and the clamp to allow sliding to occur, thereby avoiding the transmission of loads into the clamp. This is shown in Figure 18.4.2.

A less common method of riser restraint involves welding a doubler plate onto the riser and providing support by a stool or trunnion as shown by Figure 18.4.3 (overleaf). This design experiences certain shortcomings due to the fillet weld between the doubler and the riser pipe, but these can often be overcome by increasing the pipe wall-thickness locally.

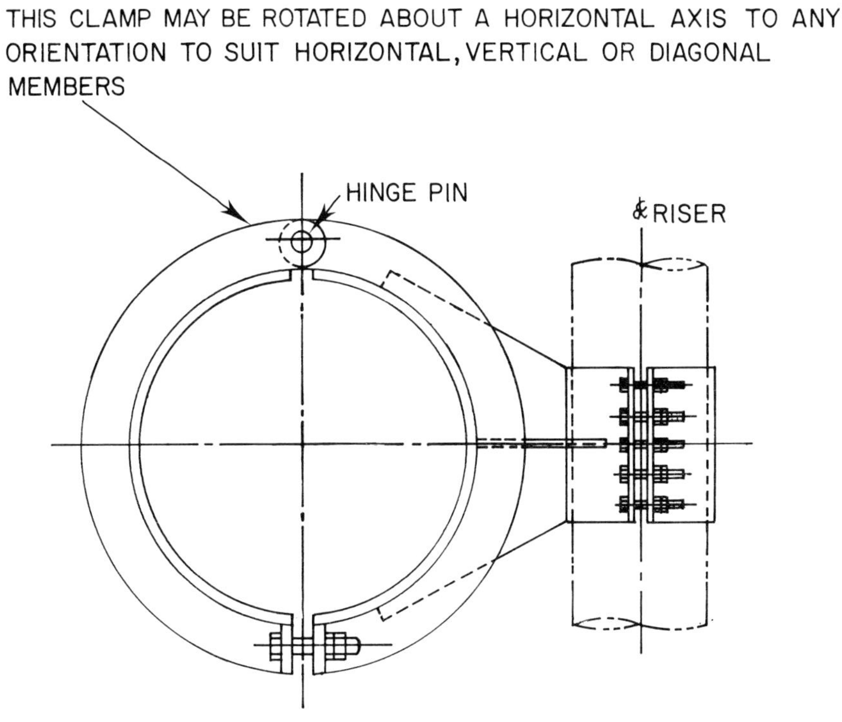

Figure 18.4.2 Bolted Type Riser Clamp

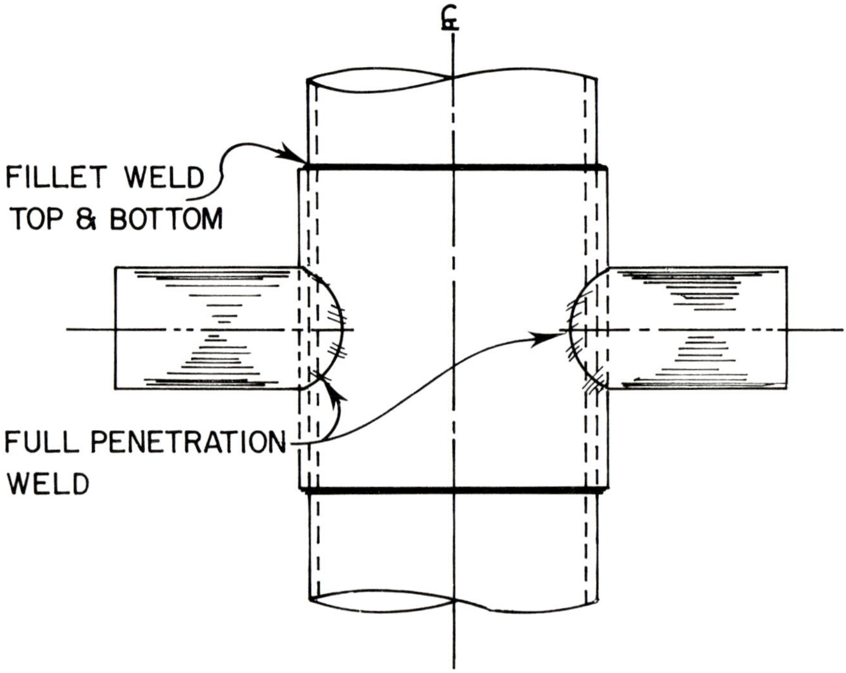

Figure 18.4.3 Trunnion Type Riser Clamp

18.5 Corrosion Protection in Riser Design

Corrosion protection of marine risers is a major concern particularly in the light of corrosion failures which have occurred on risers which were thought to be well protected.

In the splash zone where sacrificial anodes are ineffective, methods of protection are employed which physically prevent contact between the seawater and the steel surface of the riser. Although painting of the steel surface with coal tar epoxy *may* afford reasonable protection it is necessary to continually monitor the condition of the paint for breakdown or physical damage. Consequently, painting of the riser surface as the only form of protection is not currently practised, especially as severe corrosion may occur if the paint loses its bond with the steel. A recent development in the protection of steel surfaces is the use of vulcanised neoprene which affords good abrasion resistance and an excellent bond to the steel. The effects of its long-term exposure to elevated temperatures and seawater are not known but initial indications are good.

A protective sleeve similar to that shown in Figure 18.5.1, which also acts as a structural member, has several advantages. Although the sleeve itself is exposed to the harsh marine environment, it is not subjected to the same elevated temperature nor is it required to contain fluids under pressure. Consequently its corrosion is generally less severe and the stresses lower, so that its protective coating is more likely to maintain its integrity. A method of corrosion protection which is gaining in popularity is the use of non-structural protective sleeves made from monel. This method affords excellent protection of the riser surface but mechanical damage to the relatively thin monel sheathing must be avoided otherwise water ingress behind the sleeve could cause severe corrosion which would be difficult to detect.

For the remainder of the riser, from the splash zone to the seabed, primary protection is generally afforded by a cathodic protection system, which is either installed locally or as part of the overall platform protection system. The requirement for cathodic protection may be reduced by surface protection using coal tar epoxy or vulcanised neoprene.

Where the platform employs an impressed current cathodic protection system, the risers will necessitate only a small increase in power requirement. Consideration should be given to installing sufficient sacrificial anodes on main oil and gas risers to protect them until the impressed system is fully operational or at any period of time when the impressed

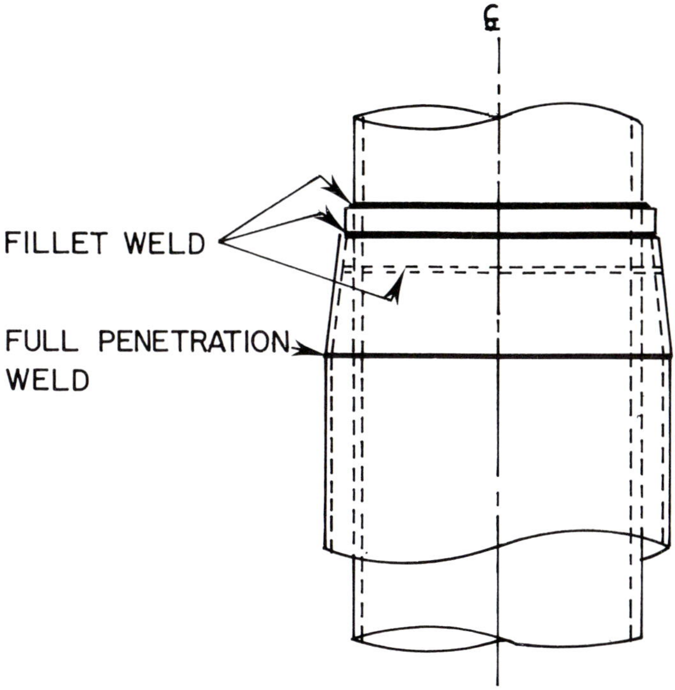

Figure 18.5.1 Protective Riser Sleeve
Note: Sleeve assembly is installed to the riser in the splash zone

system is switched off. The risers should be electrically insulated from the platform to avoid its drawing on the sacrificial anodes of the risers and consequently creating areas which are not polarised at points of contact with the structure. Furthermore, impressed current anodes should not be placed in close proximity to risers, otherwise pitting may result where current flows from the overprotected riser to the platform, particularly if the riser has insulated supports. In extreme cases the proximity of the impressed anode may cause the formation of free hydrogen on the riser surface and result in hydrogen embrittlement.

If a decision is taken that all or part of the riser's steel surface is to remain exposed and protected entirely by a cathodic protection system then it is strongly recommended that all mill scale should be removed by grit blasting before exposure to the marine environment. There is much evidence to suggest that mill scale is a dominant and contributary factor in pitting corrosion, causing the formation of local electrolytic cells even where cathodic protection is thought to be afforded.

The corrosion protection of a pipeline which is fitted in a 'J' tube is infeasible because of the damage which would be inflicted on the protective coating during the pulling operation. Consequently extra corrosion allowance must be afforded by thicker metal, particularly at the bottom of the 'J' tube bend where the pipeline is in contact with the riser.

18.6 Conclusions

This chapter provides a discussion of the basic approach which is required in the optimal development of a riser design. One of its main conclusions is the need for the riser design as an integral part of the platform design. Furthermore, the design also has to accommodate post-installation inspections for the purpose of maintaining a certificate of fitness.

Long-term maintenance requirements must also be recognised during the initial design phase and, where possible, ease of maintainability should be provided.

18.7 Bibliography

American National Standards Codes for Power Piping, ANSI B31.1/1973, *Power Piping*, ANSI B31.4/1974, *Liquid Petroleum Transportation Piping Systems* and ANSI B31.8/1975, *Gas Transmission and Distribution Piping Systems*.

Det norske Veritas *Rules for Design, Construction and Inspection of Submarine Pipeline and Submarine Riser*, Det norske Veritas, (1976).

Kaisan, D., 'Structural Analysis of Pipeline Risers', *Pipeline Industry Engineering*, (August 1977).

'Model Code of Safe Practice', *Petroleum Pipelines, Third Revised Edition, Pt 6*, (1967, reprinted 1974).

Nielsen, R., 'Simulations Help Solve Pipe Buckle Problem', *Oil, Gas Journal*, (December 1976).

Nielsen, R. and Pendered, J. W., 'Some Aspects of Marine Pipeline Analysis', *Proceedings, International Symposium on Numerical Methods in Offshore Engineering'*, (Swansea University, January 1977).

Nielsen, R. and Stephens, G., 'Computer Techniques Adapted as Riser Design Extends Existing Codes', *Offshore Engineering*, (August 1978a).

Nielsen, R. and Stephens, G., 'Riser Designers Face Significant Difference in Code Requirements', *Offshore Engineering*, (October 1978b).

Stephens, G., 'Analysis of Riser Connection', *Proceedings, International Symposium on Numerical Methods in Offshore Engineering*, (Swansea University, January 1977).

G F STEPHENS
A J GINNARD

19
SINGLE POINT MOORING SYSTEMS AS ALTERNATIVES TO LONG DISTANCE PIPELINES

Submarine pipelines are but one means of exploiting offshore oil and gas resources. Alternative means comprise fixed platform and Single Point Mooring (SPM) offshore loading facilities. The continued and varied deployment of the latter is a testimony not only of their technological capability but also their economic viability. In this respect SPMs offer a viable economic alternative to long distance pipelines and fixed offshore loading platforms in those circumstances which are described in this chapter.

19.1 Introduction

In their search for ways of meeting the inexorable increase in world demand for hydrocarbons and to realise economies in transportation, petroleum companies and host governments alike have found it necessary to pursue policies which have involved, *inter alia*:

(i) the exploration, development and exploitation of oil and gas resources located in environmentally hostile areas of the world, involving remote ocean regions and ever increasing water depths;

(ii) the rapid expansion of existing fields and the deployment of production facilities to expedite the early development of new reserves; and

(iii) the employment of supertankers of increasing size and the development of import/export terminals to accommodate their draughts.

Such policies as these have been assisted by the evolution of Single Point Mooring (SPM) technology which is providing the basis for the economical and expedient development of marginal offshore oilfields, and for safe and reliable import and export terminals throughout the world.

The marginal nature of certain offshore fields prohibits the expensive and time-consuming installation of large fixed platform complexes and, more specifically, submarine pipelines from offshore sites to distant onshore facilities. Adverse site or route conditions, which include extremely irregular seafloor bathymetry, excessive water depths, bottom instability, ice scouring, lack of suitable onshore areas for processing or terminal facilities and extremely remote field locations, can also limit pipeline applications. In these situations, the utilisation of single point mooring system technology for the development of offshore loading terminals, including on-site storage and production capabilities, provides an economically and technically viable alternative to expensive long distance submarine pipeline systems.

19.2 The Development of Single Point Mooring Systems

The first single point mooring system was installed in 1959. Up to the end of 1980 approximately 260 had been commissioned with 30 expected in 1981. SPMs have been installed in coastal and offshore areas in approximately 60 different countries. Figure 19.2.1 shows the number of SPM systems installed each year, from 1959 to 1980, and it represents the continually increasing use of SPM systems as import and export terminals, offshore loading systems, early and/or floating production facilities with integral storage and product transfer systems.

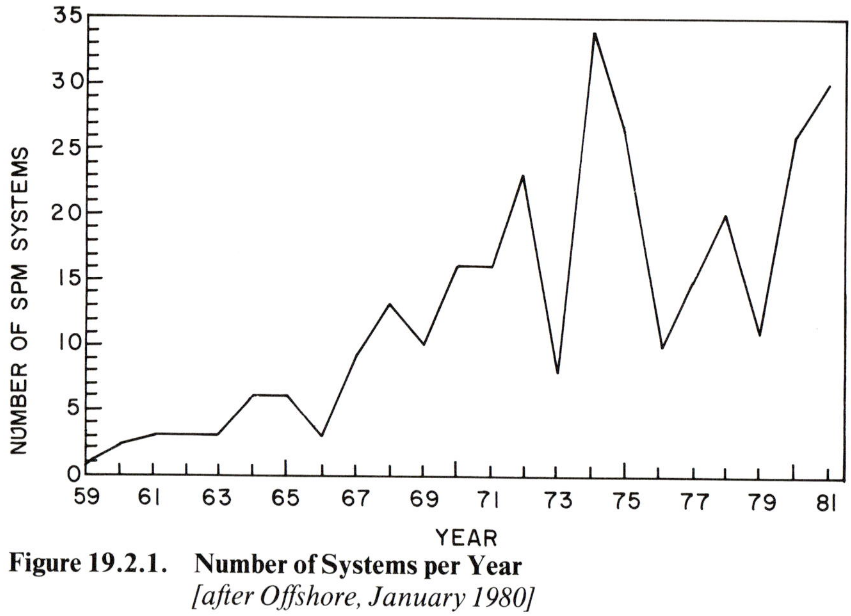

Figure 19.2.1. Number of Systems per Year
[after Offshore, January 1980]

The first SPM was designed for 3,000 dwt tankers. It employed one 4-inch diameter hose string and was located in less than 40 feet of water. However, based on accumulated operating experience, model testing programmes, improvements in manufacturing techniques and more exacting requirements, SPM technology evolved rapidly so that present systems are capable of mooring 750,000 dwt tankers in 15-foot significant wave heights with winds of 45 knots, and currents of 3 feet per second.

Design capacities have increased to 100,000 bph using two 24-inch diameter hose strings and, according to Gruy and Kiely (1977 and 1979) can be increased further to 210,000 bph with the anticipated development and use of 30-inch diameter hoses. SPM systems are now capable of operating in water depths exceeding 500 feet and, according to Knecht *et al* (1979), future designs are projected for water depths exceeding 1,000 feet. As mooring loads, throughput rates, water depths and weather capabilities have increased, so the flexibility of these systems has also improved so that several are now capable of simultaneously transferring crude oil, bunker fuel and ballast water. Hatley (1980) reports that some SPM systems are also capable of handling bulk ore slurry, coal slurry, liquefied gas and high pressure wellhead production streams consisting of crude oil, gas and/or water mixtures containing considerable amounts of H_2S, sand and paraffins.

The basic types of SPM system include the Catenary Anchor Leg Mooring (CALM), Single Anchor Leg Mooring (SALM), Articulated Loading Tower (ALT), and the Fixed Tower or Ready-Made Dolphin (RMD) system.

The CALM system, shown in Figure 19.2.2, comprises a large diameter, flat cylindrical buoy which floats on the water surface. The buoy is held on location by four or more catenary anchor legs, or mooring chains, extending from the buoy to seabed anchor points using conventional anchors or piles which are located at a distance of 750 to 1,500 feet from the buoy. The buoy supports a rotating mooring and cargo transfer

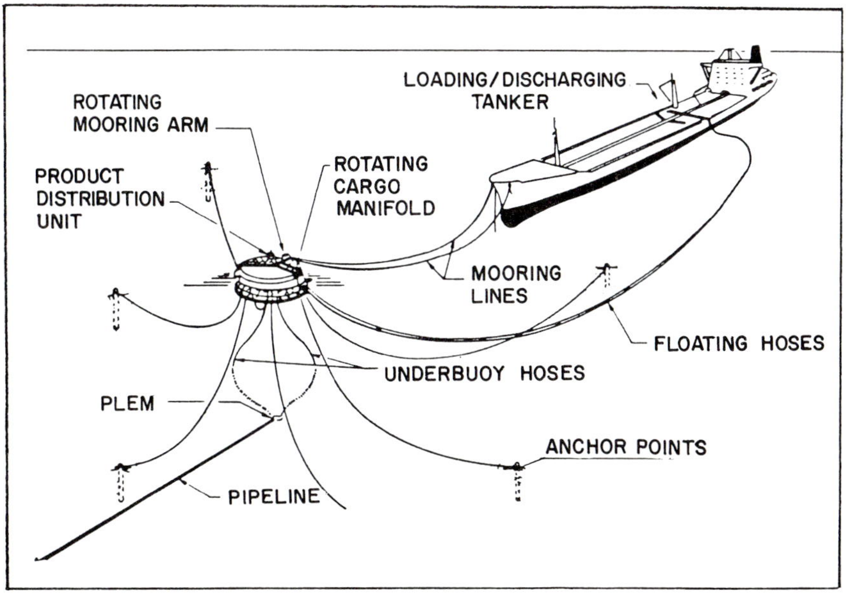

Figure 19.2.2 Catenary Anchor Leg Mooring (CALM)
[after Jansen (1980)]

system which generally includes a product swivel, cargo manifold and a mooring arm. The product swivel is connected to the pipeline end manifold (PLEM) by underbuoy or submarine hoses. The tanker is moored to the mooring arm on the buoy by one or two mooring hawsers, and floating hoses connect the SPM cargo manifold to the tanker manifold.

The SALM system shown in Figure 19.2.3, has a smaller diameter but a longer cylindrical buoy which is connected to a seafloor base by a single vertical taut-line anchor chain. Due to the positive buoyancy of the surface buoy, the anchor chain is under constant tension. To withstand the positive buoyancy of the buoy and the maximum mooring load conditions, the base is either pinned to the seafloor by piling and/or filled with ballast material. For shallow water applications, the fluid swivel and cargo piping assemblies are installed on the base or, for deepwater applications, at the top of a cylindrical riser. A series of universal joints combined with the fluid swivel allow the buoy and tanker to rotate about the base centre point. The tanker is moored to the buoy mooring bracket with mooring line(s) or hawser(s), and a combination of submarine and floating hoses connect the base fluid swivel to the tanker manifold. Short submarine hose sections or steel pipe spool pieces are used to connect the SALM base piping to the PLEM.

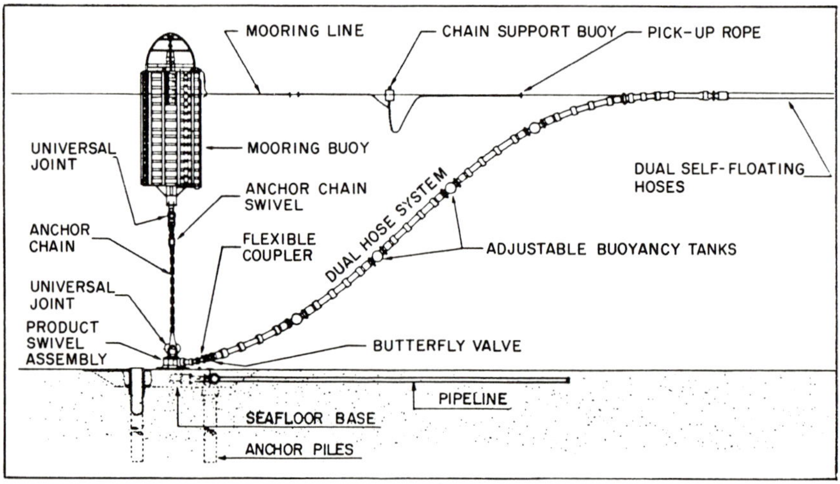

Figure 19.2.3 Single Anchor Leg Mooring (SALM)
[after Gruy and Kiely (1979)]

The ALT system, shown in Figure 19.2.4, consists of a rotating head, heliport and loading boom located on top of a vertical steel column or truss incorporating permanent ballast tanks and auxiliary floats on the lower end and a main float just below the water surface. The vertical column is attached to the seafloor base by a cast steel universal joint. The base and the column ballast tanks are flooded with water to hold the system on location. Crude oil enters the base piping and is routed through the universal joint and up the column to the swivel in the rotating head. The crude continues through the rotating head, loading boom and hose system to the receiving tanker. The ALT systems generally utilise dedicated tankers which have been modified to allow bow loading. The tanker is moored to the ALT by a single mooring hawser which is approximately 350 feet long.

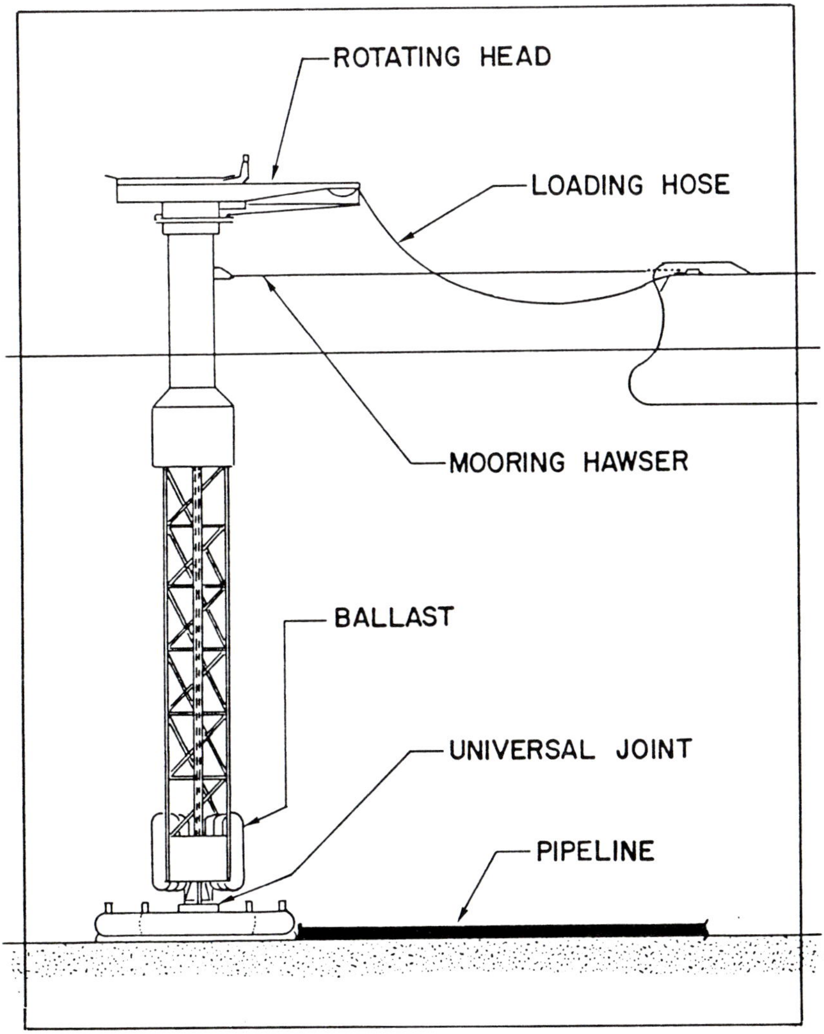

Figure 19.2.4 Articulated Loading Tower (ALT)
[after Hays et al (1979)]

The Fixed Tower or Ready-Made Dolphin system, shown in Figure 19.2.5 (overleaf), uses a tubular column or template structure normally held in place by a large area gravity base or by steel piling driven into the seafloor. A rotating turntable with mooring bracket, fluid swivel and transfer system is mounted on the tower. The fluid swivel is usually connected to the submarine pipeline by a vertical steel riser that is centrally supported by the tower. Mooring hawsers are used to moor the tanker, and marine hoses or articulated loading arms provide a cargo transfer system. Large energy absorbing fender systems are also required.

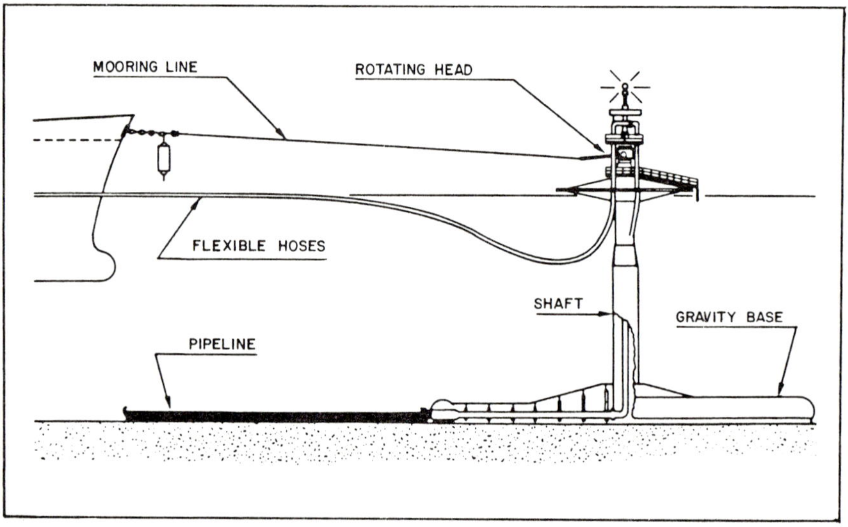

Figure 19.2.5 The Fixed Tower or Ready-Made Dolphin (RMD)

The CALM system was the basic SPM utilised throughout the 1960s. During the 1970s, however, a prototype articulated tower was developed which led to the subsequent deployment of several ALT systems and numerous SALM systems. Additionally, several hybrid types of SPM systems were installed, comprising:

(i) the Exposed Location Single Buoy Mooring (ELSBM) system, which featured a retractable hose and hawser system, as shown in Figure 19.2.6;
(ii) the Single Buoy Storage (SBS) system for crude oil and LPG, utilising a CALM-type buoy with rigid mooring arm connected to the floating storage vessel, as shown in Figure 19.2.7 (overleaf);
(iii) the moored SPAR-type floating storage and loading terminal as shown in Figure 19.2.8 (overleaf); and
(iv) the Single Anchor Leg Storage (SALS) system based on a SALM-type buoy with an arm or yoke hinged to a floating storage vessel, equipped with associated production facilities, as shown in Figure 19.2.9 (see page 289).

Table 19.2.1 (page 289) provides a generalised listing of the different types of SPM systems which have been or are currently being used, along

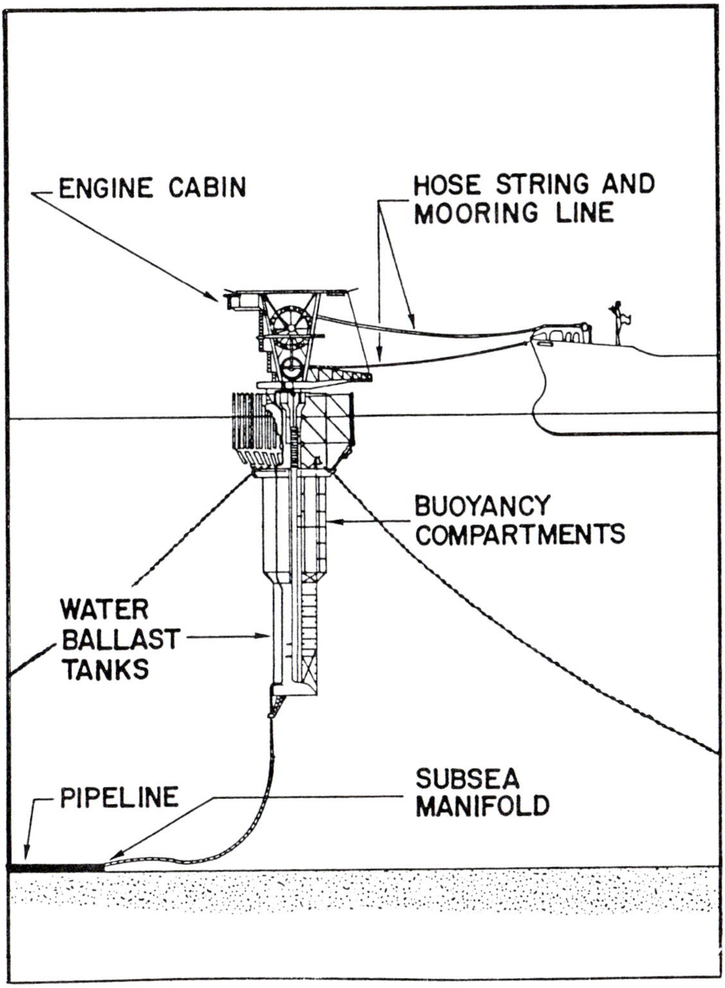

Figure 19.2.6 Exposed Location Single Buoy Mooring (ELSBM) *[after Verslius (1979)]*

with their commissioning dates and maximum water depths. By contrast, Table 19.2.2 (see page 290) provides data which is specific to their use in the North Sea.

19.3 The Advantages of Single Point Mooring Systems

As previously mentioned, SPM loading systems offer definite advantages when attempting to achieve early hydrocarbon production prior to the installation of permanent facilities and long distance pipeline sytems. They are also advantageous when a field is considered to be economically marginal, and the cost of fixed platforms and/or submarine pipelines can-

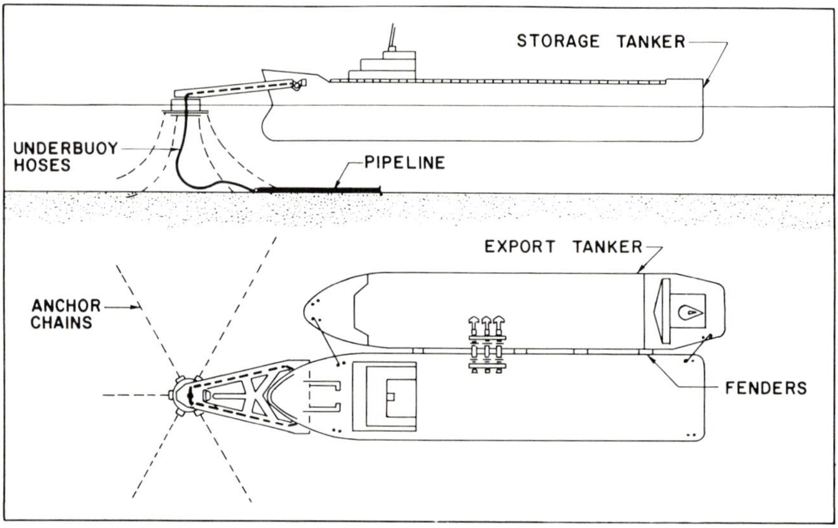

Figure 19.2.7 Single Buoy Storage (SBS) System

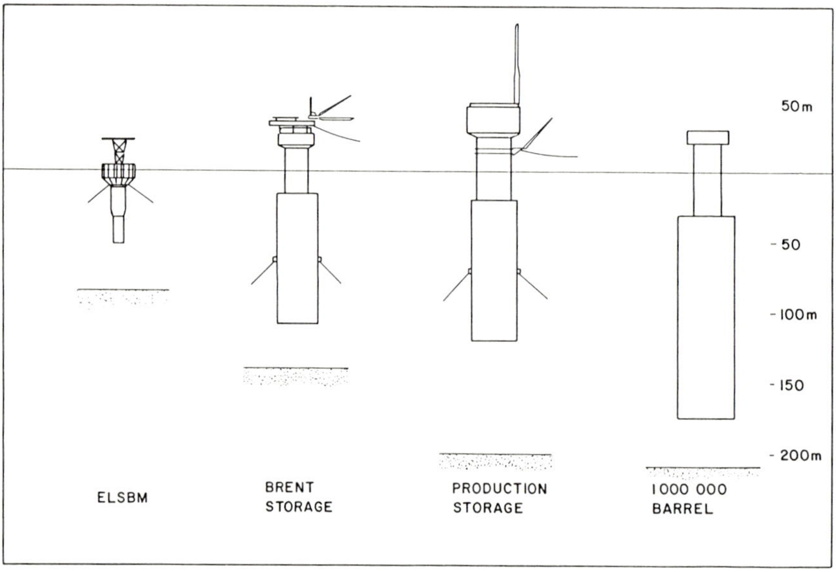

Figure 19.2.8 Typical SPAR Buoy Storage System Designs
[after van Santen and de Werk (1976)]

not be justified. Sometimes offshore loading systems using SPM facilities may be the only economic and feasible means of developing a resource if distance, seabed bathymetry, water depth or site conditions prohibit the installation of a pipeline. The development of the Beryl Field ALT and the Auk Field ELSBM, as described respectively by Hays *et al* (1979) and

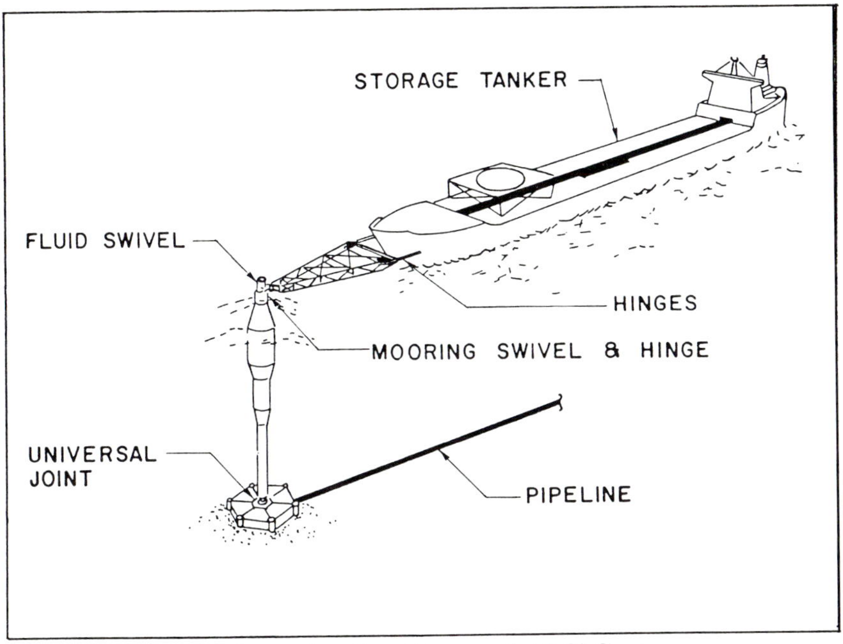

Figure 19.2.9 Single Anchor Leg Storage (SALS) System
[after Wolfram and Gunderson (1979)]

Table 19.2.1 SPM Utilisation Summary

SPM TYPE	NUMBER OF SYSTEMS	FIRST SYSTEM INSTALLATION	MAXIMUM WATER DEPTH
CALM	200	1959	369 ft
SALM	16	1969	535 ft
SBS	7	1972	302 ft
SALS	4	1976	383 ft
ALT	4	1975	480 ft
Fixed Tower	9	1962	110 ft
ELSBM	1	1974	280 ft
SPAR	1	1975	370 ft

Verslius (1979), are examples where SPMs were the preferred alternative to long distance pipelines. The Thistle Field SALM, shown below in Figure 19.3.1 (overleaf) and described by Millar *et al* (1979), is an example where the installation of an SPM allowed early production prior to the installation of the Sullom Voe pipeline. The CALM-type system, described by Hammett *et al* (1977), made possible the development of the Argyll Field, which was considered a marginal field due to its relative smallness and the unpredictable and complex nature of its reservoir characteristics.

A decisive advantage of SPM systems derives from their life-cycle costs

Table 19.2.2 North Sea Offshore Loading Systems

FIELD OPERATOR	FIELD LOCATION	SYSTEM TYPE	WATER DEPTH	YEAR INSTALLED
Phillips	Ekofisk	CALM	232 ft	1970
Phillips	Ekofisk	CALM	208 ft	1970
Danbor	Dan Field	CALM	150 ft	1971
Hamilton	Argyll	CALM	252 ft	1974
Amoco	Montrose	CALM	304 ft	1974
Shell	Anglesey	CALM	172 ft	1974
Shell/Esso	Auk	ELSBM	295 ft	1974
Mobil	Beryl	ALT	380 ft	1975
BNOC	Thistle	SALM	535 ft	1977
Shell/Esso	Brent	SPAR	360 ft	1976
Mobil/Statoil	Statfjord A	ALT	480 ft	1978
BP	Buchan	CALM	369 ft	1979
Danbor	Gorm	CALM	128 ft	1979
Mobil	Statfjord B	ALT	480 ft	1980
Shell/Esso	Fulmar	SALM	280 ft	1981
Phillips	Maureen	ALT	322 ft	1981

being 30 to 50 per cent less than for fixed berth facilities, which require piled platforms (for loading and control facilities), breasting or fendering systems, mooring dolphins, support utilities and safety systems. Several examples of these cost advantages exist currently in the North Sea, where the SPM in the Beryl Field is the preferred alternative to a 100-mile long pipeline from the offshore platform to the mainland and the Argyll floating production and offshore loading system was installed for one-third the cost of fixed platforms.

Besides their life-cycle cost advantage over fixed berth facilities, SPM systems can usually be constructed in six to 12 months due to their minimal fabrication requirements, the ready availability of anchor chain, hawsers and hose components and, in the case of SBS or SALS systems, medium sized tankers can be acquired readily. Hatley (1980) demonstrates this advantage by citing an example taken from a South-east Asian project where oil was produced 18 months after its discovery, and only 11 months after the decision was made to develop the field. In the North Sea, the Argyll system was commissioned and production commenced in less than half the time it would have taken to build a fixed platform complex and pipeline to shore.

A further decisive advantage to be realised by the use of SPMs is that their suppliers currently build and stock CALM-type systems for immediate installation in emergency or rapid deployment situations.

Several SPM systems have been developed which combine operational objectives by providing an export berth (*via* direct, alongside or tandem berthing), on-site storage capacity for up to one million barrels and space for production equipment and quarters, as incorporated on the SBS, SALS and SPT systems. An associated advantage of multi-functional SPM systems is that the SPM fabrication, ship-board modifications and

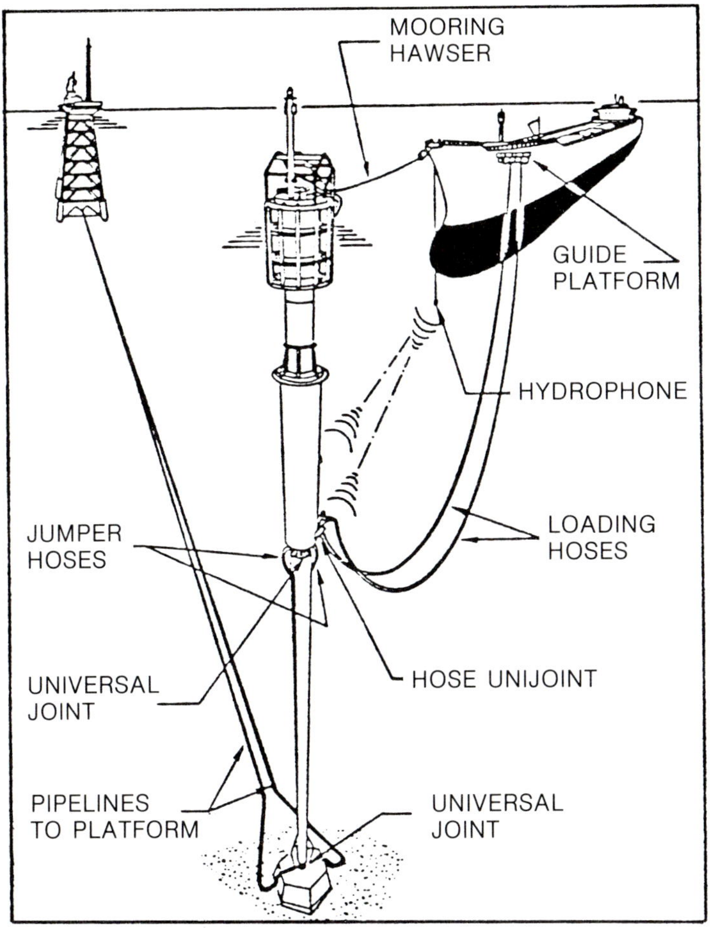

Figure 19.3.1 Deepwater SALM System
[after Millar et al (1979)]

support systems can be completed onshore, thereby minimising expensive offshore fabrication, installation, hook-up, commissioning and start-up activities.

The utilisation of SPM systems is basically unlimited by water depth, environmental conditions, location, seabed conditions, vessel size, throughput, type of product or the nature of their combined operational functions. Because of their tremendous operating flexibility, SPM systems are found in the Gulf of Mexico which is subject to hurricanes, in the North Sea with its extreme weather conditions, and in Far East Asian waters which are subjected to typhoons.

Another feature of the versatility of SPM systems, when used for offshore terminals or floating storage/production facilities, is that they can easily be moved from one similar location to another with minor modification. This requirement obtains when a field becomes depleted or permanent fixed facilities and pipelines are installed as replacements to SPM facilities.

Due to the exposed and uncongested location of offshore SPM terminals, combined with the fact that an SPM can be approached from the most favoured direction depending on environmental conditions, berth approach and departure times are reduced significantly. Mooring and de-mooring time is also reduced by the use of only one or two mooring lines compared with four or more which are required at fixed berths. Furthermore, offshore SPM terminals generally require only one or two mooring launches with small crews to handle the mooring hawsers and hose strings, unless the system is especially designed to eliminate them altogether. In effect, this represents a significant reduction in number and cost of tugs, support vessels and associated crews required to tend and moor supertankers at fixed berths.

The distinct advantage of SPM systems over fixed berthing facilities derives from their elasticity or flexibility of mooring and their weather vaning capabilities. The elasticity of these systems dampens relative motion between ship and buoy and peak mooring loads and reduces damage between the buoy and tanker if the tanker should accidentally overrun or collide with the buoy. In addition, the vessel can rotate or 'weather vane' freely about the mooring point depending on the changing directions of the prevailing wind, wave and current forces; so the vessel always assumes a bow-on orientation which significantly reduces system mooring loads. These combined factors allow vessels to remain moored to SPM loading systems in sea states that would not be possible with fixed berths and to some extent, fixed towers with rotating loading systems. Furthermore, the downtime performance of SPMs in rough weather conditions is remarkably good. Considerable weather downtime data has been collected, analysed and reported by such authors as Kolpak *et al* (1980) for offshore loading systems in the North Sea, especially for the Montrose, Argyll and Auk systems. A review of these data, operational procedures and design criteria indicates that most systems have been designed to allow tankers to remain moored at the SPM during 50-knot winds and maximum wave heights of 25 feet. As reported by Nichols and Westby (1980), however, the concrete articulated loading column (ALC) being built for the Maureen development is designed to accommodate tankers in 20-foot significant wave heights, which generally equate to 34-foot maximum waves. With regard to mooring operations, it appears that the general trend is to design such systems to allow tankers to approach the SPM and moor safely in 12 to 15-foot maximum waves with 30 to 50-knot winds. Based on actual field experience, annual weather downtime for such systems approximates 15 to 20 per cent, but projected system designs, such as the Maureen ALT, should be able to reduce this further.

19.4 Flow-line Connections to SPMs

Although offshore loading systems are viable alternatives to long distance pipelines, nevertheless submarine pipelines of shorter length are an important and integral part of any SPM terminal. Most offshore loading systems are located approximately 6,000 to 8,000 feet from the production or storage facility to provide adequate clearance for vessel manoeuvring and mooring operations. Consequently submarine flow-lines must be installed to connect the production and/or storage platform to the loading terminal. Flow-lines vary in diameter up to 56 inches, as used by the Louisiana Offshore Oil Port, and their short lengths make them ideally suited for bottom-tow installation methods.

Deepwater import or export terminals requiring water depths in excess of 100 feet, to accommodate very large tankers, are often located at considerable distances offshore, so that pipelines of considerable design are required to connect the marine terminal to the onshore storage facilities and interconnecting pipeline systems. For example, the Louisiana Offshore Oil Port (LOOP) marine terminal is connected to the onshore pipeline system by an 18-mile long, 48-inch diameter submarine trunkline.

19.5 Conclusion

Field experience has demonstrated that systems utilising single point mooring technology are an economically and technically viable alternative to long distance pipelines. They are also applicable to situations requiring early production, the development of marginal fields, rapid expansion of existing fields, reliable backup or stand-by facilities, and for use as deepwater import and export terminals. Because of their life-cycle cost advantage, coupled with their low installation time, multiple operational capability, increased recovery value and unlimited application with regard to site location or type of product, the trend toward the development and deployment of additional SPM systems will continue to be a major factor in solving problems associated with the production and transportation of new oil and gas reserves.

19.6 Bibliography

Gruy, R. M. and Kiely, W. C., 'The World's Largest Single Point Mooring Terminals: Design and Construction of the SALM System for 750,000 dwt Tankers', Paper OTC 2826, *Proceedings, Ninth Annual Offshore Technology Conference,* (Houston, 1977).

Gruy, R. M., Kiely, W. C., Maresh, L. A. and Pedersen, K. I., 'The Loop Deepwater Port: Design and Construction of the Single Anchor Leg Mooring (SALM) Tanker Terminals', Paper OTC 3562, *Proceedings, Eleventh Annual Offshore Technology Conference,* (Houston, 1979).

Hammett, D. S., Johnson, J. S. and White, J. L., 'First Floating Production Facility – Argyll', Paper OTC 2821, *Proceedings, Ninth Annual Offshore Conference,* (Houston, 1977).

Hatley, A. G., 'Offshore Oil Field Put Onstream in 18 Months', *World Oil,* (December 1980).

Hays, D. L., McSwiggan, M. and Vilain, R., 'Operation of an Articulated Oil Loading Column at the Beryl Field in the North Sea', Paper OTC 3563, *Proceedings, Eleventh Annual Offshore Technology Conference,* (Houston, 1979).

Knecht, H. L., Stubbs, S. B. and Erb, P. R., 'Floating Caisson Vessel System for Oil Production in Deep Water', Paper OTC 3637, *Proceedings, Eleventh Annual Offshore Technology Conference,* (Houston, 1979).

Kolpak, M. M., Fairbrother, N. J. and Beynet, P. A., 'Reducing Tanker – SPM Loading Downtime', Paper 210, *EUR,* (October 1980).

Jansen, M. B., 'Shipping of Slurries using Offshore Single Point Mooring Terminals', *Skillings Mining Review,* (February 1980).

Millar, J. L., Hughes, H. and Dyer, R. C., 'First Year's Operation Experience of the Deepest SPM in the World', Paper OTC 3561, *Proceedings, Eleventh Annual Offshore Technology Conference,* (Houston, 1979).

Nichols, J. H. and Westby, K. A., 'Innovative Engineering Makes Maureen Development a Reality', Paper 231, *EUR,* (October 1980).

van Santen, J. A. and de Werk, K., 'On the Typical Qualities of Spar Type Structures for Initial or Permanent Field Development', Paper OTC 2716, *Proceedings, Eighth Annual Offshore Technology Conference,* (Houston, 1976).

Verslius, J., 'Exposed Location Single Buoy Mooring', Paper OTC 3805, *Proceedings, Seventh Annual Offshore Technology Conference,* (Houston, 1975).

Wolfram, W. R. and Gunderson, R. H., 'Structural Design of Production Risers and Offshore Production Terminals', Paper OTC 3535, *Proceedings, Eleventh Annual Offshore Technology Conference,* (Houston, 1979).

A R SCHULTZ

APPENDIX A

Modelling Subgrade Reaction

The forces due to soil-pipeline interaction are generally represented by a model comprising three components using X, Y, Z coordinates, as

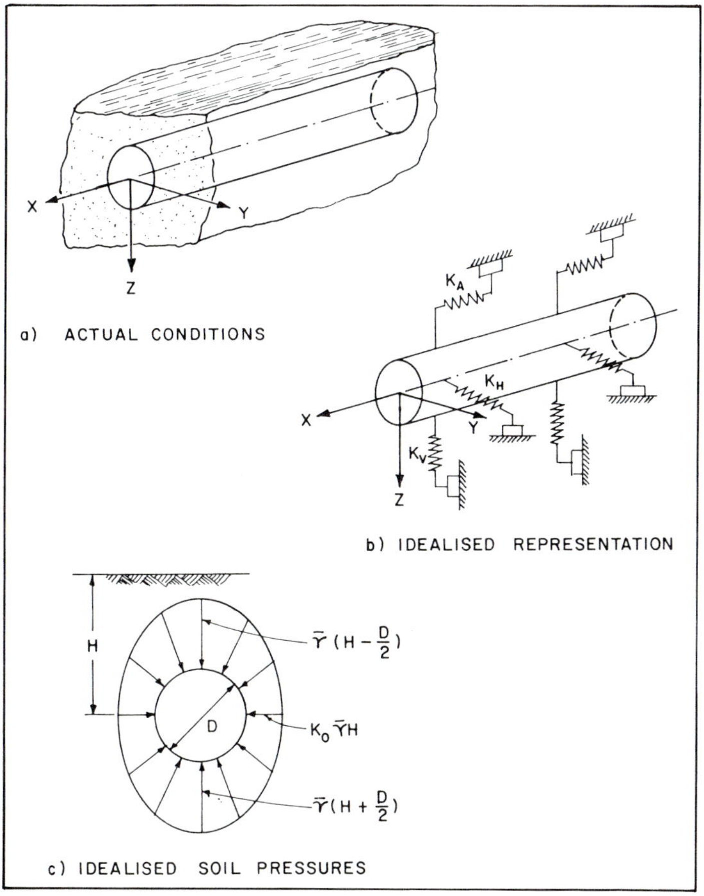

Figure A.1 Subgrade Reaction Idealisation

shown by Figure A.1. Depending on the relative motion between the soil and the pipeline, the seabed soils may either load or restrain the pipeline. The pipeline is usually represented by a simple structural beam with model forces t, p and q representing the loads, and discrete springs representing the restraints from the seabed. The amount of this restraint is usually a nonlinear function of the relative motion between the soil and the pipeline as illustrated by Figure A.2, where t–x, p–y and q–z refer to the axial, horizontal and vertical reactions and displacements respectively. If displacements greater than x_u, y_u and z_u occur then the soil loads may reach constant ultimate values of t_u, p_u and q_u respectively. This concept has been used for over a century and is known as 'subgrade reaction' characterised by coefficients of subgrade reaction.

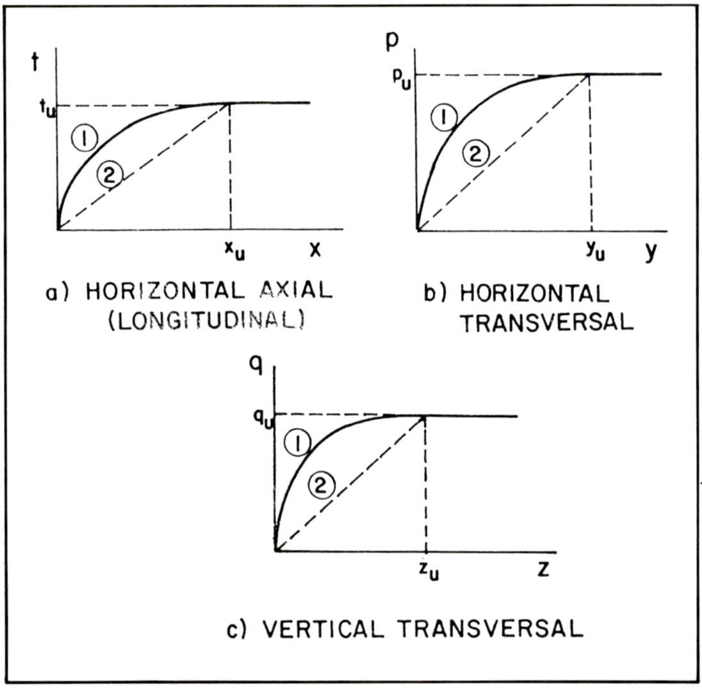

Figure A.2 Load Deformation Relationships for Soil Restraints or Loadings (① = Actual nonlinear relationship, or hyperbolic approximation, ② = Simplified bilinear representation).

Figure A.3 represents the structure of the model used to analyse the resultant forces on a pipeline which is subjected to mudflows and other similar seabed movements. For calculation purposes, the analysis assumes that a mudflow subjects the line to a uniform load distributed over the width of the mudflow, while a faulting process subjects the line to

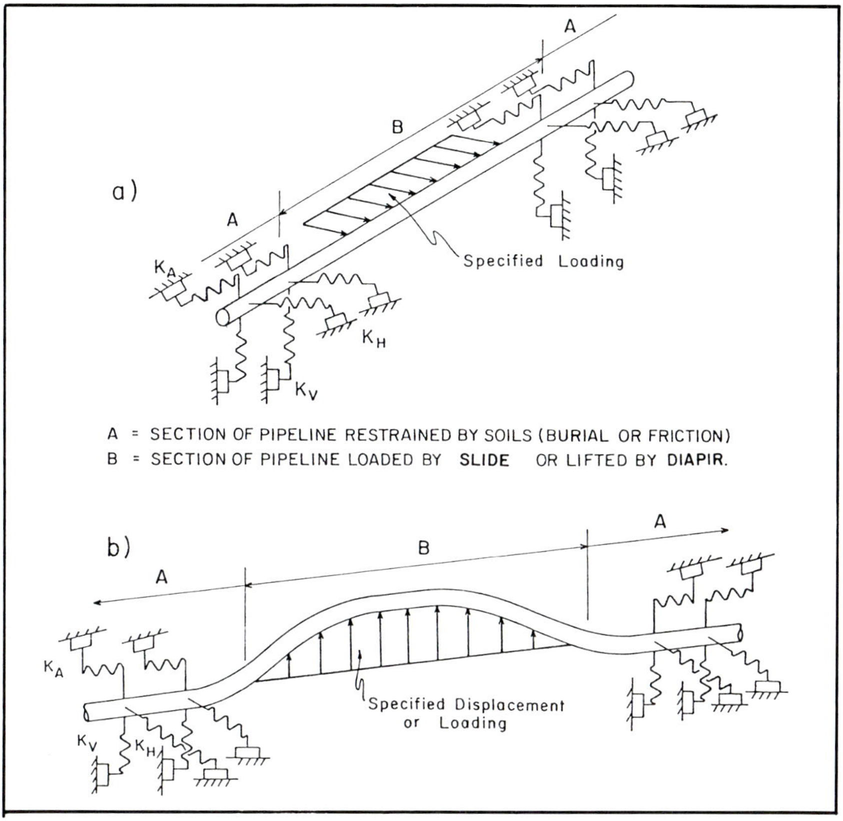

Figure A.3 Structural Modelling of Pipeline Subjected to: (a) Slide, (b) Fault Displacements

a specified displacement. Various algorithms are available for calculating these forces and displacements. These are described below and they are reviewed in Table A.1 (overleaf).

A.1 Evaluation of the Axial or Longitudinal (T–X) Component Force on a Pipeline

This component force represents the side friction along a pipeline. Due to its similarity with the unit skin friction caused by shaft resistance to piles, the load-deformation relationship developed for load-transfer along pile shafts is used. For sands this gives:

$$t = \sigma_n \tan \delta$$

where:

σ_n = effective normal stress along the periphery of the pipeline
δ = interface angle of friction between soil and pipeline

Table A.1 Soil Loading and Restraint Algorithms *[adapted from Audibert et al, 1978]*

COMPONENTS AND RELATIONSHIPS	PROPOSED LOAD/RESTRAINT DEFORMATION RELATIONSHIPS	REMARKS
Axial Component t–x Curve	$T_u = \begin{cases} \alpha C A_c & \text{for clays} \\ \iint_{A_c} \bar{\sigma}_n \tan \delta \, dA & \text{for sands} \end{cases}$ $x_u = 0.1$ to 0.2 inches (2.5 to 5 millimetres) where: A_c = soil-pipelines contact area/unit length C = undrained shear strength α = load transfer coefficient $\bar{\sigma}_n$ = effective normal stress at soil-pipeline interface δ = interface friction angle dA = incremental contact area	Inferred from pile shaft load-transfer technology
Horizontal Component p–y Curve	Surface Lines: (Cases i and ii below) $P_u = F(W_b - F_v)$ $y_u = \Delta H \frac{D}{Z}$ where: F = coefficient of lateral stability W_b = buoyant weight of pipeline/unit length F_v = hydrodynamic lift force/unit length ΔH = lateral displacement	After Karal (1977)
	Buried Lines (Cases iii and iv below) $P_u = \begin{cases} CN_cD & \text{for clays} \\ \bar{\gamma} Z_c N_q D & \text{for sands} \end{cases}$ $y_u = \begin{cases} 0.00 \text{ to } 0.06\,(Z_c + D/2) & \text{in clays} \\ 0.015 \text{ to } 0.02\,(Z_c + D/2) & \text{in sands} \end{cases}$ where: N_c, N_q = bearing capacity factors (horizontal direction) $\bar{\gamma}$ = effective soil unit weight Z_c = depth to centre line of equivalent footing or anchor plate D = pipeline diameter	Inferred from footing and pull-out capacity technology of vertical anchor plates (after Audibert *et al*, 1977)

Since the vertical and horizontal soil stresses are considered the principal stresses in a soil mass at rest, the following relation holds:

$$\sigma_n = \frac{\bar{\gamma} Z}{2} [1 + \cos 2\theta + K_o(1 - \cos 2\theta)]$$

Table A.1 *(continued)*

Vertical Component q–z Curve	<u>Downward Direction</u> $Q_u = \begin{cases} C\,N_c\,B & \text{in clays} \\ (\bar{\gamma} H\,N_q + \frac{1}{2}\bar{\gamma} B\,N\gamma)\,B & \text{in sands} \end{cases}$ $Z_u = \begin{cases} 2\,C_F\,B & \text{in clays} \\ 0.1\,B & \text{in sands} \end{cases}$ where: $N_c, N_q, N\gamma$ = bearing capacity factors (vertical direction) H = depth of embedment of equivalent footing of width B B = width of equivalent footing $\leq$ diameter of pipeline C_F = axial strain to failure in triaxial compression test	Inferred from bearing capacity technology for strip footings
	<u>Upward Direction</u> $Q_u = (C\,F_c' + \bar{\gamma} Z_c\,F'_q)\,B$ all soils $Z_u = 0.01H$ all soils where F_c', F_q' = breakthrough factors for long cylinders of diameter B	Inferred from pull-out capacity technology for horizontal anchor plates & cylinders

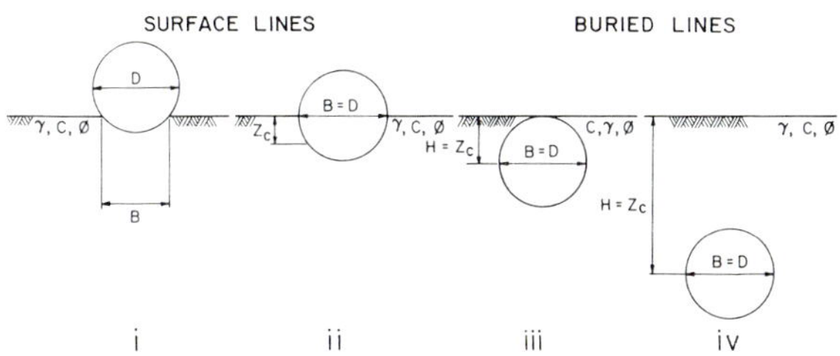

<u>Notes:</u> In the above formulations, it is assumed that undrained conditions prevail in clays while drained conditions prevail in sands.
T_u, P_u and Q_u are ultimate soil loads or restraints per unit length of pipelines.

where:

θ = angular coordinate indicating the position of the point under consideration along the periphery of the pipeline
$\bar{\gamma}$ = submerged unit weight of soil
Z = depth from mudline to pipeline axis
K_o = coefficient of 'at rest lateral earth pressure' (ratio of horizontal to vertical earth pressure)

Since the soil spring representations of previous figures consider only the overall or gross soil structure interaction between the pipeline and the soil, the resulting maximum axial load per unit length of pipeline (T) is obtained from the following relationship:

$$T = 2 \int_{-\pi/2}^{\pi/2} \bar{\gamma} Z \tan \delta \, [1 + \cos 2\theta + K_o (1 - \cos 2\theta)] \, ds$$

which simplifies to:

$$T = \frac{\pi}{2} \bar{\gamma} H D (1 + K_o) \tan \gamma$$

where:

D = pipeline diameter
H = depth of burial of pipeline (mudline to pipeline axis)

This represents the maximum or ultimate load transfer available, denoted by t_u in Figure A.2.

For clays the following relationship applies:

$$t = c_a$$
$$= \alpha S_u$$

where:

c_a = force of adhesion
S_u = undrained soil shear strength
α = empirical coefficient varying with S_u

The resulting maximum axial load per unit length of pipeline (T) is obtained from the equation:

$$T = \pi D \alpha S_u$$

From pile technology, it is generally accepted that the maximum load transfer occurs at displacements X_u of the order of 0.1 to 0.2 inches only. The load-deformation relationship between zero and the ultimate value is known to be nonlinear and can be approximated by a hyperbolic relationship, or more simply by a bilinear relationship.

A.2 Evaluation of the Horizontal or Lateral Component (P–Y) Force on a Pipeline

When the line is exposed at the mudline, the lateral stability of the line under hydrodynamic lift and drag forces must be considered. The lateral resistance of the soil to horizontal motion of a submarine pipeline is given by:

$$P = f(W_b - F_v)$$

where:

f = friction coefficient
W_b = buoyant weight of pipeline
F_v = vertical component of hydrodynamic force (lift)

Such an approach assumes that the supporting medium is rigid, the pipe is sliding parallel to the surface of the supporting medium, and the sliding resistance P is independent of the lateral displacement. Karal (1977) has argued that these assumptions are inadequate for most practical purposes because the soil is not rigid, sliding does not develop along the soil surface and the soil resistance depends on the displacement of the pipe. Instead he proposed that the lateral stability of a pipe is similar to bearing capacity analysis of footings loaded by an inclined load, and he suggested a solution based on two prescribed failure mechanisms observed in model tests. Initially, the penetration depth of the pipe into the soil is determined from the first failure mechanism, then the lateral soil resistance is calculated from the second failure mechanism by minimising the equation of energy equilibrium. The lateral soil resistance is calculated for successively increasing amounts of horizontal displacements until either the maximum horizontal force is found (further displacements result in lower horizontal force), or some prescribed failure criterion is reached.

According to Karal (1977) the lateral soil resistance is given by:

$$P = F(W_b - F_v)$$

where:

F = a coefficient of lateral stability of an unburied pipeline

In the following analyses a similar approach of wider applicability is proposed. Based on experimental evidence and results, Audibert and Nyman (1975, 1977) developed an analytical method to determine the restraint offered by granular soils and sands to the transversal/horizontal movement of buried pipelines. Their method is based on the horizontal bearing capacity of vertical strip footings and the nonlinear p–y relationship is given by the hyperbolic relationship:

$$P = \frac{y}{A' + B'y}$$

where:

A' = $0.145\, y_u/q_u$
B' = $0.855/q_u$
q_u = $\bar{\gamma} H N_q$
y_u = $0.02\,(H + D/2)$ for loose sand
= $0.015\,(H + D/2)$ for dense sand
$\bar{\gamma}$ = submerged unit weight of soil

H = depth to centreline of pipeline
N_q = bearing capacity factor for vertical strip footing, horizontally loaded, (after Hansen (1961))

The same approach can also be used for cohesive clays, in which case the ultimate soil restraint is given by:

$$P_u = cN_c$$

where:

c = undrained shear strength or cohesion
N_c = bearing capacity factor for vertical strip footings, horizontally loaded, (after Hansen (1961))

It is also expected that the p–y relationship would also be similar, with y_u possibly larger than for sands (cohesive soils are generally more ductile than sands). A value of y_u of about three to five per cent of H would seem reasonable.

A.3 Evaluation of the Vertical-Lateral Component (Q–Z) Force on a Pipeline

While the previous two components have a symmetrical response (same response regardless of direction of motion), the vertical component is unsymmetrical.

For downward motions, the pipeline is acting as a cylindrical shape strip footing and the q–z, load-deformation relationships are given by conventional bearing capacity theory:

$$q_u = cN_c + \bar{q}N_q + 1/2\,\bar{\gamma}D\,N_\gamma$$

where:

c = soil undrained strength, or cohesion
N_c, N_q, N_γ = bearing capacity factors
$\bar{q}$ = effective overburden pressure at pipeline axis
γ = submerged unit weight of soil
D = pipeline diameter.

For the particular case of undrained loading in clays this equation becomes:

$$q_u = cN_c$$

In accordance with general soil mechanics principles, Reese and Casbarian (1968) considered that N_c would increase from 5.7 at the ground surface (very little to no embedment) to possibly 12 for depth of embedment ratios of 8 or more.

The ultimate displacement is generally considered to be of the order of ten to 15 per cent of the foundation width and bilinear elastic-plastic relationship can be easily obtained. Alternatively, the method proposed by Skempton (1951) may be used. In this method, he suggested that a q–z curve for a strip footing could be directly extrapolated from the stress–strain curve ($\sigma - \epsilon$) obtained in a triaxial test using the following relationship:

$$z = 2\,\epsilon D$$

with z and ϵ corresponding to the same stress level and:

$$q/q_u = \sigma_1 - \sigma_3/(\sigma_1 - \sigma_3)_f$$

where:

σ_1 = major principal stress in triaxial compression test
σ_3 = minor principal stress in triaxial compression test
$(\sigma_1 - \sigma_3)_f$ = the difference in major and minor principal stresses at failure.

For upward motions, Reese and Casbarian (1968) reported results from pull-out tests on model pipelines, and suggested that the maximum uplift resistance of a model pipeline section of length, L, and diameter, D, is given by:

$$F = K_c\, cLD \quad \text{in soft clay}$$

and:

$$F = K_s\, \gamma HLD \quad \text{in loose sand}$$

where K_c and K_s represent empirical factors for soft clay and loose sand, respectively.

Alternatively, a pipeline can be considered to act as a buried anchor and the q–z relationship can be inferred directly from the results of buried anchor plates, as reported, for example, by Das *et al* (1975), Vesić (1969), and others.

Vesić (1969) proposed an analytical approach to the problem of break-out resistance of objects embedded in the ocean bottom based on solutions for the problem of expansion of cylindrical cavities close to the surface of a semi-infinite, rigid-plastic solid. According to Vesić *et al* (1965), the ultimate radial pressure is:

$$q_o = cF'_c + \gamma HF'_q$$

where F'_c, F'_q = cavity breakthrough factors dependent on shape, relative depth of cavity and angle of internal friction of soil.

Vesić (1969) extended the solution to long rectangular plates as follows:

$$q_o = c\overline{F}'_c + \gamma HD\overline{F}'_q$$

where $\overline{F'}_c$, $\overline{F'}_q$ = plate breakthrough factors with:

$$\overline{F'}_q = F'_q + \frac{\pi D}{8 H}$$

$$\overline{F'}_c = F'_c$$

Numerical values of the breakthrough factors for long buried cylinders and rectangular plates have been published by Vesić *et al* (1965).

As far as displacements are concerned, very little information is available in the literature inasmuch as researchers have reported on the ultimate pull-out resistance of anchors rather than on the load-deformation characteristics of an anchor-soil system. From tests reported by Esquivel-Diaz (1967) on the pull-out resistance of 3-inch circular model anchors buried in sand, the following results pertain:

(i) In dry dense sand, the ultimate pull-out resistance occurred at 0.1 inches and 0.6 inches of vertical displacement for H/D = 4.5 and 10, respectively.

(ii) In dry loose sand, the ultimate pull-out resistance occurred at 0.8 inches and 0.5 inches of vertical displacement for H/D = 4.5 and 10, respectively.

Similarly, Ali (1968) reported on the pull-out resistance of 3-inch diameter circular anchor plates buried in soft bentonite clay. The ultimate vertical displacement was determined to vary between 0.4 inches and 0.8 inches, for embedment ratios varying from 0.12 to 3, respectively.

It is not known to what degree scale and shape effects may affect such results, but it is believed that ultimate displacement may increase proportionately with the size of the anchor plate. It is thus tentatively proposed that the ultimate displacements obtained by Esquivel-Diaz and by Ali may be scaled in direct proportion with the ratio of the diameter of the pipeline, expressed in inches and the 3-inch diameter of the model plates accordingly:

$$(Z_u) \text{ uplift on pipeline} = (Z_u) \text{ model plate} \times \frac{D \text{ (inches)}}{3 \text{ (inches)}}$$

Alternatively, an ultimate displacement equal to four per cent of the depth of embedment (0.04H) has been used in engineering practice.

A.4 Bibliography

Ali, M., 'Pullout Resistance of Anchor Plates and Anchor Piles in Soft Bentonite Clay', Thesis submitted to Duke University, Durham, North Carolina in partial fulfilment of the requirements for the degree of Master of Science, (1968).

Audibert, J. M. E. and Nyman, K. J., 'Coefficients of Subgrade Reaction for the Design of Buried Piping', *Proceedings, Second ASCE Speciality Conference on Structural Design of Nuclear Plant Facilities,* 1A, 109–141, (New Orleans, Louisiana, December 1975).

Audibert, J. M. E. and Nyman, K. J., 'Coefficients of Subgrade Reaction for the Design of Buried Piping', *Proceedings, Second ASCE Speciality Conference on Structural Design of Nuclear Plant Facilities,* 1A, 109–141, (New Orleans, Louisiana, December 1975).

Audibert, J. M. E. and Nyman, K. J., 'Soil Restraint against Horizontal Motion of Pipes', *ASCE J. of the Geotechnical Engineering Division,* 103, GT10, (October 1977).

Das, B. M. and Seeley, G. R., 'Breakout Resistance of Shallow Horizontal Anchors', *ASCE J. of the Geotechnical Engineering Division,* 101, GT9, (September 1975).

Hansen, J. B., *'The Ultimate Resistance of Rigid Piles against Transversal Forces',* Bulletin 12, Danish Geotechnical Institute, (1961).

Karal, K., 'Lateral Stability of Submarine Pipelines', Paper OTC 2967, *Proceedings, Offshore Technology Conference,* (Houston, 1977).

Reese, L. C. and Casbarian, A. O. P., 'Pipe Soil Interaction for a Buried Offshore Pipeline', Paper SPE 2343, *Proceedings 43rd Annual Fall Meeting of the Society of Petroleum Engineers of AIME,* (Houston, 29 September to 2 October, 1968).

Skempton, A. W., 'The Bearing Capacity of Clays', *Collected Papers, Building Research Congress,* 180–189, (London, September 1951).

Vesić, A. S., Wilson, W. E., Clough, G. W. and Tai, T. L., *'Theoretical Studies of Cratering Mechanisms Affecting the Stability of Crated Slopes',* Final Report, Phase II, Project No A-655, Eng Exp Sta, Georgia Institute of Technology, (1965).

Vesić, A. S., 'Breakout Resistance of Objects Embedded in Ocean Bottom', *Proceedings, ASCE Ocean Engineering Conference,* (Miami Beach, 10–12 December, 1968). Also available as Duke University School of Engineering *Soil Mechanics Series,* 20, (1969).

R G BEA

APPENDIX B

The Reduction of Residual Longitudinal Stress during Hydrostatic Tests

A hydrostatic test partially erases residual longitudinal stresses left by plastic bending during reeling. A simple model from elementary plasticity theory shows how this happens.

A pipeline can be idealised as a thin-walled tube. In the absence of torsion and shear, the principal directions are longitudinal and circumferential, and the corresponding principal stresses are the longtitudinal stress and the hoop stress. The longitudinal stress generally varies around the circumference. The third principal stress is radial, and is negligibly small by comparison with the other two. A state of biaxial stress can be described by a point on a diagram such as Figure B.1 (overleaf), in which the axes are longitudinal stress and hoop stress. A yield condition is described by a closed curve on this diagram: points within the curve are below yield, and points on the curve are at yield. If the steel can be idealised as a von Mises material, the yield locus is an ellipse.

During construction, the hoop stress is small by comparison with the longitudinal stresses induced by plastic bending. Consider a section of the pipe whose residual longitudinal stress after construction is represented by point a, a compressive stress about two-thirds of yield. If the pipeline is now pressurised, the hoop stress increases, and – if there is sufficient axial constraint for the longitudinal strain to be zero – the longitudinal stress increases in such a way that the change in longitudinal stress is the change in hoop stress multiplied by Poisson's ratio. This is true as long as the response to pressurisation is elastic, that is, as long as the stress point remains within the yield locus. As the pressure increases, the stress point follows the path ab. When it reaches c, the material begins to yield. The stress point then moves along the yield locus, in such a way that negative longitudinal plastic strain increments balance positive longitudinal elastic strain increments and the resultant longitudinal strain is zero. This can continue until e is reached. If the maximum pressure corresponds to d, and the pressure is reduced, the stress point moves back along a line df parallel to ab. When the internal pressure is again zero, the stress state is f and the longitudinal stress has been reduced from its initial value. Sub-

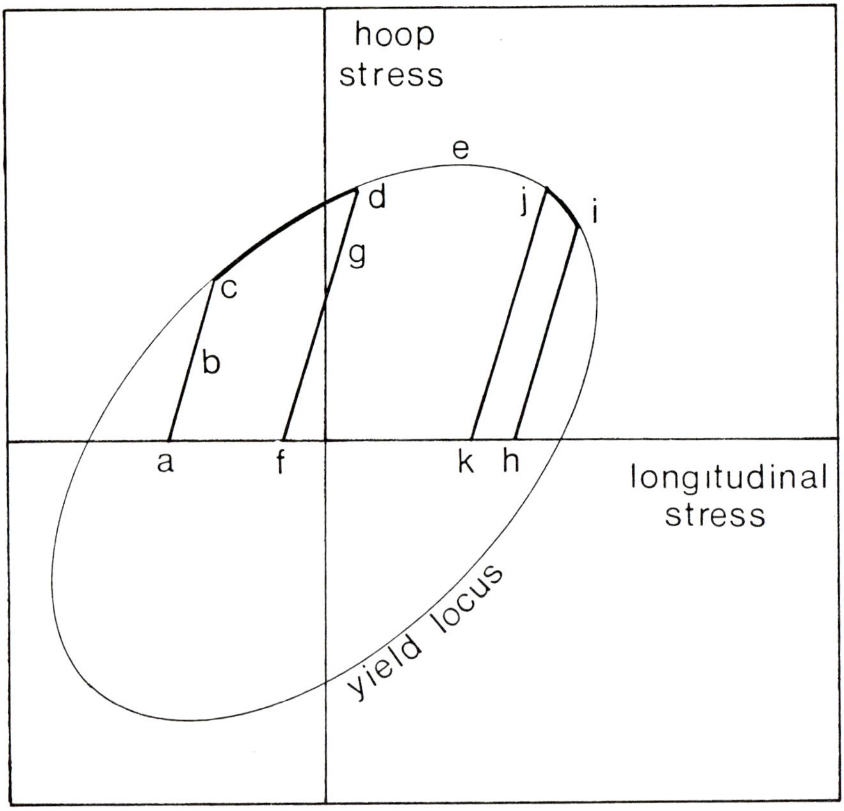

Figure B.1 The Effect of Hydrostatic Tests on Longitudinal Stresses

sequent pressurisation for operation, at a lower pressure than the maximum test pressure, induces an elastic response, and the stress point moves along fd to g.

If the initial longitudinal stress had been positive, the stress point would have followed the path hijk.

This idealisation has assumed that there is no strain hardening. A real material does strain-harden, and the yield locus moves and changes in shape as plastic deformation occurs. The response is qualitatively similar, but the degree of reduction of residual stress depends on the hardening rule. The analysis can be made quantitative, but requires a detailed knowledge of the plastic stress-strain relations of the material, and this is rarely available in practice.

R J BROWN
A C PALMER

APPENDIX C

Vortex Shedding

This appendix incorporates the theories which apply to vortex shedding from model cylinders under steady current conditions and real pipelines under transient wave conditions. Design criteria are cited along with the need for further R&D to permit an even better understanding of vortex shedding problems under complex real world conditions.

C.1 Forces on Horizontal Cylinders in Steady Currents

Extensive studies over many years have established a reasonably good understanding of the flow patterns around rigid cylinders and the forces on those cylinders under steady flow conditions. Some of the more important aspects are summarised in Figure C.1.1 (overleaf). The drag and lift forces are given by:

$$F_L = \tfrac{1}{2} V^2 C_L \rho D \qquad \text{(C.1.1)}$$

$$F_D = \tfrac{1}{2} V^2 C_D \rho D \qquad \text{(C.1.2)}$$

where:

F_L = lift force

F_D = drag force

V = undisturbed normal fluid velocity

C_L = coefficient of lift

C_D = coefficient of drag

ρ = fluid density

D = cylinder outside diameter

A reasonable estimate of the time varying forces, the fluctuating drag coefficient C_D' and the fluctuating lift coefficient C_L' can be found from the literature if they are required for design calculations. However, the extrapolation of results established mainly at low Reynolds numbers becomes uncertain under prototype conditions. Some progress with the effects of roughness at high Reynolds number has been made recently for the time average drag.

Other factors requiring more data at high Reynolds numbers are the effects of distorted velocity profiles.

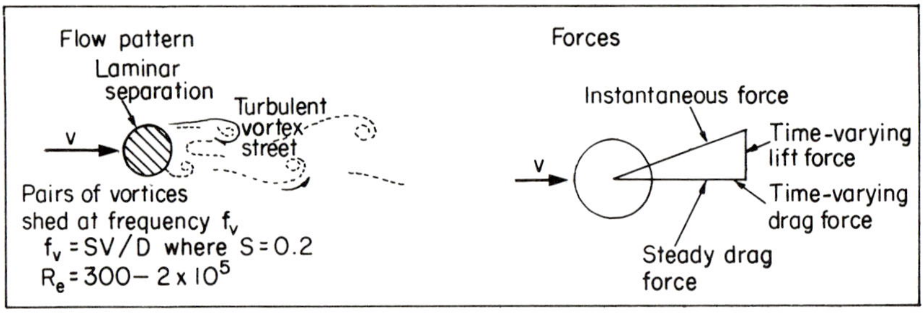

Figure C.1.1 Flow about a Rigid Cylinder. (The forces are functions of the Reynolds number (R_e), surface roughness, turbulence and L/D ratio. L = length of cylinder; D = cylinder diameter; S = Strouhal number)

The main effect of structural flexibility is that the movement of the cylinder, produced by the time varying components of the forces, can modify the flow patterns around the cylinder and hence the actual drag and lift forces themselves. There is thus a fluid-elastic effect, which causes the 'lock-in' phenomenon shown diagrammatically in Figure C.1.2 for cross-flow ('lift' direction) vibration.

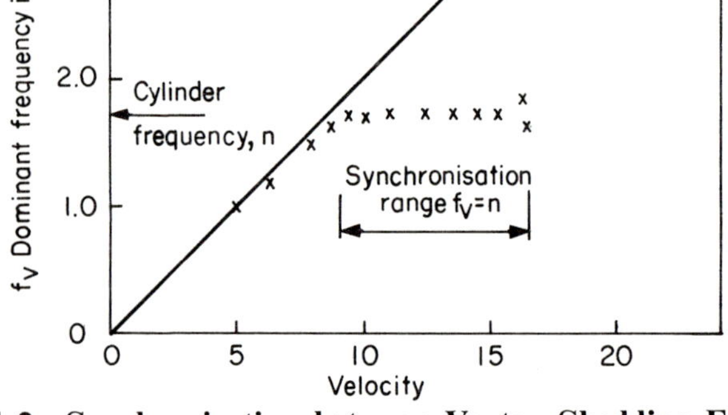

Figure C.1.2 Synchronisation between Vortex Shedding Frequency f_v and Flexible Cylinder Frequency n. ($S = f_v D/V$)

Since the fluctuating drag component C_D' has a frequency twice the Strouhal frequency, a flexible cylinder may be expected to vibrate in-line ('drag' direction) at a velocity of half the value required for cross-flow vibrations. The 'lock-in' effect in fact extends the velocity range above and below the expected value and the response of a flexible cylinder, measured in terms of a non-dimensional deflection, will be similar to Figure C.1.3. The important point to note is that the cylinder is stable if $V/nD < 1.2$.

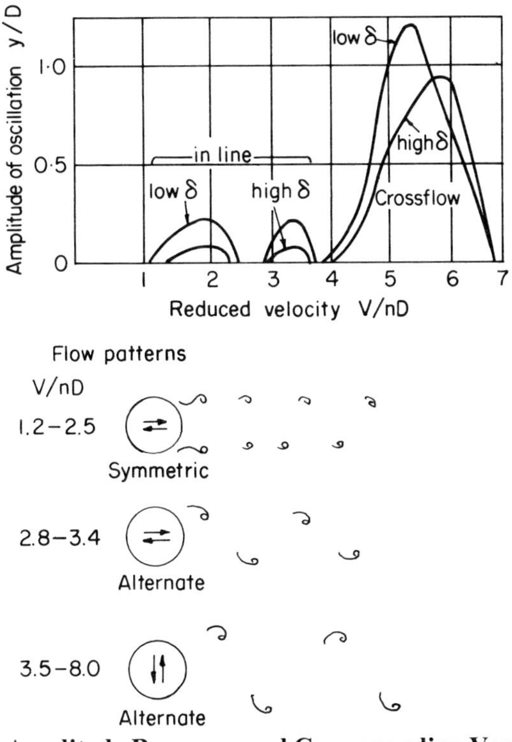

Figure C.1.3 Amplitude Response and Corresponding Vortex Shedding Patterns

Structural and fluid damping, together with the ratio of structural density to fluid density, also influence the actual stability and the resulting response if the structure does vibrate. The effect of damping on the response is indicated in Figure C.1.3, but damping and the mass ratio may be combined to form a parameter known as the combined stability parameter K_S, where:

$$K_S = \frac{2Me\delta}{\rho D^2} \qquad \text{(C.1.1)}$$

where:

$$Me = Ms + Mo + Ma$$

$$\delta = 2\pi\,\xi$$

ξ = damping ratio in air (structural damping)

Me = equivalent mass per unit length

Ms = structural mass of pipeline per unit length

Mo = mass of content in the pipeline per unit length

Ma = added mass per unit length

δ = logarithmic decrement of damping

The response is a function of K_S, with a differing critical value for each structural mode and direction of vibration. Again, the importance of the stability parameter K_S is that, to avoid sustained vibrations, K_S must be greater than 1.2 for the in-line direction and greater than about 16 for the cross-flow direction.

Figures C.1.4 and C.1.5 show that the hydrodynamics are self regulated in that amplitudes of cross oscillations (y) are not greater than approximately one pipe diameter. Note that $y/D > 0.5$ only for $4.5 < V/nD < 6.5$.

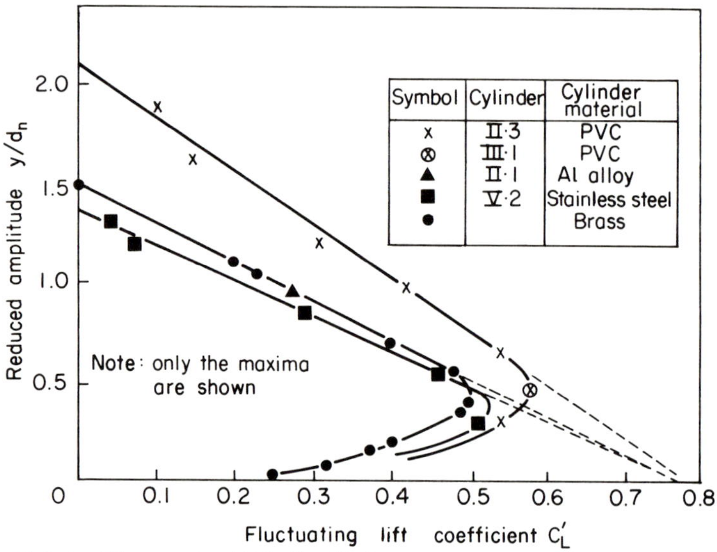

Figure C.1.4 Fluctuating Lift Coefficients *versus* Amplitudes of Oscillation Cross-flow. *[after King (1977)]*

C.2 Wave Forces on Vertical Cylinders

Forces on rigid cylinders due to waves of various heights and periods are fairly well established for engineering purposes. A further discussion is included here because vortex-induced oscillations require a deeper description of the oscillating part of drag and lift and the frequencies involved. There are adequate methods for calculating forces in the reflec-

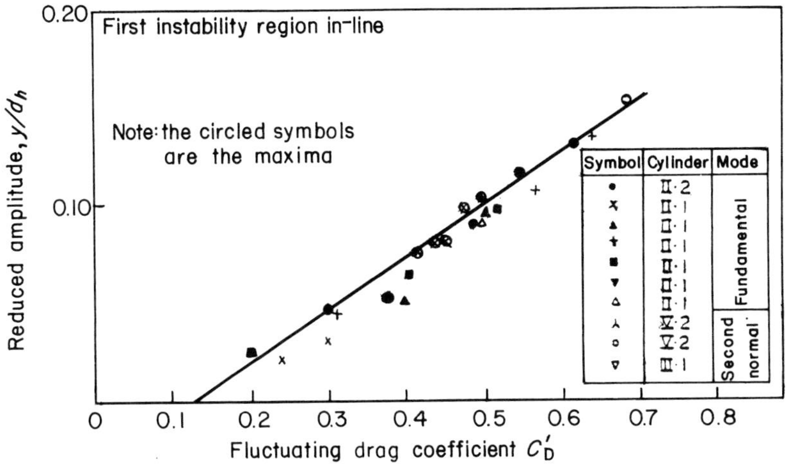

Figure C.1.5 Fluctuating Drag Coefficients *versus* Amplitudes of Oscillation In-line. (First instability region). *[after King (1977)]*

tion and diffraction regimes, and data exist for the Morison equation in the drag and inertia regimes. However, when more precise design data are required, there are many difficulties with the Morison equation, particularly in the drag regime. The basic equation does not account for 'lift' forces which are perpendicular to the wave direction, and although the values of the drag and inertia coefficients may be allowed to vary with flow parameters related to the maximum velocity (say) in a wave cycle, no variation is permitted through a wave cycle. The equation thus implies a far simpler force-time history than may actually occur. Figure C.2.1 (overleaf) shows the measured stresses in a vertical cylinder in a laboratory wave tank.

In steady currents, the value of the drag coefficient is very much related to the separation point, the type of separation, and the resultant wake width, which are in turn primarily a function of Reynolds number.

$$Re = \frac{V_{max} D}{\nu} \qquad (C.2.1)$$

where:

v_{max} = maximum partial velocity (also denoted by V_m)

υ = kinematic viscosity

In waves there is a time factor in that a wake must form in a half wave cycle before flow reversal occurs and the Keulegan-Carpenter number

$$K_c = \frac{V_{max} . T}{D} \qquad (C.2.2)$$

where T = the period of the cyclic flow

expresses this temporal constraint. For low values there may be insufficient time for a wake to form, notwithstanding the Reynolds number achieved. Thus in waves, drag and inertia coefficients are related to the wake, and one might expect the Reynolds number and the Keulegan-Carpenter number to be important. Considerable research has been undertaken to determine these coefficients against Reynolds number and Keulegan-Carpenter number, or some combination of these. The most extensive to date is that of Sarpkaya (1976) and (1977) on oscillating flow.

The following information is taken from CIRIA (1977).

For shallow-water waves the Keulegan-Carpenter number can be expressed as:

$$K_c = \frac{\pi\lambda}{D} \qquad \text{(C.2.3)}$$

The depth parameter:

$$k_d = \frac{2\pi d}{\lambda} \qquad \text{(C.2.4)}$$

where d = the water depth; and

λ = the wavelength, is also important.

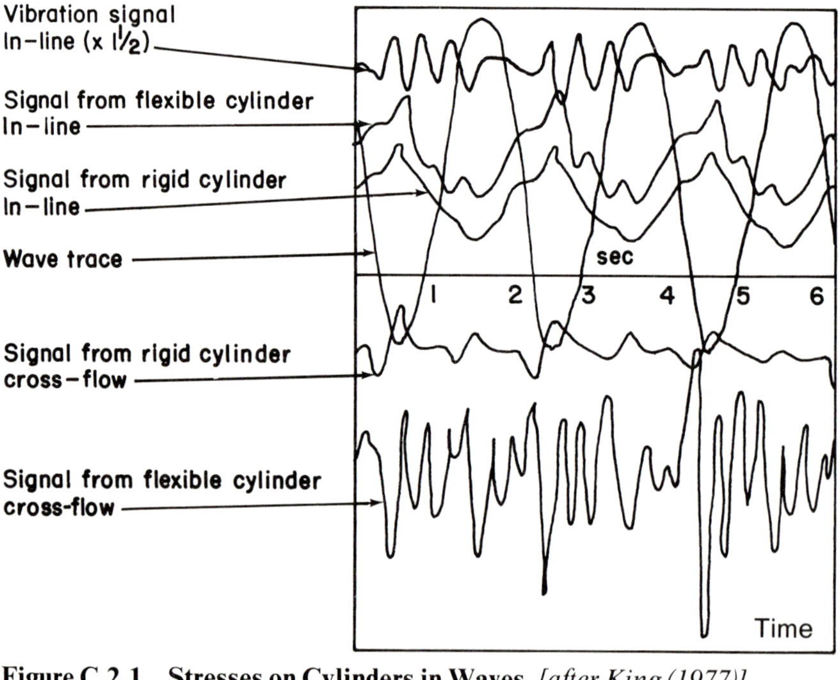

Figure C.2.1 Stresses on Cylinders in Waves. *[after King (1977)]*

The Keulegan-Carpenter number is therefore proportional to the number of diameters per wavelength, in shallow-water waves.

The lift coefficient C_L' is given by:

$$C_L' = \frac{F_L'}{\frac{1}{2}\rho (V_{max})^2 D \cdot d} \qquad (C.2.5)$$

Because the Reynolds number involves V_{max}, which must be calculated for each wave condition associated with a cyclinder, it is convenient to replace the Reynolds number with a parameter that depends on the wave period. This parameter is given by:

$$\text{frequency parameter } F_p = D^2/\upsilon T \qquad (C.2.6)$$

Hence the Keulegan-Carpenter number, the Reynolds number and the frequency parameter are related by:

$$Re = F_p K_c$$

The Keulegan-Carpenter number is the parameter which determines where lift occurs. Experimental work with waves and some 'starting-flow' situations strongly suggest that for vortex shedding to promote lift $K_c > 5$.

The lift frequency ratio f_L'/f_w is an integer value primarily dependent on K_c and to a lesser extent on R_e, as can be seen from Figures C.2.2 and C.2.3 (overleaf). The lowest value occurs for K_c between 5 and 16 where f_L'/f_w is 2 and as K_c increases above 16, f_L'/f_w increases, although integral multiples are maintained for ranges of K_c. The frequency ratios are found to lie between two lines corresponding to:

$$f_L' D/V_{max} = 0.125 \text{ and } f_L' D/V_{max} = 0.2$$

for Reynolds numbers of around 10^3 and 10^6 respectively. The latter value should be used for most full-scale situations.

Figures C.2.2 and C.2.3 show that, for a particular value of K_c, the value of the integer frequency ratio is likely to be 2 less for vertical cylinders than for two-dimensional flow.

The depth parameter k_d determines the variation of lift along the length of the vertical cylinders. For horizontal cylinders, normal to the wave direction, two-dimensional criteria based on the local flow conditions may be applied. For $k_d < 0.3$, corresponding to shallow-water waves, the flow is effectively two-dimensional and the lift is identical at all positions. For intermediate-depth waves, ($0.3 < k_d < 3$) experimental results indicate that vortex shedding is still well correlated along the length. A quasi-two-dimensional summation of the lift forces, based on local flow conditions along the length, however, gives a magnitude which is less than that measured directly. For deepwater waves, $k_d > 3$, the flow along the length is highly three-dimensional with relatively small lift forces. This is shown schematically in Figure C.2.3.

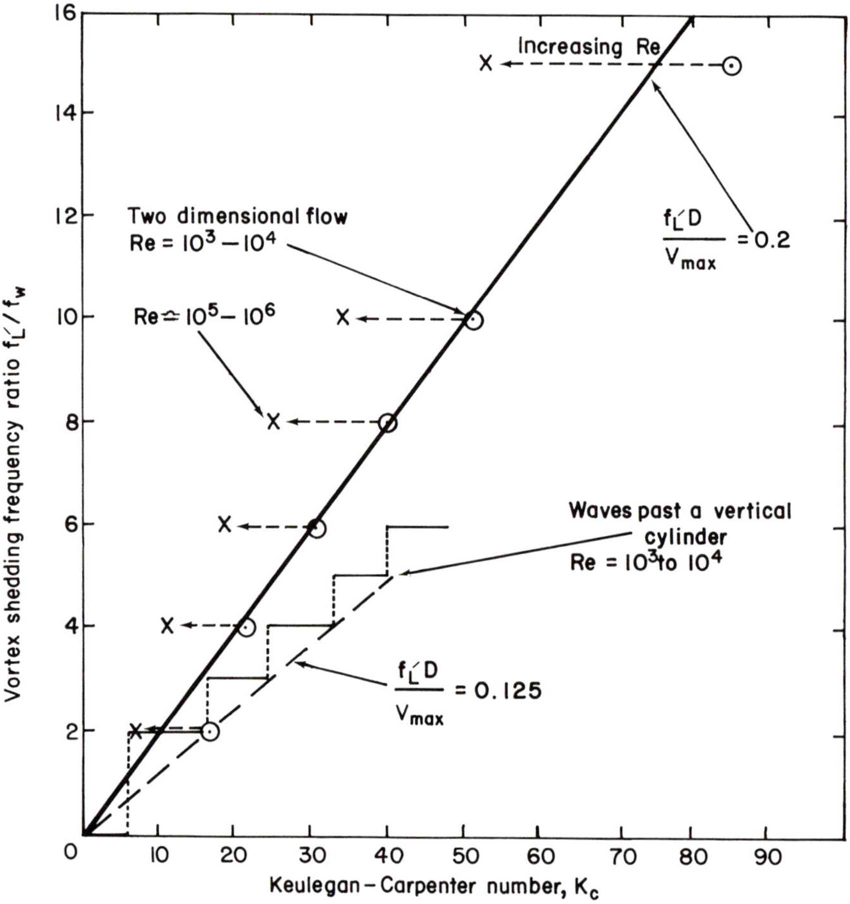

Figure C.2.2 The Frequency of Vortex Shedding in Waves. *[after CIRIA (1977)]*

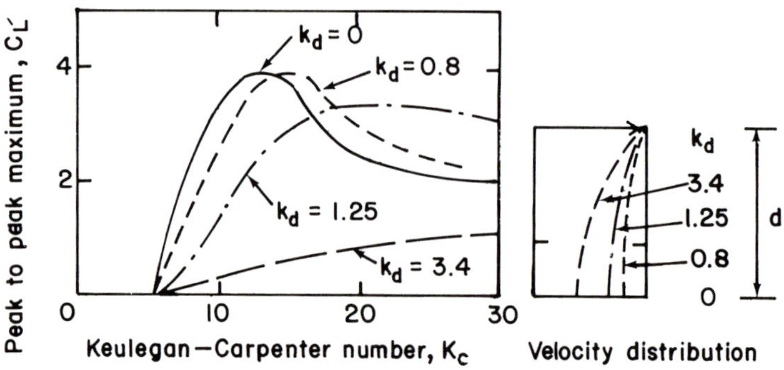

Figure C.2.3 An Indication of the Effect of Three-dimensionality of Wave Flow on the Lift Forces on a Vertical Cylinder. *[after CIRIA (1977)]*

The value of the lift coefficient as a function of Reynold's number and frequency parameter is shown in Figures C.2.4 and C.2.5. These graphs were obtained from two-dimensional periodic flow and, therefore, relate to horizontal cylinders in waves and to vertical cylinders where the

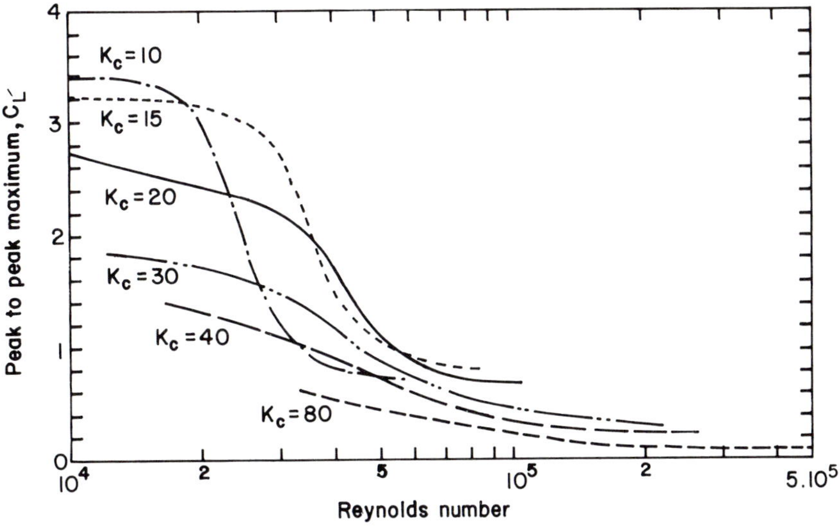

Figure C.2.4 Lift Generated in Periodic Flow Plotted against Reynolds Number. *[after CIRIA (1977)]*

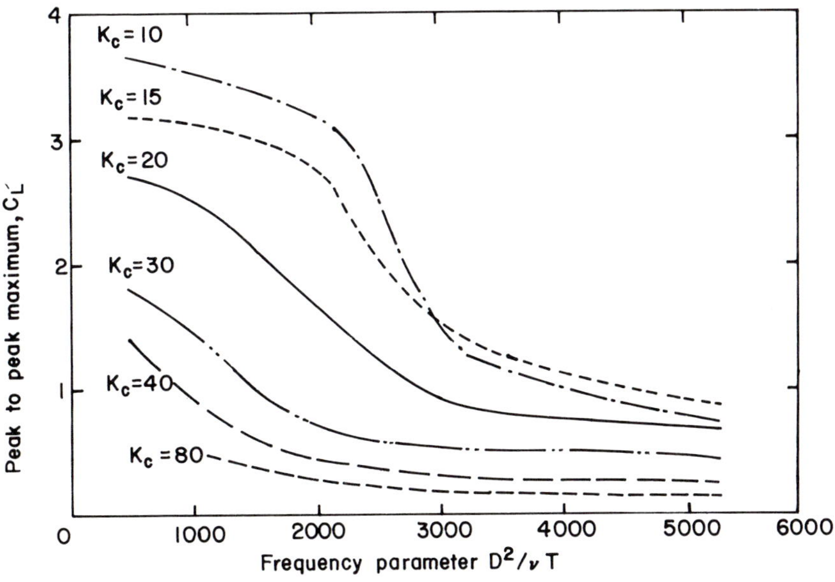

Figure C.2.5 Lift Generated in Periodic Flow Plotted against the Wave Frequency Parameter. (Data is identical to that plotted in Figure C.2.4). *[after CIRIA (1977)]*

velocity is uniform over their length. The values of C_L' presented are the maximum that occur during the passage of the wave. In practice, the largest lift force tends to occur at the time of maximum velocity and is, therefore, coincident with the largest in-line force for a cylinder in which drag predominates over the inertial force. The change of C_L' with Reynolds number is similar to that of steady current. The effect of response of a cylinder on the vortex-induced forces, which occurs in steady flow, is not understood in the case of waves. It has been suggested that the magnification of the forces is much less in wave flow, but this is not proven.

A flexible cylinder in waves will, in practical applications, be slender and in the drag regime of wave loading. One would not expect the flexible cylinder to experience the 'lock-in' phenomena which occurs in a steady flow as the motion is continually changing in magnitude and direction. However, as the eddy shedding forces are strong and may be at a frequency several times higher than the wave frequency, a dynamic magnification may occur between the eddy shedding frequency and the flexible cylinder's natural frequency. The flexibility alters the eddy shedding and spanwise correlation of vortices in oscillatory flow and in steady current flows, but this effect is not so great in waves.

Recent experimental work at BHRA (Verley (1977)) has studied the response of flexible cylinders in waves. In this work the stressing of flexible cylinders was compared directly with rigid cylinders. The vibrational component of the flexible cylinder near its natural frequency is shown in Figure C.2.6, the additional stresses being due to a dynamic response to eddy shedding. The vibration was a function of Vmax/nD, which can be interpreted as the ratio between the eddy shedding frequency and the natural frequency. No vibration occurred below Vmax/nD = 1, for both the in-line and cross-flow directions. The response was nearly independent of structural damping, in marked contrast with steady currents, where K_S, which involves damping, is of prime importance, as shown by Figure C.2.7 (overleaf).

According to Sarpkaya (1978b), the amplitudes of the lift coefficient and the relative displacement may be shown to be governed by:

$$\{C_{LM}, Y_M/D\} = f_i\left\{V_m T/D,\ D^2/\nu T,\ V_m/f_n D,\ k/D,\ \xi,\ M/\rho L D^2\right\} \qquad \text{(C.2.7)}$$

in which $C_{LM} = F_L/(0.5\rho LDV^2{}_m)$ and Y_m represents the amplitude of velocity oscillation; f_n, the in-water natural frequency of the oscillating system; D, the diameter of the cylinder; ν, the kinematic viscosity; T, the period of flow oscillations; k, the average roughness height; ξ, the damping ratio in air; M, the mass of the oscillating system; ρ, the density of fluid; and L, the length of the cylinder. The last two parameters in this equation may be combined as $M\xi/(\rho LD^2)$ on the basis of experience with hydroelastic oscillations in steady flow.

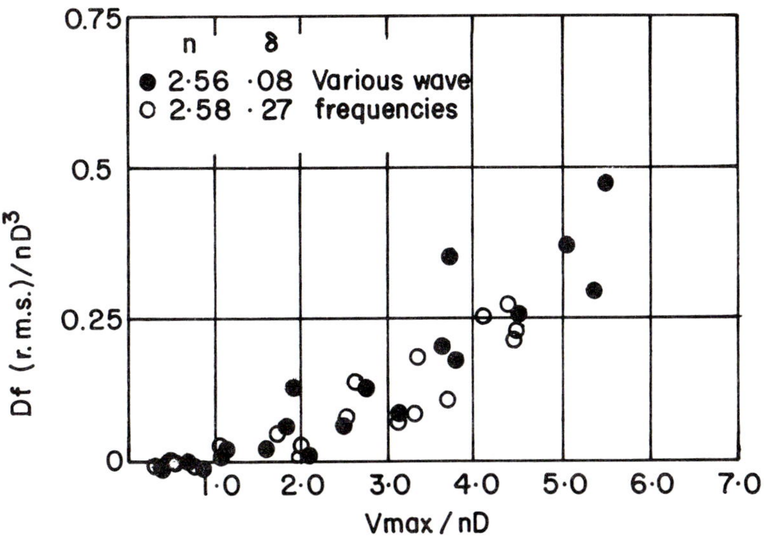

Figure C.2.6 Effect of Damping on Vibration in Waves. *[after Verley (1977)]*

The lift coefficient and the Strouhal number for a non-oscillating cylinder are governed by:

$$\left\{S^{o}, C^{o}_{LM}\right\} = g_i\{V_m T/D, D^2/\nu T, k/D\} \qquad \text{(C.2.8)}$$

In fact, the three parameters in Equation C.2.8 determine uniquely all of the charcteristics of the base flow. Missing from the above relationship is the effect of the correlation length.

Experiments with bodies with fixed separation lines (and suitable after-bodies) have shown that the increase in correlation length can account for only a small fraction of the large increase in the transverse force near lock-in. The fluid response to the body motion or the interaction betwen the fluid motion and the body motion is the primary reason for the lock-in and the increase in the transverse force. It is, therefore, natural to expect the dynamic response of the body to depend not only on V_m/f_nD and the mass parameter $M\xi/(\rho LD^2)$ but also on C^{o}_{LM} as defined in Equation C.2.8.

The equation of motion for an elastically-mounted rigid cylinder of mass M, damping coefficient ξ, circular natural frequency ω_n, and spring constant k, is given by:

$$M\ddot{y} + 2M\xi\omega_n\dot{y} + ky = F(t) \qquad \text{(C.2.9)}$$

where y represents the transverse displacement of the cylinder and F(t) the exciting force or the lift force. In general F(t) depends on the parameters given by Equation C.2.7 or simply on $V_r = V_m/f_nD$, $M\ /(\rho LD^2)$, and C^o_{LM}. For the sake of simplicity, F(t) may be assumed to be given by:

$$F(t) = 0.5\rho LDV_m^2 C_{LM} \sin(\omega_v t + \theta) \qquad \text{(C.2.10)}$$

in which $\omega_V = 2\pi f_V$ and θ is the phase angle. However, even during the period of nearly perfect synchronisation, the lift force does not retain a constant amplitude and frequency. Furthermore, the velocity V varies as $V = V_m \sin 2\pi t/T$ during a given cycle. In general, it is necessary to

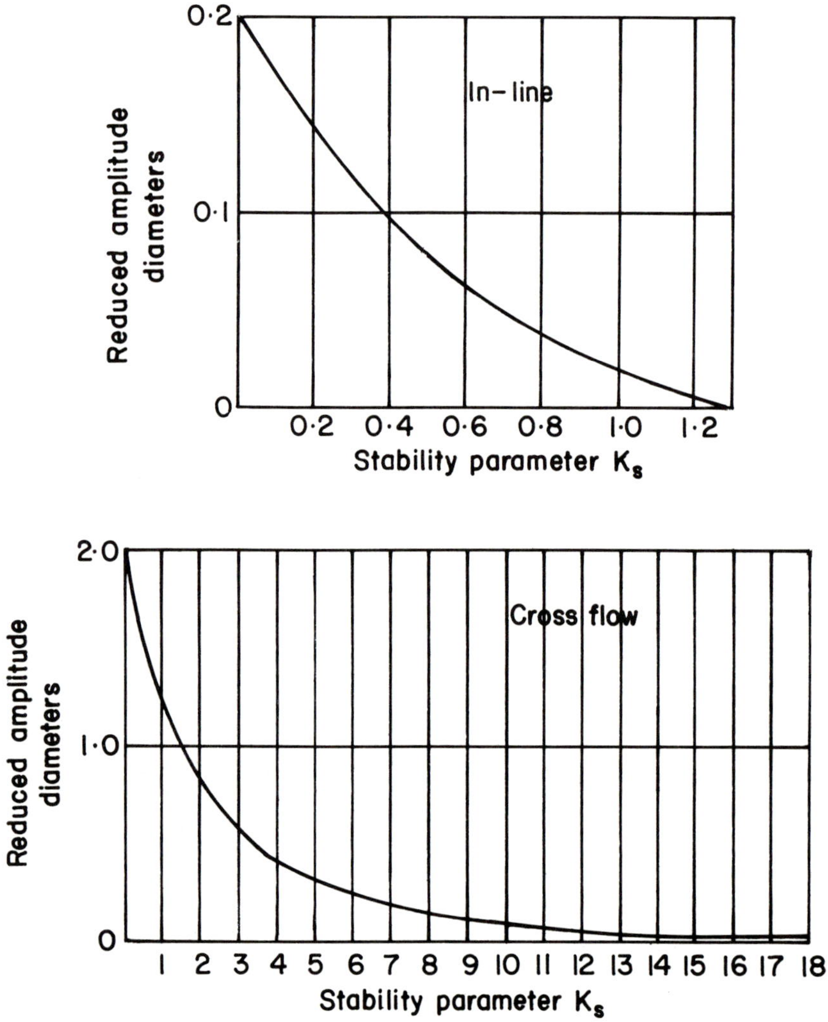

Figure C.2.7 Limiting Amplitudes of Oscillation In-line and Cross-flow as Functions of K_s. *[after Verley (1977)]*

consider the harmonics of the lift force together with their spectral energy. Prior to such a detailed analysis, it will be hypothesised, and subsequently verified, by the present results that the rate of energy transfer from the fluid to the oscillating body is proportional to the rms value of the velocity. Then Equation C.2.9 becomes:

$$F(t) = 0.25\rho LDV_m^2 C_{LM} \sin(\omega_v t + \theta) \qquad (C.2.11)$$

or:

$$F(t) = 0.25\rho LDV_m^2 C^o_{LM} (C_{LM}/C^o_{LM}) \sin(\omega_v t + \theta) \qquad (C.2.12)$$

in which C_{LM}/C^o_{LM} represents the amplification of the lift force at or in the neighbourhood of perfect synchronisation.

Equation C.2.9 coupled with Equation C.2.12 may be solved only for nearly constant values of C_{LM}/C^o_{LM}. At synchronisation the result reduces to:

$$Y_M/D = \frac{(V_m/f_n D)^2 \, C_{LM}/C^o_{LM}}{32\,\pi^2 M/(\rho LD^2 C^o_{LM})} \qquad (C.2.13)$$

This surprisingly simple equation shows that a unique relationship should exist between Y_M and $R_p = M\xi/(\rho LD^2 C^o_{LM})$ for both smooth and rough cylinders if it can be demonstrated that perfect synchronisation occurs for unique values of V_r and C_{LM}/C^o_{LM}. The parameter R_p will be referred to herein as the 'response parameter'.

Figure C.2.8 shows the variation of f_v/f_n as a function of the normalised time t/T for various values of V_r in the vicinity of perfect synchronisation. Clearly, at perfect synchronisation the frequency ratio remains nearly constant throughout the cycle and the vortex shedding frequency

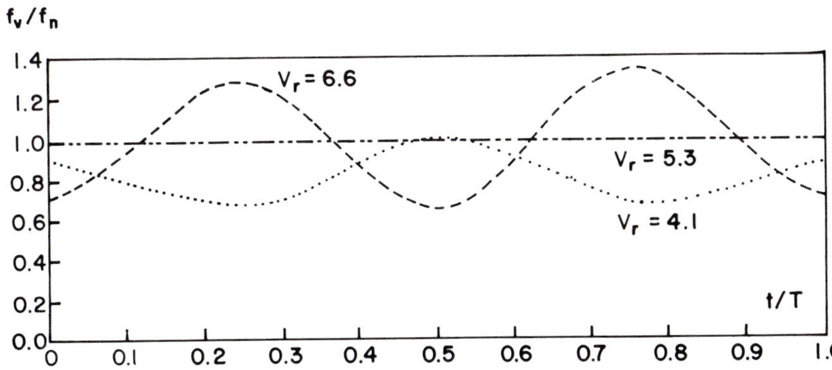

Figure C.2.8 Variation of the Vortex Shedding Frequency with Time during a Cycle for Various Values of V_r. *[after Sarpkaya (1978b)]*

locks into the in-water natural frequency of the cylinder. For values of V_r smaller or larger than that corresponding to perfect synchronisation (V_r^*), f_v/f_n oscillates about a mean value, ie, $f_v = \bar{f}_v + \Delta f \sin \omega t$. It is not therefore correct to assume that f_v varies as $f_v = (V_m S^o/D)|\sin \omega t|$.

Figure C.2.9 shows the variation of C_{LM}, C_{LM}/C^o_{LM}, C^o_{LM}, and Y_M/D as a function of V_r for a sand-roughened cylinder (k/D = 1/100). The C^o_{LM} values for the corresponding Keulegan-Carpenter numbers and $\beta = D^2/\nu T$ values were taken from Sarpkaya (1977c). The data shown in Figure C.2.9 are quite representative of all the data obtained with smooth and rough cylinders.

Apparently, the maximum lift amplification occurs for $V_r < V_r^*$ and $f_v/f_n < 1$. At $V_r = V_r^*$, Y_M/D reaches its maximum value and the lift amplification drops to about 2. For values of V_r larger than about 8, the cylinder is subjected to forced oscillations. These observations and measurements are in conformity with those previously made in connection with self-excited oscillations in steady flows.

The relative amplitude of oscillation at synchronisation, for all smooth and rough cylinders, is shown in Figure C.2.10 as a function of the response parameter R_p. The correlation is quite remarkable if one considers the fact that, M, ξ, D, $K = V_m T/D$, and $Re = V_m D/\upsilon$ and hence C^o_{LM} underwent wide variations during the course of the investigation.

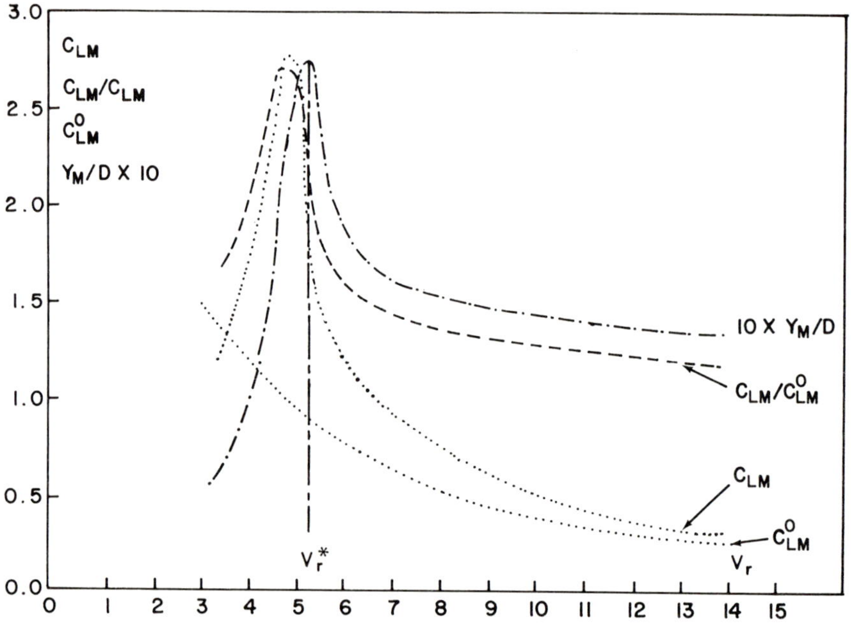

Figure C.2.9 Various Lift Coefficients and Relative Displacement *versus* V_r for a Sand-roughened Cylinder in Harmonic Flow. (k/D = 1/100). *[after Sarpkaya (1978b)]*

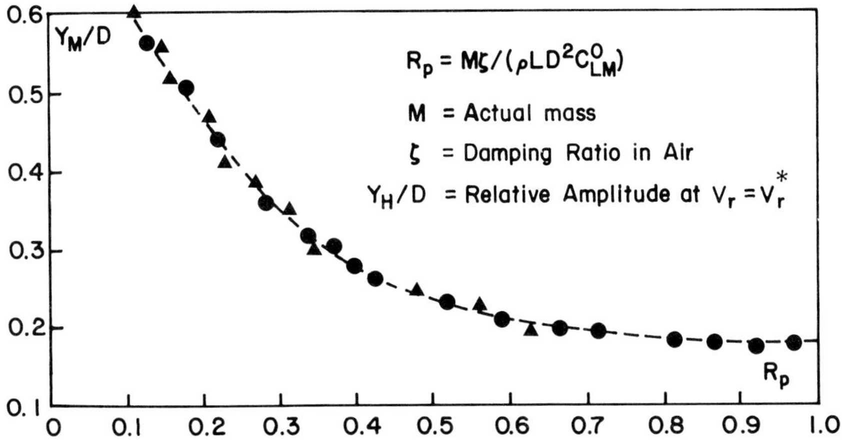

Figure C.2.10 Relative Amplitude *versus* Response Parameter at Synchronisation, ($V_r = V_r^*$) for all Smooth and Rough Cylinders

In Figure C.2.10, M stands for the actual mass of the oscillating body and does not include the displaced mass of water. Also, ξ is the damping ratio measured in air. Engineers, in plotting such data as a function of $M\xi/\rho LD^2$ (or slightly modified versions of it), prefer to use an equivalent mass M_e which includes the displaced mass and a damping ratio ξ_w which includes the effects of the surrounding fluid in which the body is to be subjected to self-excited oscillations. The equivalent mass approach can also be used. Accordingly, the damping ratio in (Sarpkaya 1977c), Figure C.2.10 has been replotted in Figure C.2.11 replacing M by M_e and ξ by

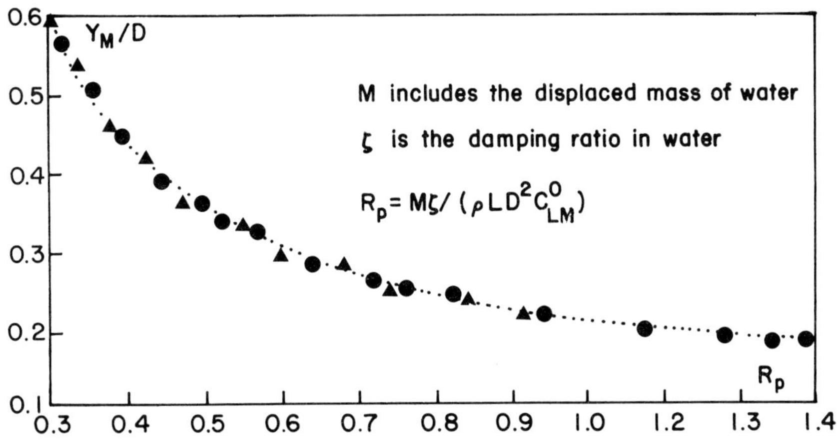

Figure C.2.11 Relative Amplitude *versus* Response Parameter at Perfect Synchronisation for all Smooth and Sand-roughened Cylinders

ξ_W (obtained in water by slightly displacing the cylinder from its equilibrium position and then letting it to undergo gradually-damping free oscillations).

In either plot the correlation of Y_m/D with R_p is quite good. The ramifications of these plots may be summarised as follows:

(i) in steady flows one obtains similar variations between Y_m/D and $M\xi/\rho LD^2$ because the range of Reynolds numbers has been quite limited and C^o_{LM} did not significantly vary (aside from the effects of the correlation length). In fact, the variations in C^o_{LM} and the test conditions may have been the cause of large scatter in the said plots;

(ii) the larger the lift coefficient of a stationary body, the more susceptible the body is to large transverse oscillations;

(iii) the most effective way of reducing the oscillation amplitude (aside from increasing M and ξ and decreasing D) is to decrease the lift coefficient in the nascent state (eg, the use of a splitter plate, spoilers, etc).

It is recommended that Figures C.2.10 and C.2.11 should be plotted in terms of the Root Mean Squared (rms) values of the lift coefficient and the relative displacement to reduce the scatter.

Figure C.2.12 shows a plot of the rms values of the measured lift coefficient as a function of the rms value of the calculated lift coefficient. The

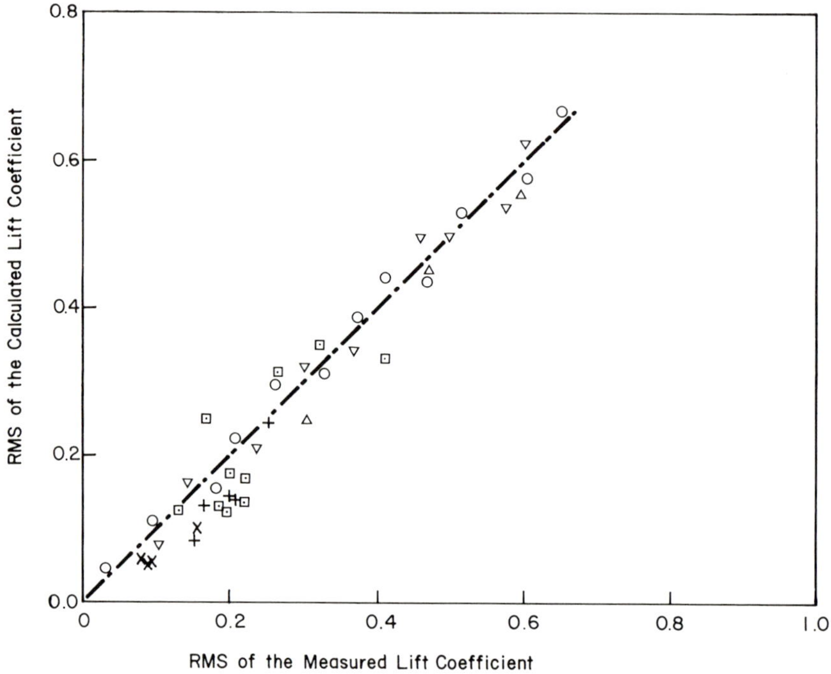

Figure C.2.12 Comparison of the rms Values of the Measured and Calculated Lift Coefficients

latter has been obtained from Equation C.2.13 through the use of the rms values of the measured relative displacements at perfect synchronisation. Evidently, the correlation between the two coefficients is quite good and shows that the assumptions leading to the derivation of Equation C.2.13 are justified.

Finally, Figure C.2.13 shows a comparison of the measured and calculated values of the instantaneous displacement as a function of time. The measured force was used as input in the Duhamel's integral. The close agreement between the measured and calculated displacements is not surprising. It simply proves that the various parameters such as mass, damping ratio, etc, have been properly measured.

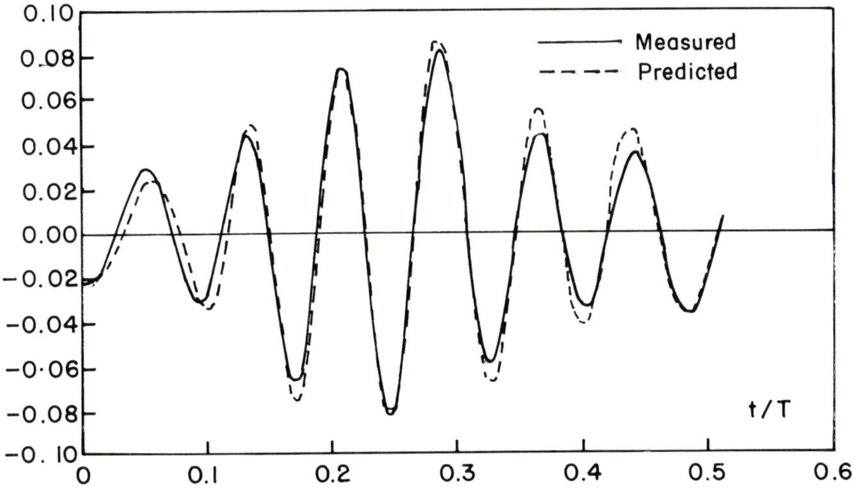

Figure C.2.13 Measured and Predicted Relative Displacements *versus* Time for $V_r = 4.3$

To conclude this section: the hydroelastic oscillations of smooth and rough circular cylinders in harmonic flow have been investigated for a wide range of parameters. It has been shown that the perfect synchronisation occurs at $V_r^* = 5.4$; the maximum lift amplification occurs for V_r values slightly lower than V_r^*; and the frequency ratio f_v/f_n remains nearly equal to unity only at perfect synchronisation. A response parameter R_p has been identified and its power in correlating the relative displacement for smooth and rough cylinders for a wide range of Reynolds numbers and Keulegan-Carpenter numbers has been demonstrated. The response parameter R_p uniquely incorporates all of the physical characteristics of the body, including its afterbody shape, and governs the forced or self-excited transverse oscillations of the body. For in-line oscillations, C_{LM}^o in R_p must be replaced by the normalised amplitude of the fluctuating component of the in-line force.

C.3 Classical Approach to Defining Criteria for Resonant Vortex Shedding on Pipelines

The lowest resonance frequency f_n for a pipe span may be calculated for specified span length (l), type of supportings, weight (m = mass per unit length including added mass), and stiffness of pipe (EI = bending-stiffness):

$$f_n = \frac{k\pi}{2l^2}\sqrt{\frac{EI}{m}} \qquad \text{(C.3.1)}$$

k is a factor depending upon the support

k = 1.0 for a normally supported pipe

k = 2.25 for fully fixed pipe ends

The maximal free span length with no resonance oscillations is:

$$l_m^2 = d^2 \frac{k\pi}{2\,SVn}\sqrt{\frac{E}{2\,(d/t)\,\delta}} \qquad \text{(C.3.2)}$$

where δ = mass of pipe + added mass per unit volume

n = safety factor on resonance frequency

S = Strouhal number

t = wall-thickness of pipe

Table C.3.1 gives typical span lengths.

Table C.3.1 Span Lengths which Give Lowest Resonance Frequency 20 per cent above Limiting Exitation Frequency

Supporting	Current velocity	
	0.25 m/s	1 m/s
Fixed ends	250 d	125 d
Free supportings	168 d	84 d

$S=0.2$, $d/t=30$, $\delta=3.0\ t/m^3$, $E=2.1 \cdot 10^8\ kN/m^2$

Example:

For a 34-inch diameter line, a free span of length 50 metres may have a natural frequency of 0.46 Hz, and will be subjected to in-line resonant vibrations due to vortex shedding for current velocities in the range of 0.46 to 0.91 m/seconds.

Stresses due to resonant vortex shedding induced vibrations of a free span on a pipeline may be calculated using the following approach.

The amplitude of the motion in the middle of the free span may be calculated from:

$$a_n = F \cdot DLF \cdot \frac{1}{\omega_n^2} \cdot \frac{\int_0^\lambda \mu_n(x)\, dx}{\int_0^\lambda Me \cdot \mu_n^2(x)\, dx} \qquad (C.3.3)$$

where:

$F = \frac{1}{2} \cdot \rho \cdot C \cdot D \cdot V_{cr}^2$: static force per unit length

$DLF = \frac{1}{2\xi}$: dynamic load factor

$\mu_n(x)$ = normalised mode shape of the fundamental mode

Me = equivalent mass per unit length

$\omega_n = 2\pi f_n$: angular frequency of the oscillation

V_{cr} = critical velocity, corresponding to a reduced velocity of 2.5 for in-line and 5.0 for cross-flow motion

ξ = damping ratio

C = in-line or cross-flow fluctuating force coefficients

Values of the fluctuating drag and lift coefficients are not uniquely given in available literature but the figures in Section C.2 may be used as guidance for determining these coefficients.

Assuming the deflection of the pipeline calculated as outlined above to be equal to the deflection caused by a uniform static load along the free span of the pipeline, the stresses may be calculated when the relation between the curvature and the deflection of the pipeline is known.

Example:

Free span length: 50 metres
Natural frequency: 0.46 Hz
$C_{LRMS} = 0.06$, $C_{DRMS} = 0.02$ } from CIRIA (1977)
Bending stress δ_b = 77 kp/cm² when subjected to in-line resonant vibration.

Current-induced vortex shedding of submarine pipelines in connection with free spans or long unsupported lengths of risers may be regarded as a phenomenon that seldom occurs in practice. However, if it occurs it may lead to a short fatigue life. A conservative fatigue evaluation may be carried out as follows:

(i) if no measured data exists, assume current variation over a 12-hour cycle;

(ii) simulate the current as a tidal current with period of 12 hours with a given velocity amplitude at the sea bottom;

(iii) calculate natural frequency from pipe dimension and given free span;
(iv) find critical particle velocity (Vr);
(v) calculate the time interval within each tide cyclus when the current velocity is in the critical area for resonant shedding;
(vi) calculate a number of 'stress blocks' for use in the Miner-Palmgren cumulative fatigue rule. The stress range is taken as two times the bending stress;
(vii) as fatigue criteria (S-N curve) use AWS (C or D curve).

Example:
Pipeline data as given above.

Free span length: 50 metres
Natural frequency of oscillation: 0.46 Hz

The free span is subjected to resonant in-line oscillation for current velocities in the range of 0.46 to 0.91 m/seconds.

Bending stress due to resonant in-line vortex shedding induced oscillation: $\sigma_b = 77$ kp/cm^2
Stress range: $\Delta\sigma = 2\sigma_b = 154$ kp/cm^2

Assuming a current velocity variation with time as shown in Figure C.3.1 the number of stress cycles in each 12-hour current cycle is determined to be 13,168, and in one year $9.61 \cdot 10^6$. Fatigue calculation for $l = 50$ m:

C-curve: $\Delta\sigma = 15.36$ N/mm$^2 \Rightarrow N = 2.94 \cdot 10^9$

Fatigue life: N/n = <u>305 years</u>

D-curve: $\Delta\sigma = 15.36$ N/mm$^2 \Rightarrow N = 4.13 \cdot 10^8$

Fatigue life: N/n = <u>43 years</u>

One may conclude that with a proper inspection of the pipeline during operation and remedial action to reduce longer free span, current-induced vortex shedding can be regarded as a minor problem for most submarine pipelines.

Critical length for free spans are, at any rate, determined to prevent in-line resonance rather than to determine fatigue life. The main reason for this is that any cyclic loading may cause propagation of cracktype defects.

C.4 A Modified Approach to Vortex Shedding Problems which Takes Account of Actual Pipeline Conditions

In analysing the structural response of the pipeline, decisions must be made regarding:

(i) the excitation;

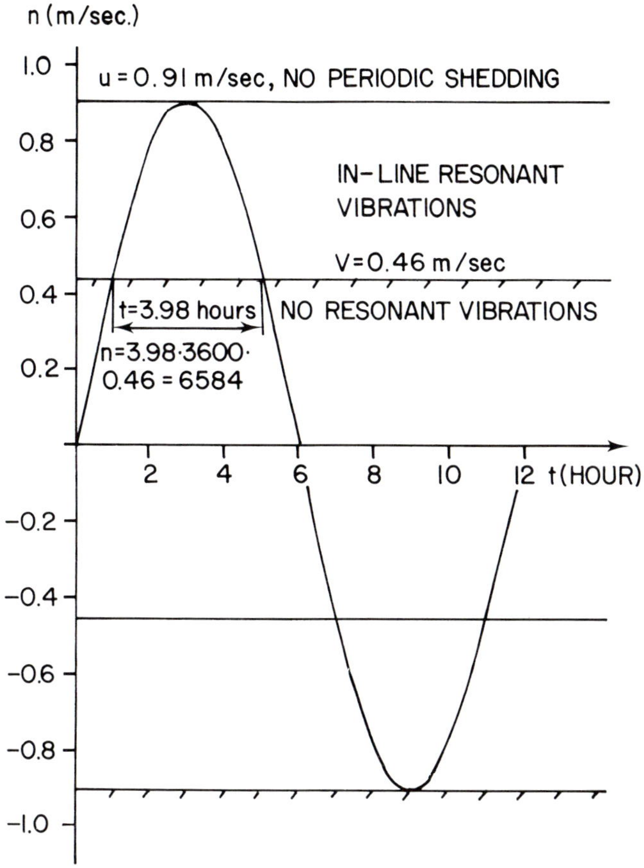

Figure C.3.1 Current Velocity Vibration Vortex Shedding

(ii) the value of damping;
(iii) the pipeline and conditions and stress calculations.

The actual dynamic motion of the pipeline is highly damped due to:

(i) change of the threshold level of current required for initiation of vortex shedding;
(ii) irregular wave action;
(iii) nonlinear hydrodynamic damping;
(iv) change of tension with deflection;
(v) damping due to contiguous spans.

All these factors will contribute in the suppression of dynamic motion. However, the classical approach of the preceding section must be extended to include:

(i) an unrestrained pipeline as:
  a) a simple beam;
  b) a simple beam with tension due to pressure; and

(ii) an unrestrained pipeline analysed as;
 a) a pipeline with sag tension;
 b) a pipeline with both sag and pressure tension.

Such an analysis could show that the pipeline vibrations can be tolerated due to the damping involved and due to the occurrence of the dynamic load at a shorter span length than the static load.

Structural damping has a value between 0.02 and 0.05 which, at the critical frequencies, corresponds to amplification factors in the order of 10–25. This means that it will normally be the hydrodynamics which control the damping.

Hydrodynamic damping is described as applying to the case of small amplitudes, a vibration where y/d is less than 0.2 (CIRIA (1977)). In this case for in-line motion:

$$\delta = \frac{\rho V D C_b}{2\, M f_n} \qquad (C.4.1)$$

where:

C_b = drag coefficient

V = particle velocity

f_n = natural frequency of pipeline

ρ = mass density of water

For cross oscillation δ_c is half the value for in-line motion.

Pipelines spans would have mass damping parameters ranging from δ = 0.12 to 0.5. This means that the time scale $T_L = T_o/\delta$ for the oscillation (following a linear description of the oscillation) amounts to from 8 to 2 times the resonance period.

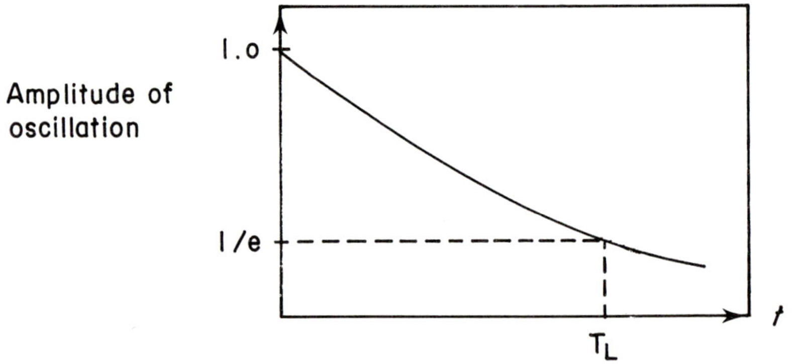

Figure C.4.1 Time Scale of Oscillation. *[after Gravesen (1975)]*

The ratio $T_L/T_O = \frac{1}{\delta}$ describes, in other words, how many force oscillations of a certain fixed amplitude are necessary to obtain an oscillation amplitude equal to the asymptotic one times $(1 - \frac{1}{e}.)$

We have thereby a rough determination of the number of oscillations needed to develop an asymptotic resonance amplitude. The length of time which is required for the oscillatory currents to make a significant contribution to the steady-state current, without diminishing the vortex shedding regime, is approximately equal to T_L.

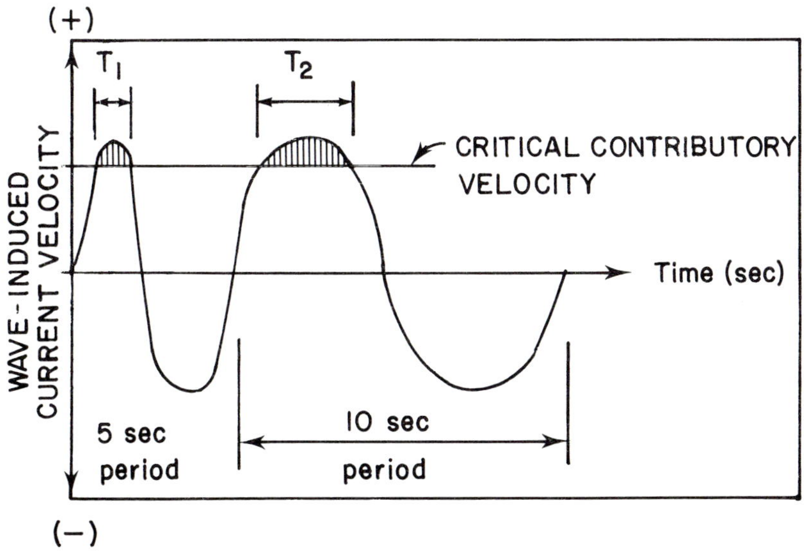

Figure C.4.2 The Effect of Wave-induced Current on Pipeline Oscillation

According to the analysis illustrated in Figure C.4.2, for the same wave height (H), the short duration T_1 may be less effective than the longer duration T_2 depending on the damping ratio δ.

To this comes the nonlinear effects caused by axial load induced in the pipeline due to sag, deflection or pressure, These will stiffen the pipeline, upset the threshold frequency at which vortex shedding occurs and will thereby reduce the amplitude (depending on the particular span length).

The nonlinear hydrodynamic damping of the pipeline will also contribute to reduce the amplitude of response. This latter effect is a matter of oscillations close to the seabed and the influence of end foundations in the bed material.

Much of what is known about marine structures was developed with offshore platform structures in mind. The theory which applies to the beam elements of an offshore structure, such as platform members and piling, does not necessary apply to a pipeline structure. In comparison the pipeline structure is extremely long and does not behave completely as a beam element or as a beam on contiguous supports. Figure C.4.3 illustrates some typical spans in a pipeline.

In the classical small deflections theory, end supports of the beam element are free to move inwards as the beam deflects. The pipeline supports, however, remain in place and the pipeline slides over the support.

In addition, the initial slope of the supports are not known with any accuracy nor indeed the height misalignment.

The usual method of analysis is to assume the ends to be fixed, but this need not be the case unless the spans are equally spaced. Except in the most unusual combination of span, the assumption of fixed ends is mildly conservative. A number of three-moment calculations carried out indicate a fixity factor between 0.8–0.9 of the fixed end conditions. Obviously, out of line supports and non zero end slopes change the actual conditions of fixity.

It is in the area of vibrations of the pipeline where the question of end fixity has the most significance. At present, the usual design procedure is to assume a pinned condition to determine the maximum span length which avoids vortex shedding. The case of fixed ends would give a higher threshold current and longer spans. Figure C.4.4 shows an example of such a calculation for a 10.75-inch, 0.5-inch wall pipeline. These curves have been determined from the method described in Nielsen (1979).

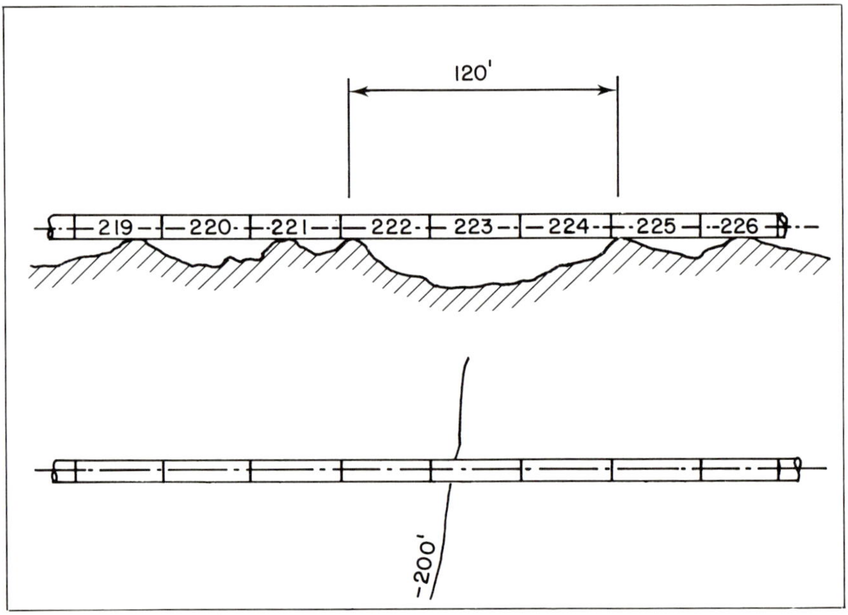

Figure C.4.3 Typical Pipeline Span for 10.75-inch Pipeline

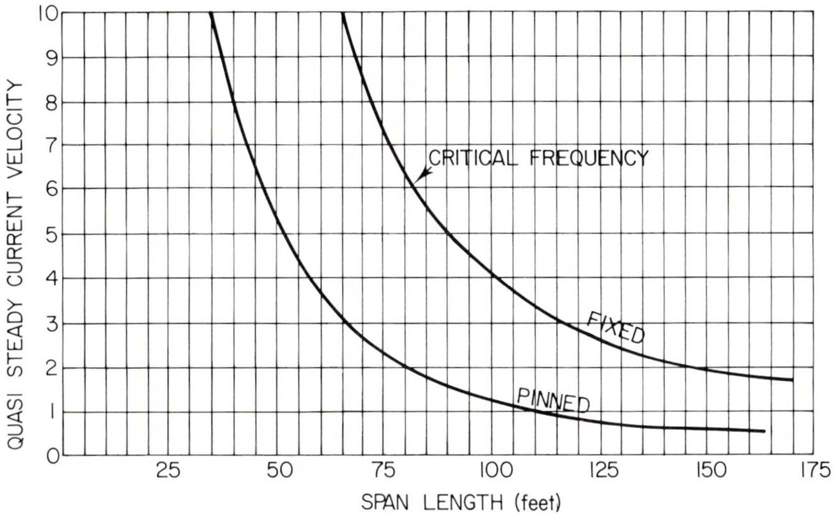

Figure C.4.4 Allowable Span Length *versus* Current for 10.75-inch Pipeline. (0.5-inch wall)

C.4.1 *Analysis Linear Theory (Assumes no Axial Restraint)*
A continuous beam with many equal spans which is supported on rollers may be considered as shown below:

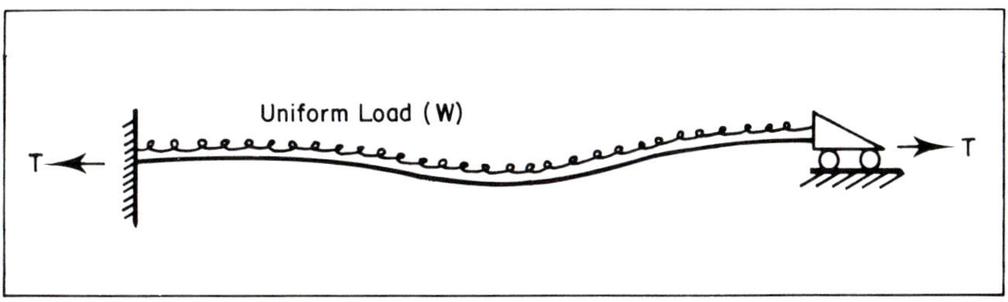

Figure C.4.1.1 Model of a Typical Span

$$\text{end moment } ME = \frac{WL^2}{12} \qquad \text{(C.4.1.1)}$$

$$\text{mid-span moment } M_M = \frac{WL^2}{24} \qquad \text{(C.4.1.2)}$$

The application of internal pressure causes a hoop stress in the pipe which shortens the pipe by a strain equal to Poisson's ratio times the hoop strain. The pressure at the end of the pipe exerts a force equal to the internal pressure times the area of the endcap. This endcap force induces an axial stress in the pipe.

Maximum bending stress:

$$\sigma b = \frac{WL^2R}{12\,I}\,LMF \qquad (C.4.1.3)$$

where: R = pipe radius
I = pipe moment of inertia
LMF = load magnification factor

The load Magnification factor (LMF) is in this case demagnified due to tension.

The bending stress values for the various loadings over a range of span lengths may thereby be calculated.

For the mathematical model shown above it is valid to add the axial pressure stress to the bending stress to obtain the maximum axial stress. For a typical span of a continuous pipe, however, the endcap will be many spans away and so the endcap force will be resisted by friction between the pipe and the farthest supports, rather than strain along the entire length of the pipe.

If this is the case, however, it follows that the frictional restraint offered by these farthest supports will also cause the axial strain inducted by hoop stress to become stress rather than movement as was the case for the unrestrained pipe.

For small deflections, it is the axial pressure stress that should be used for the continuous pipeline supported on equally spaced supports. The axial pressure-induced stress should be added to the bending stress to obtain the total axial stress.

Assuming a hoop stress which is one half the allowable stress for an X-42 specification pipe gives:

$$\sigma_H = \frac{PD}{2t} = 36\ \sigma y = 15{,}120\ \text{lb/in}^2 \qquad (C.4.1.4)$$

where:

P = operating pressure 1,406 psi

D = nominal outside diameter = 10.75 inches

t = nominal wall-thickness = 0.50 inches

If: μ = Poisson's ratio = 0.30

And: σp = longitudinal stress in restrained line due to internal pressure

$$\sigma p = \mu \sigma_H$$

$$\sigma p = 0.30 \times 15{,}120\ \text{lb/in}^2 \qquad (C.4.1.5)$$

$$= 4{,}536\ \text{lb/in}^2$$

The stresses may thereby be calculated for various loading cases and span lengths.

The elastic section modulus used in the calculation of bending stress is:

$$Z = 2\frac{I}{D} = 39.4 \text{ in}^3 \tag{C.4.1.6}$$

C.4.2 *Nonlinear Theory (Accommodates Sag Tension)*

When the deflection of a beam is large, the effects of restraint to axial movement and applied axial load become important. A continuous pipeline resting on supports offering friction and subject to internal pressure is under tension (caused both by flexure and by the axial strain induced by the hoop strain caused by pressure). This tension reduces the bending moment, hence the total stress due to induced tension and bending is less than the simple bending stress for an unrestrained support analysis. Such a situation is depicted by Figure C.4.2.1 below.

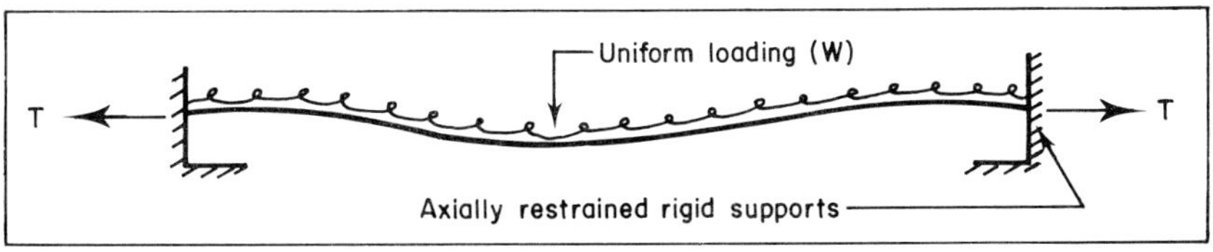

Figure C.4.2.1 The Tension Induced in a Pipeline Subject to Sagging

The analysis uses a method outlined in Roark and Young (1975). The first step in this method is the solution of the following equations.

$$Y + \frac{A}{16\,I}\,Y^3 = \frac{WL^4}{4\pi^4 EI} \tag{C.4.2.1}$$

where:

L = the span length

A = the cross sectional area of the pipe

I = the moment of inertia of the pipe

E = Young's modulus

Y = the mid-span deflection

The solution of this equation enables the calculation of the flexurally induced tension, given by:

$$T_1 = \frac{\pi^2 EA}{4L^2} Y^2 \tag{C.4.2.2}$$

The pressure-induced tension is then calculated:

$$T_2 = \mu \sigma_H A = \sigma_p A \qquad \text{(C.4.2.3)}$$

where μ is Poisson's ratio.

The total tension is then used in the evaluation of a term kL which is required for the use of a table which provides moment demagnification factors (DMF).

$$kL = \sqrt{\frac{TL^2}{EI}}$$

The actual moments resulting from the nonlinear analysis are the product of this DMF and the moments from the linear analysis of the same span.

The total longitudinal stress is the summation of the average stress caused by tension and the maximum bending stress.

The DMF values for various values for kL can be calculated based on Roark (1975).

C.4.3 *Discussion of Results*

The discussion to follow will suggest that the worst case for design is the fixed case. This comes about for two reasons. Firstly, the occurrence of vibrations at a shorter span length means that the pipeline has more static stress reserve to absorb the dynamic stress imposed at a shorter span than the fixed case where dynamic loads occur at a longer span and/or high currents.

Secondly, the axial end condition of restrained or unrestrained ends will also determine the stresses. A succession of spans will develop high sag tensions and these sag tensions will be balanced by the contiguous spans causing a circuit restrained conditions with stationary supports. These tensions will not only cause a reduction in bending stress but the tensions will also serve to raise the critical vibration frequency of the pipeline.

Figure C.4.3.1 shows the results for a fixed end restrained pipeline under various loadings and Figure C.4.3.2 shows similar results for a pinned ended beam. It is striking that after a span of about 100 feet the stress doesn't increase for the pin ended case. This is one of the principle reasons that the fixed case for vibrations is the limiting case.

Figures C.4.3.3 and C.4.3.4 (overleaf) show the result of tension on lifting the threshold current for vortex shedding. The frequency shift factor is given on Table C.4.3.1 (see over) together with the sag tensions used in the calculations of Figure C.4.3.3 and C.4.3.4. The shift factor:

$$S_F = \sqrt{\frac{L^2 T}{\pi^2 EI} + 1} \qquad \text{(C.4.3.1)}$$

is applicable to the pinned case but is used here for the fixed case as well, since the tension T is developed from the nonlinear sag deflection equation which take into effect the fixed end induced shape.

Figure C.4.3.5 (see page 340) gives the results of adding the dynamic

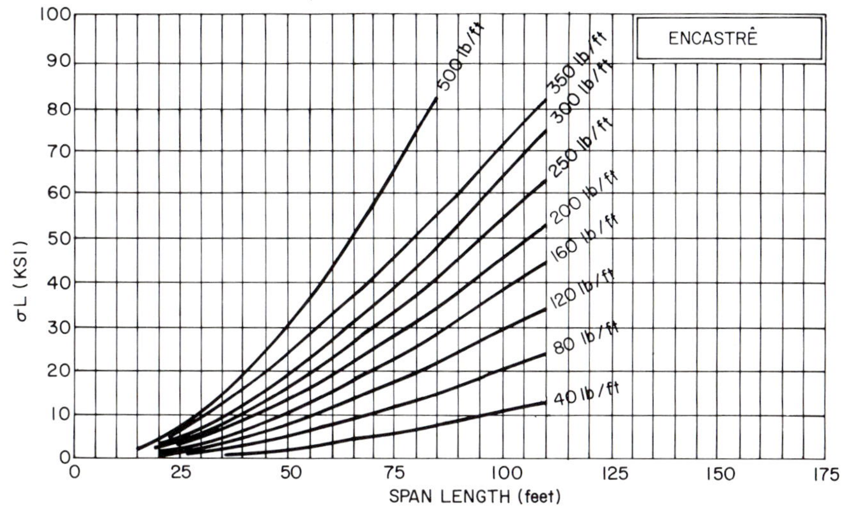

Figure C.4.3.1 Maximum Total Longitudinal Stress *versus* Span for 10.75-inch Pipeline. (0.5-inch wall)

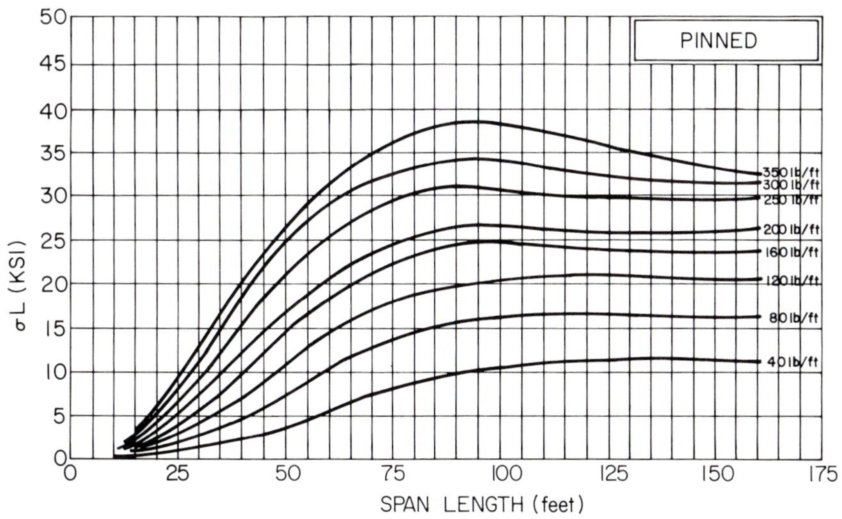

Figure C.4.3.2 Maximum End Stress *versus* Span for 10.75-inch Pipeline. (0.5-inch wall)

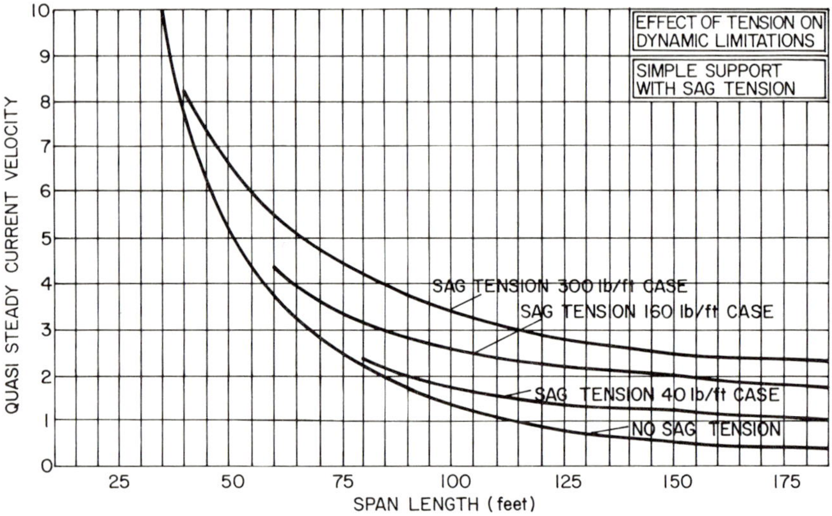

Figure C.4.3.3 Allowable Span Length *versus* Current for 10.75-inch Pipeline. (0.5-inch wall)

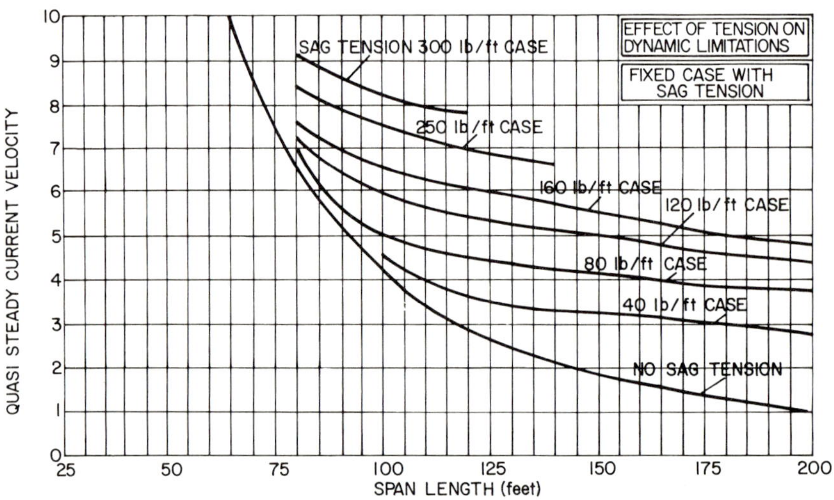

Figure C.4.3.4 Allowable Span Length *versus* Current for 10.75-inch Pipeline. (0.5-inch wall)

load to the steady force load on the pipeline. The results of the stress analysis for a DMF = 25 shows that the pinned case is immune to the dynamic loading while the fixed case is limited to a span of about 100 feet for X-42 (0.96 of SMYS permitted).

Table C.4.3.1 Self Weight Sag of 10.75-inch Pipeline for Various Loadings

	FIXED CASE – TENSION (KIPS)						PINNED CASE – TENSION (KIPS)			
SPAN	40#/ft	80#/ft	120#/ft	160#/ft	300#/ft	500#/ft	40#/ft	80#/ft	120#/ft	160#/ft
60	0.3	0.12	3.7	4.2	12.5	40.1	6.8	21.7	38.8	58.2
80	1.7	6.5	13.8	23	62.8	124	23.4	53.2	80	105
100	6.0	20.6	38.9	58	124	209	41.9	80	113	142
120	15.2	42.6	69.4	96	178		57.3	101	139	172
140	28.0	64.9	98.7	129			70.1	119	161	198
160	41.6	85.2	123	157			81.1	135	180	221
180	54.2	103	145	180			90.9	149	198	242
200	65.6	119	164	204			99.7	162	215	262

$S_F = \sqrt{1 + \frac{T}{EI}\left(\frac{l}{\pi}\right)^2}$ $I = 212 \text{ in.}^4$ $E = 28 \times 10^6/\text{in.}^2$

C.5 Conclusion

More information is needed about damping of pipeline systems, particularly the action of unexcited contiguous spans which react with the seabed or contribute hydrodynamic damping to the system. This factor, together with tension, must explain why a number of pipeline spans which theoretically must be excited to high amplitudes survive nevertheless.

Surely the major contribution of the tension is to raise the span threshold frequency above steady current excitation. The high currents from waves are not regular and hence not as prone to cause regular vortex shedding.

The pipeline operating case where pressure is present will have less static stress due to the demagnification action of the tension. The limiting stress case is at zero pressure and this condition will most likely exist at the design storm conditions.

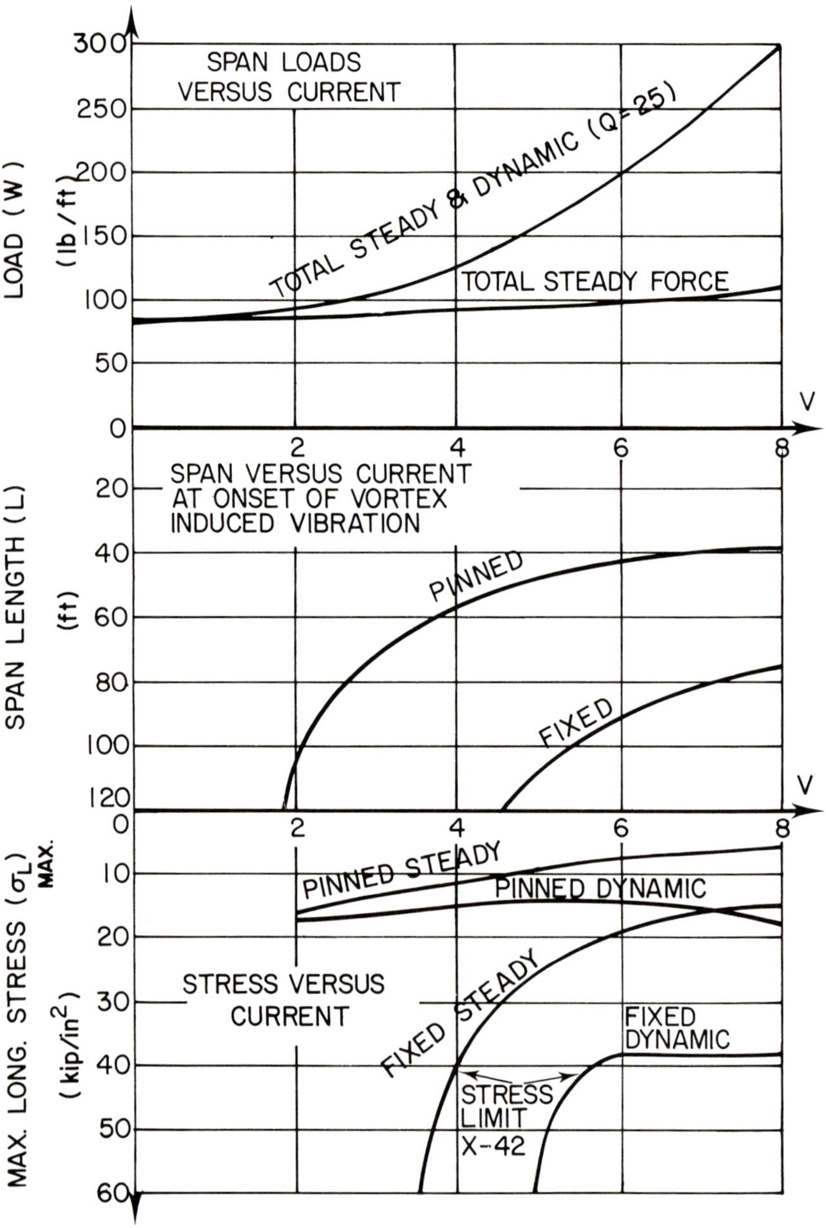

Figure C.4.3.5 Dynamic Span Calculation Table. (1.5-inch cover)

C.6 Bibliography

Abbott, M. B., *Computational Hydraulics, Elements of the Theory of Free Surface Flows,* Pitman Publ. Ltd, (London, 1979).

CIRIA, *Dynamics of Marine Structures*, Underwater Engineering Group, Report VR8, (London, 1977).

Det nørske Veritas, *Rules for the Design, Construction and Inspection of Submarine Pipelines and Pipeline Risers,* Det nørske Veritas, *Report* (1976).

Forristall, G. Z. *et al*, 'The Directional Spectra and Kinematics of Surface Gravity Waves in Tropical Storm Delia', *Journal of Physical Ocean,* 8, 5, (1978).

Gravesen, H., *Analysis and Reproduction of a Natural Three-dimensional Wave Field,* Report to the Danish Council of Technology, Danish Hydraulic Institute, (1979, not published).

Gravesen, H. *et al.*, 'Problems of Physical Model Tests with Harbours', *Proceedings, ASCE Meeting, Modelling 75,* (San Francisco, 1975).

Hallam, *et al.*, 'Dynamic Design of Conductor Tube Arrays', Paper OTC 2264, *Proceedings, Offshore Technology Conference*, (Houston, 1975).

King, *et al.*, 'The Suppression of Structural Vibrations Induced by Current and Waves', *Boss 76,* (Trondheim 1976).

King, R., 'A Review of Vortex Shedding Research and its Application'. *Ocean Engineering,* 4, 141–171, (1977).

Longuet-Higgins, *et al.*, 'Observations of the Directional Spectrum of Sea Waves, Using the Motion of a Floating Buoy', *Proceedings, Conference on Ocean Wave Spectra,* National Academy of Science, (1963).

Lundgren, H., *et al.*, 'Waveloads on Pipelines on the Seafloor', *Boss 76*, (Trondheim 1976).

Lundgren, H., *et al.*, 'Natural Wave Trains: Description and Reproduction', *Proceedings, 16th Coastal Engineering Conference,* (Hamburg, 1978).

Lundgren, H., *et al.*, 'Improved Physical Basis of Wave Forces', *Civil Engineering in the Oceans IV*, (San Francisco, 1979).

Naudascher, E., 'Flow-induced Vibration – a Unified Approach', *IAHR*, (Baden Baden, 1977).

Nielsen, R., 'Spanning, Some Aspects of Statics and Dynamics', *Proceedings, 1979 European Seminar, Offshore Oil & Gas Pipeline Technology,* (London 1979).

NOU, 40, *Pipelines in Deep Water*, with Appendixes (in Norwegian), (1975).

Olbjorn, E. H., *Some Submarine Pipeline Engineering Problems*, Norske Veritas, (Oslo, 1978).

Pendered, J., 'Vortex Shedding Problems of Subsea Pipelines', *Proceedings, 1978 European Seminar, Offshore Oil & Gas Pipeline Technology,* (London, 1978).

Roark and Young, *Formula for Stress and Strain,* 5th ed., McGraw-Hill, (New York, 1975).

Sand, S. E., 'Three-dimensional Deterministic Structure of Ocean Waves', Technical University of Denmark, PhD Thesis, (1979).

Sarpkaya, T., 'In-line and Transverse Forces on Cylinders in Oscillatory Flow at High Reynolds Number', Paper OTC 2553, *Proceedings, Offshore Technology Conference,* (Houston, 1976).

Sarpkaya, T., 'Hydroelastic Response of Flexibly-mounted Cylinders in Harmonic Flow', Paper OTC 2897, *Proceedings, Offshore Technology Conference,* (Houston, 1977a).

Sarpkaya, T. 'Wave Forces on Rough-walled Cylinders at High Reynolds Numbers', Paper OTC 2901, *Proceedings, Offshore Technology Conference,* (Houston, 1977b).

Sarpkaya, T., 'In-line and Transverse Forces on Cylinders near a Wall in Oscillatory Flow at High Reynolds Numbers', Paper OTC 2898, *Proceedings, Offshore Technology Conference,* (Houston, 1977c).

Sarpkaya, T., 'A Theoretical and Experimental Investigation of the Hydroelastic Response of Circular Cylinder in Uniform Flow', *IAHR,* (Baden Baden, 1977d).

Sarpkaya, T., 'Hydroelastic Response of Circular Cylinders in Harmonic Flow', *IAHR*, (Baden Baden, 1977e).

Sarpkaya, T., 'Fluid Forces on Oscillating Cylinders', *ASCE WW4,* (1978a).

Sarpkaya, T., 'Lateral Oscillations of Smooth and Rough Cylinders in Harmonic Flow', *Proceedings, Symposium Wave induced Forces on Cylinders,* (Bristol, 1978b).

Verley, R. L. P., *et al*, 'Wave-induced Vibrations of Flexible Cylinders', Paper OTC 2899, *Proceedings, Offshore Technology Conference,* (Houston, 1977).

R NIELSEN
H GRAVESEN

APPENDIX D

Calculation of the Probability of Pipeline Damage Due to Ship Anchoring following Machinery Failure (The Danish Great Belt)

25,000 ships complete the 60 kilometre Storebælt passage (Røsnæs-Omø) each year, one half travelling in each direction. The average vessel speed of 22.2 kilometre/hour gives an average passage time of 2.7 hours.

According to DnV Report 80–0129, the incidence of serious machinery damage for international merchant shipping is 6.9×10^{-5} per hour. Damage events of this type are assumed to involve stoppage of the main engine and lead to anchoring. This assumption is probably on the conservative side.

The probable number of such events in Storebælt each year is thus:

$$25{,}000 \times 2.7 \times 6.9 \times 10^{-5} = 4.7$$

The anchor dragging distances indicated in Appendix E for those vessels in the Storebælt traffic population which can inflict pipeline damage (ie, those in excess of about 5,000 to 15,000 dwt) range from 30 metres to 140 metres and are typically of the order of 70 metres (in mud) when the vessel speed at 'letting go' is 2 knots, ie, when the vessel is drifting with a fast current or proceeding with a speed somewhat less than steerage way (steerage way is typically 3 knots). If the anchor is let go at the steerage way threshold of 3 knots, the typical dragging distance will be of the order of 150 metres (in mud). The most representative dragging distance for all speeds in mud is thus of the order of 100 metres. The corresponding dragging distances in clay are only some 25 per cent of those in mud. The mud figure is the one adopted here as shown by Figure D1 (overleaf).

The proportion of all machine stoppage anchorings occurring within just that range of positions which will ultimately lead to letting go the anchor within the 100 metre zone preceding the pipepine is 0.1/60 or 1.67×10^{-3}.

Some 60 per cent of the Storebælt shipping traffic is less than 15,000 dwt, and it is shown in Appendix E that anchors of ships up to this size are unlikely to cause serious damage to a 30-inch pipeline. Even if ships down to 5,000 dwt can cause serious damage, the proportion of ships involved is only 50 per cent of the total population.

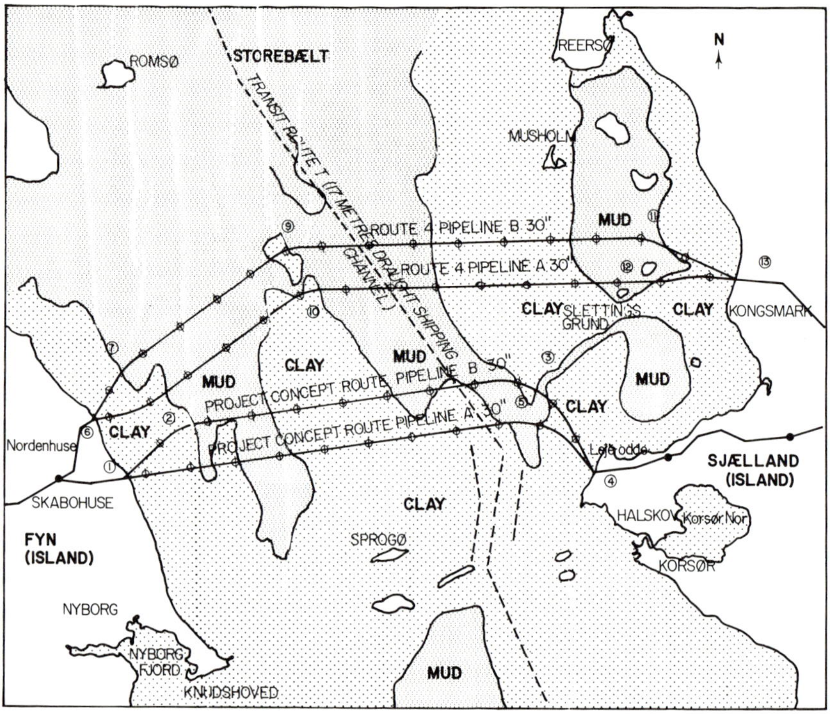

Figure D.1 Distribution of Seabed Soils and Storebælt Crossing Pipeline Location Plan

The annual probability of pipeline damage due to anchoring following engine trouble is therefore no more than:

$$50\% \times 1.67 \times 10^{-3} \times 4.7$$
$$= 3.9 \times 10^{-3}$$

It is an implicit assumption in the derivation of this probability figure that the navigating officer does not take any precautions to avoid letting go anchor in the vicinity of the pipeline. This is likely to be the case for a ship near the Project Concept Route, whereas the navigator of a ship near Route 4 will be under much less pressure and will have far greater scope for precautionary considerations. Captain Heimdal considered that the factor relating the probabilities at the two locations was between two and five.

APPENDIX E

Calculation of the Probability of Pipeline Damage due to Ship Anchoring following Collision (The Danish Great Belt)

With N vessels/hour proceeding at velocity V kilometres/hour through a channel of length L kilometres, the number of vessels n within the channel at any point in time is NL/V.

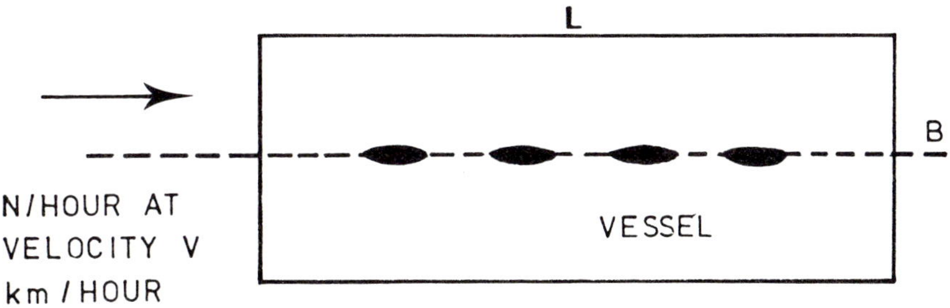

Figure E.1 Illustration of the Number of Vessels Passing through a Channel.

If the width of the channel is B, the average channel area per vessel is BL/n or VB/N. The average distance D between adjacent vessels is thus:

$$\sqrt{\frac{VB}{N}}$$

on the assumption that the vessels follow different courses all parallel to the channel centreline but spaced at random across the channel width, and that $B >> D$.

If the vessels sail in single file along the same line, the heading distance is V/N.

If B is equal to V/N, then the single-file heading distance and the average ship-to-ship distance on the basis of random lateral positioning within the channel are the same. In Storebælt the single file vessel frequency N is:

$$\frac{50\% \times 25{,}000}{365 \times 24} = 1.43 \text{ vessels/hour/direction}$$

so the single file heading distance H is:

$$\frac{V}{N}=\frac{22.2}{1.43}=15.6 \text{ kilometres}$$

The average distance between vessels where there are two parallel lines of vessels relatively close together and moving in opposite directions is just over half this, ie, 7.8 kilometres.

The average effective channel width between Røsnæs and Omø is taken as 4 kilometres, so that the average ship-to-ship distance on the basis of a random lateral distribution of vessels is:

$$D=\sqrt{\frac{VB}{N}}=\sqrt{\frac{22.2\times 4}{2.63}}=5.8 \text{ kilometres}$$

This figure is the more conservative, is probably the more correct one, and is the appropriate one in the context of the theory used below, in spite of the fact that the formula is not strictly valid when B and D have comparable magnitudes.

The models usually used to deduce the probability of ship collisions at sea are modifications of the 'molecule theory' and involve the concept of 'mean free path', as used by MacDuff (1974).

The mean free path F of a vessel of dimension L travelling through a planar continuum containing similar vessels of dimension L distributed with a density ($1/D^2$) per unit area, ie, with an average spacing of D, is equal to D^2/L.

If all the vessels in the continuum have a parallel vector velocity V, then the mean free path (measured against a stationary grid) of an aberrant vessel with velocity V at an angle θ to the general direction of motion can be shown by vector analysis to be:

$$F=\frac{D^2}{L}\times\frac{1}{2\sin\theta/2}$$

The geometric probability p_o of a collision within a radius R of the starting point of the aberrant vessel is thus R/F or:

$$p_o=\frac{2\,RL\sin\theta}{D^2}$$

If θ is regarded as a variable which can have any random value between O and 2π radians, then the average value of p_o is given by:

$$\frac{1}{2\pi}\int_0^{2\pi}\frac{2\,RL\sin\theta/2}{D^2}\,.\,d\theta \text{ or } \frac{4\,RL}{\pi D^2}$$

If the moving continuum of non-aberrant vessels has a finite width B perpendicular to the general direction of motion (ie, the effective channel width), then that proportion P of collisions within the radius R which occur within the channel of dimensions $2R \times B$ (all the remainder being *ipso facto* substituted by groundings) is:

$$\frac{2\,RB}{\pi\,R^2}$$

according to Figure E.2 below, so that the geometrical probability of a collision within the channel is given by:

$$Pg = P \times {}_{po} = \frac{2\,RB}{\pi\,R^2} \times \frac{4\,RL}{\pi\,D^2} = \frac{8\,BL}{\pi^2\,D^2} \text{ per aberrant vessel event}$$

Substituting values as follows:

B = 4 kilometres 2R = 60 kilometres (Røsnæs-Omø)
L = 0.15 kilometres
D = 5.8 kilometres

We get:

$$Pg = \frac{8 \times 4 \times 0.15}{\pi^2 \times 5.8^2} = 1.45 \times 10^{-2}$$

and the proportion of such collisions occurring within a transverse pipeline's defined risk corridor of width s (See Appendix D) is s/2R or 0.10/60

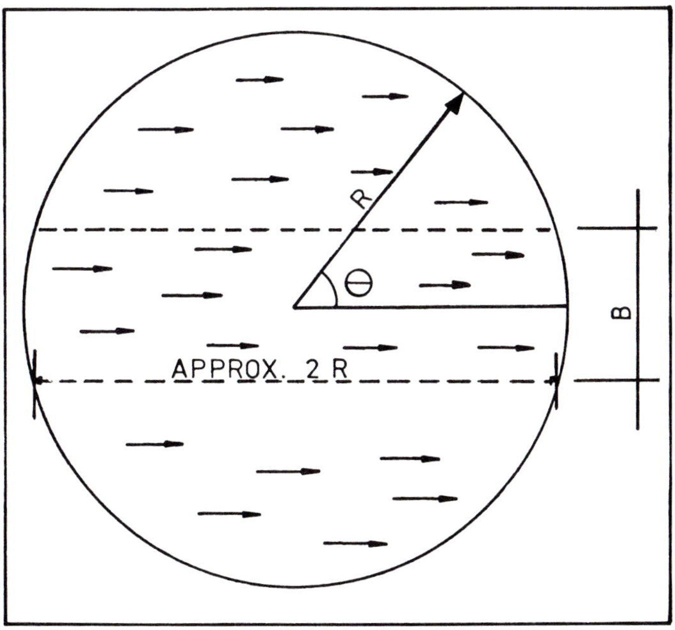

Figure E.2 The Geometry of the Collision of an Aberrant Vessel

or 1.67×10^{-3}, so that the probability of any aberrant vessel causing a collision involving the pipeline is:

$$1.45 \times 10^{-2} \times 1.67 \times 10^{-3} = 2.42 \times 10^{-5}$$

It should be noted that if 50 per cent of the continuum of non-aberrant vessels move in one direction and 50 per cent in the opposite direction instead of all moving in the same direction, the mathematical value of the integrated probability is unaltered. According to Fujii (1974) the incidence of aberrant vessels is one per 10,000 passages of a channel, where special attention is expected. The traffic in Storebælt is 25,000 vessels/year (both directions together), so that we may expect 2.5 aberrant vessels per year. The annual probability p_a of ship collision involving a single pipeline is thus:

$$2.5 \times 2.42 \times 10^{-5} = 6.05 \times 10^{-5}$$

Some 50 per cent of the ship population is under 5,000 dwt, so the proportion of collisions involving at least one ship in excess of 5,000 dwt is 0.75. If the anchors of ships under 5,000 to 15,000 dwt cannot damage the pipeline [MPPO 30462 (1980)] the annual probability p_a of a collision which can lead to anchor damage to the pipeline is:

$$0.75 \times 6.05 \times 10^{-5} = 4.54 \times 10^{-5}$$

the risk of accident in general is significantly lower in Route 4 than in the TC Route. On this basis it is to be expected that the event probability in the PC Route is somewhat higher than the theoretical figure derived above, while that for Route 4 is somewhat lower.

It should be noted that these calculations involve the very conservative assumption that the non-existent vessel involved in the collision takes no evasive action. In fact evasive action is normally taken and is normally successful, and for this reason the true probability of collision is several orders of magnitude lower than that calculated.

Ship sinkings only occur in a very small fraction of collision events, and there are few sinking risks in Storebælt other than collision (ie, there are no hard rock reefs and no likelihood of weather heavy enough to cause a ship to founder except in the most abnormal circumstances).

The incidence of ship sinking is therefore one or two orders of magnitude below that of ship collision, which itself is low, and ship sinking may therefore be neglected.

APPENDIX F

Calculation of the Probability of Pipeline Damage due to Ship Grounding (The Danish Great Belt)

Of all aberrant vessels, the proportion involved in collisions is 1.45×10^{-2} (see Appendix E) and the proportion exiting the channel ends can be shown by simple geometry to be $2[\arctan (B/2R)]/\pi$ where B and 2R are the breadth and length of the (rectangular) channel respectively. With B = 4 kilometres and 2R = 60 kilometres this proportion becomes 4.24×10^{-2}.

The remainder, or in other words some $94\frac{1}{2}$ per cent of all aberrant vessels, will go aground rather than collide with other ships or exit the channel. The annual probability of ship grounding is therefore slightly less than the annual incidence of aberrant vessels. According to Fujii (1971) this incidence is one per 10,000 passages of a channel where special attention is expected and the traffic in Storebælt is 25,000 vessels/year (both directions together), so the annual probability of ship grounding (including vessels of all sizes over 500 brt) is simply:

$$p_l = 94\tfrac{1}{2}\% \times 25{,}000/10{,}000 = 2.36$$

(Note: This tallies reasonably closely with the observed frequency of grounding in the Storebælt area.

This agreement serves to confirm the appropriateness of the statistics which are the basis of both the collision frequency and grounding frequency calculations).

The geometrical probability of such a grounding straddling the pipeline can be derived as follows:

Channel dimensions are $2R \times B$.

Aberrant vessel sails at random angle θ to proper course and grounds outside channel.

Terminal depth of bite into seabed h_G can be determined from the values of h_{GT} obtained from the DHI program for various combinations of ship size, speed, seabed material and seabed slope.

k is the cotangent of the angle of slop of the seabed in the grounding area (perpendicular to the channel).

G is the distance traversed between the time the bow strikes the seabed and the time the ship is brought to rest.

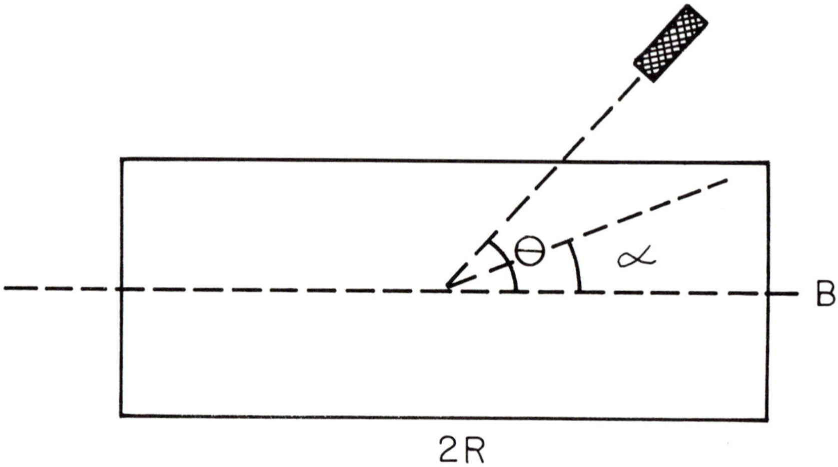

Figure F.1 The Geometry for Calculating the Probability of a Ship Grounding

Then $G = h_G k/\sin\theta$

The geometric probability p_2 of a grounding straddling the pipeline is thus:

$$p_2 = \frac{G\cos}{2R}\theta$$

The grounding distance G is given by:

$$G = h_G k/\sin\theta$$

The value of the ultimate bite depth h_G for any given value of θ can be related to h_{GT}, the ultimate depth of bite when the ship strikes perpendicular to the shore (ie, $\theta = \pi/2$). For the seabed soil in the grounding depths in Storebælt (clay) and the ship sizes prevailing there, the resistance force during grounding is directly proportional to the instantaneous depth of bite, and under these circumstances it can be shown that:

$$h_G = h_{GT} \times \sin^{\frac{1}{2}}\theta$$

whence:

$$p_2 = \frac{h_G k\cos\theta}{2R\sin\theta} = \frac{h_{GT}\cos\ \sin^{\frac{1}{2}}\theta}{2R\sin\theta}$$

$$= \frac{h_{GT}k}{2R}\cos\theta\sin^{-\frac{1}{2}}\theta$$

The geometric probability p_3 of the ship actually striking the pipe in any specific event of a grounding straddling the pipe is given by:

$$p_3 = \frac{h_G - x_t}{h_G} = 1 - \frac{x_t}{h_G}$$

where x_t is the depth from seabed to top of pipe.

Whence:

$$p_3 = 1 - \frac{x_t}{h_{GT}\sin^{\frac{1}{2}}\theta}$$

Thus:

$$p_2 \times p_3 = \frac{h_{GT}k\cos\theta\sin^{-\frac{1}{2}}\theta}{2R}\left(1 - \frac{x_t}{h_{GT}\sin^{\frac{1}{2}}\theta}\right)$$

$$= \frac{h_{GT}k}{2R}\left(\cos\theta\sin^{-\frac{1}{2}}\theta - \frac{x_t}{h_{GT}}\cot\theta\right)$$

valid for $\theta \nless \arcsin({}^{x}t/h_{GT})^2 = ß \ (h_{GT} > x_t > o)$

while for $\theta < ß \ p_2 \times p_3 = o$

The overall geometric probability p_G of a pipe strike per grounding event is the average value of $(p_2 \times p_3)$, ie, the full circle integral of the function with respect to θ.

Thus:

$$p_G = \frac{4}{2\pi}\cdot\frac{h_{GT}k}{2R}\cdot\int_{\beta}^{\pi/2}(\cos\theta\sin^{-\frac{1}{2}}\theta - \frac{x_t}{h_{GT}}\cot\theta)\,d\theta$$

$$= \frac{h_{GT}k}{\pi R}\left[{}_{\beta}^{\pi/2}\ 2.\sin^{\frac{1}{2}}\theta - \frac{x_t}{h_{GT}}\ln|\sin\theta|\right]$$

But note that when $x_t < o$ (ie, pipeline projecting above seabed) the lower limit ß is no longer applicable and must be replaced by the limit α = arctan B/2R in order to exclude the aberrant vessels exiting the channel.

Since h_{GT} varies with ship size, the function must be evaluated for each ship size group individually. This is done in Table F.1 and Table F.2 (see pages 354 and 355) for the ship size distribution specified by MPPO 30462, the values of h_{GT} obtained from the DHI programme, and a ship speed of 12 knots.

If the seabed were mud at the critical depth contours the risk would be greatly increased. The justification for adopting the bite depths applicable to clay is demonstrated in Figure F.1.

It will be noted that the geometric probability is related to the seabed slope perpendicular to the channel axis. The slope of 1:100 assumed in the calculation gives a conservative value of p_G for positive values of x_t. The slope in the vicinity of the pipeline routes is considerably flatter. Figure F.2 indicates how the value of P_G falls off with decreasing slope, and is in fact reduced to zero at a slope flatter than about 1:400.

The results are not particularly sensitive to the assumed value of the effective channel width, and even less sensitive to the assumed value of channel length.

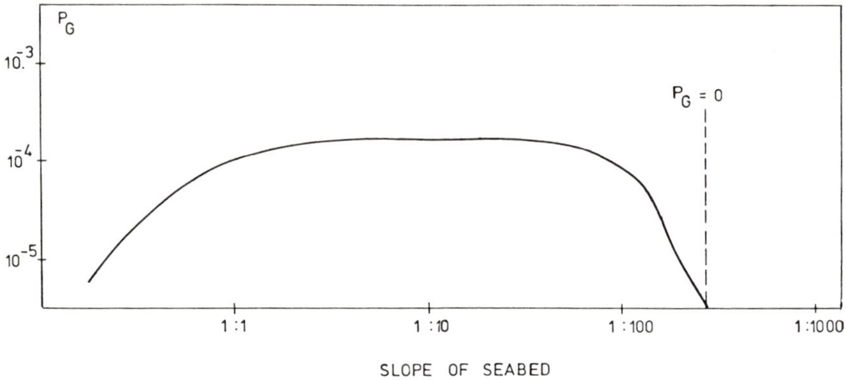

Figure F.2 Variation of Geometric Probability P_G with Seabed Slope when $x_t = 1$ metre

Nor are the results seriously affected if the channel, instead of being a simple rectangle, bends back and forth, unless the pipeline actually crosses in a bend, in which case the inside bend is a risk area. This is not the case with Route 4 or with the Project Concept Route.

Strictly speaking, the formula for p_G is only valid when the vessel becomes aberrant at the geometric centre of the channel, and a rigorous calculation involves integrations with respect to lateral and longitudinal positions as well as the integration with respect θ. It can be shown, however, that such finesse only alters the results by small amounts, and they are therefore omitted for simplicity.

On the other hand, the results are relatively sensitive to trenching depth in the critical grounding zones, ie, between the 8 and 12 metres depth contours.

The annual probability of pipeline damage due to ship grounding with the trenching depths defined in Table F1 is:

$$p_l \times p_G = 2.36 \times 1.16 \times 10^{-4} = 2.74 \times 10^{-4}/\text{year}.$$

The corresponding figure for the trenching depths defined in Table F.2 is:

$$2.36 \times 3.37 \times 10^{-4} = 79.5 \times 10^{-4}/\text{year}.$$

The derivation of these probability figures does not take into account any local variation of risk level along the passage. It does not, for example, take account of the fact that a significant proportion of aberrant vessel situations are associated with course changes. For this reason there is in theory a concentration of risk in a corridor to the north of the northbound lane of the traffic separation zone; this risk concentration is associated with aberrant northbound vessels which fail to change course at turnpoint No 19. This gives rise to a special situation in the area of Slettings Grund and Musholm Bank.

There is a similar concentration of risk on the Sjælland shore approach of the PC Route. This is associated with aberrant southbound vessels setting a wrong course from turnpoint No 18 and failing to turn in a timely manner at turnpoint No 19. It is impossible to quantify the risk concentration in either of these cases.

On the other hand, the observed locations of groundings, as shown by Figure F.3, in Storebælt are mainly south of Korsør and there have been no reported groundings of vessels of significant size either in the area of Route 4 or in the area of the Project Concept Route. The supposed risk concentrates are thus theoretical rather than real. Indeed, both of the pipeline crossings under consideration seem to be in a zone of risk deconcentration. Adoption of the theoretical figures for the channel as a whole is therefore considered to be satisfactorily conservative.

It will be noticed from Table F.2 that the risk of pipeline damage from ship grounding is some 30 times greater if the pipeline lies untrenched on the seabed.

Figure F.3 Locations of Observed Ship Groundings in Storebælt

Table F.1 Geometric Probability of Pipeline Strike per Ship Grounding Event in Storebælt. (1 metre from seabed to top of pipe throughout)

Proportion of ships (r)	Tonnage dwt	Typical draught m	Depth of water m	Predom, seabed soil	Depth of bite in metres (h_{GT}) m	Depth from seabed to pipe top (x_t) m	Computation of geometric probability P_G		Weighted probability rxp_G
							$\beta = \arcsin \frac{1}{2} (x_L/h_{GT})^{\frac{1}{2}}$	$p_G = \frac{h_{GT}k}{\pi R}\left[2\sin^{1/2}\theta - \frac{x_t}{h_{GT}}\ln.\sin\theta\right]_{\beta}^{\pi/2}$	
0.30	< 500	3	0–4.5	clay	0.4	1.0	$\pi/2$	0	0
0.20	500 to 5,000	5	3.5–6.5	clay	0.8	1.0	$\pi/2$	0	0
0.10	5,000 to 15,000	8	5.5–10	clay	1.0	1.0	$\pi/2$	0	0
0.38	15,000 to 85,000	12	8–14	clay	1.6	1.0	0.40	2.72×10^{-4}	1.03×10^{-4}
0.02	> 85,000	15	12–17	clay	2.0	1.0	0.25	6.58×10^{-4}	0.13×10^{-4}
1.00							Overall geometric probability		1.16×10^{-4}

Bite depths are taken from DHI Ship Grounding Computer Program Result.
Seabed slope perpendicular to channel typically 1:100

Table F.2 Geometric Probability of Pipeline Strike per Ship Grounding Event in Storebælt. (Pipe resting on seabed; ie, projecting approx 70 cm above bottom)

Proportion of ships (r)	Tonnage dwt	Typical draught m	Depth of water m	Predom, seabed soil	Depth of bite in metres (h_{GT}) m	Depth from seabed to pipe top (x_t) m	Computation of geometric probability P_G		Weighted probability rxp_G
							$\alpha =$ arctan b/2R	$p_G = \frac{h_{GT}k}{\pi R}\left[\frac{\pi/2}{2\sin \frac{1}{2}\theta} - \frac{x_t}{h_{GT}} \ln.\sin\theta\right]$	
0.30	< 500	3	0–4.5	clay	0.4	−0.7	0.067 rads	1.79×10^{-3}	0.54×10^{-3}
0.20	500 to 5,000	5	3.5–6.5	clay	0.8	−0.7	0.067 rads	3.27×10^{-3}	0.65×10^{-3}
0.10	5,000 to 15,000	8	5.5–10	clay	1.0	−0.7	0.067 rads	3.57×10^{-3}	0.36×10^{-3}
0.38	15,000 to 85,000	12	8–14	clay	1.6	−0.7	0.067 rads	4.52×10^{-3}	1.72×10^{-3}
0.02	> 85,000	15	12–17	clay	2.0	−0.7	0.067 rads	5.16×10^{-3}	0.10×10^{-3}
1.00							Overall geometric probability		3.37×10^{-3}

Bite depths are taken from DHI Ship Grounding Computer Program Result.
Seabed slope perpendicular to channel typically 1:100
When $\theta <$ ß vessels exit the channel ends

APPENDIX G

Calculation of the Probability of Damage to the DONG A/S 30-inch Diameter Export Pipeline by High-holding Anchors Associated with Offshore Installation and Repair Activities in the South East Approach Quadrant at Tyra East within 1 Kilometre of the TE–E Platform

Assumptions

(i) High-holding anchors are only used by vessels involved in the laying and trenching of offshore pipelines and in the repair of rupture or serious damage. Vessels associated with routine maintenance (eg, diver support vessels, survey vessels, and RCV mother ships) use only standard anchors, and these vessels are of such a size (normally less than 10,000 dwt) that their anchors cannot damage a 30-inch diameter pipe.

(ii) The degree of negligence will never be so great that the anchor spread of a vessel working on the installation or repair of a pipeline will damage that pipeline, but the degree of negligence or positioning error may be such that damage is inflicted on other pipelines in the vicinity.

(iii) The length L_N of new pipeline expected to be installed within anchor spread reach of the DONG A/S export pipeline during the course of 30 years is 3 kilometres.

(iv) Installation of a pipeline within 1 kilometre of a platform complex involves the passage of a laybarge during which there will be 20 anchor drops per kilometre run (ie, a full anchor-spread relocation for every 500 metres of advance), and the passage of a trenching barge during which there will be a further 20 anchor drops per kilometre run. The width of the spread is 2 kilometres, yielding in all 20 drops/square kilometre (N_I).

(v) The length L_E of existing pipeline (excluding the DONG A/S pipeline) within anchor spread reach of the DONG A/S pipeline at the date of commissioning of the DONG A/S pipeline is expected to be 3 kilometres.

(vi) Repair of a pipeline, typically involving the cutting out and replacement of a section of pipe, will involve 40 anchor drops in an area 2 × 2 kilometres, ie, 10 drops/square kilometre (N_R).

(vii) The frequency of repair events per kilometre of pipeline in the vicinity of the DONG A/S pipeline is equal to the probability P_{QD} of rupture or serious damage per kilometre/year.

(viii) The value P_{QD} for a 30-inch diameter pipe trenched 1 kilometre below seabed is given by DnV as 2×10^{-2} events/kilometre/year and is restricted to damage from high-holding anchors (see (i) above). However, interfield lines of substantially smaller diameter and lying on or close to the seabed will also be susceptible to anchor damage by routine maintenance vessels.

(ix) Maintenance and inspection operations on the interfield lines involve 40 standard-anchor drops per year per kilometre of pipeline maintained and inspected (N_M).

(x) The proportion of all anchor drops which are negligent or subject to position error (P_N) is ten per cent.

(xi) The design life T of the DONG A/S pipeline is 30 years.

(xii) The average effective dragging distance S_H of a 30-ton high-holding anchor dropped by tug and winch-tensioned from the barge is of the order of 15 metres in sand (ie, total length of impact and hooking interference zone).

(xiii) The average dragging distance S_S of a 4-ton standard anchor dropped from a 10,000 dwt vessel proceeding at slow speed is of the order of 20 metres in sand.

Figure G.1 depicts the 'random' (ie, negligent or erroneously positioned) anchor drop circles (ie, the circles whose diameter(s) is the effective anchor dragging distance) within a 1 square kilometre area, and also shows a 'victim' pipeline. The interference width is $2 \times S$.

Since the actual anchor drag trace can have any random orientation (Ø relative to the victim pipeline) within the indicated circle, the probability of interference with the pipeline is less than unity and depends on the position of the circle relative to the pipeline, as shown by Figure G.2.

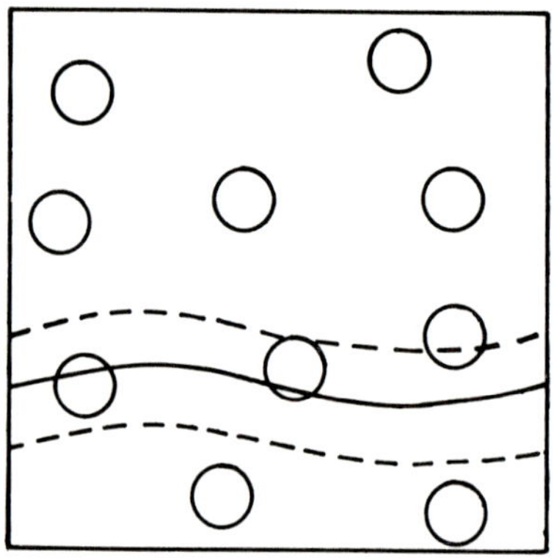

Figure G.1 An Illustration of Anchors Dropped Erroneously onto a Pipeline

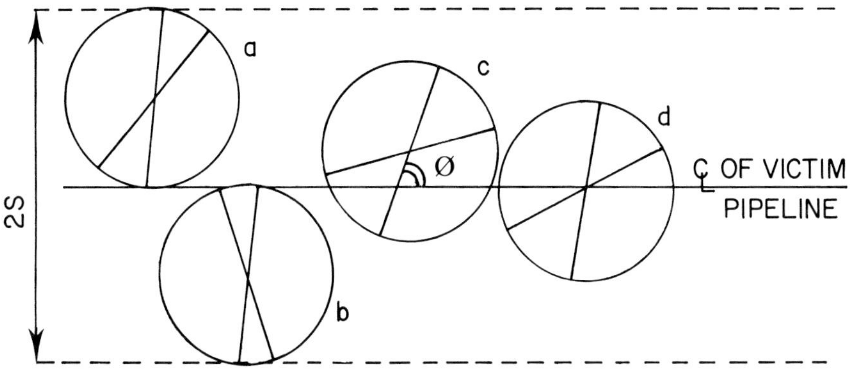

Figure G.2 An Illustration of the Probability of an Anchor Striking a Pipeline

The probability of a strike in circle positions a and b is clearly zero. The probability in position d is unity. The probability in position c is intermediate. The average (f) of the probability for all circle positions totally within the corridor width 2s (ie, whose centres lie within a corridor width s) is assumed to be approximately $2/\pi = 0.64$.

The first step in the calculation is to evaluate P_{QD} for the initial inter-field lines:

Number of maintenance and inspection work anchor drops	$= L_E \times N_M$/year/sq km
Number of such drops subject to negligence or position error	$= L_E \times N_M \times P_N$/year/sq km.
Number of these 'random' drops which will interfere with a victim pipeline (assuming 2 km of victim pipeline per sq km)	$= L_E \times N_M \times P_N \times \frac{S_S}{1{,}000} \times f \times 2$ per year/km of affected pipeline $= P_{QD1}$

$$\text{whence } P_{QD1} = 3 \times 40 \times 0.1 \times \frac{20}{1{,}000} \times 0.64 \times 2$$

$$= 30 \times 10^{-2} \text{ events/km/year}$$

The corresponding figure for damage to high-holding anchors associated with offshore installation and repair work in accordance with assumption (viii) above is:

$$P_{QD2} = 2 \times 10^{-2} \text{ events/km/year}$$

Thus $P_{QD} = P_{QD1} + P_{QD2} = 32 \times 10^{-2}$ events/km/year.

In the next step we turn our attention to the DONG A/S pipeline:

Number of high-holding anchor drops associated with the installation of subsequent pipelines (averaged over field development of, say, four years) $= \frac{L_N \times N_I}{T_{dev}}$

$= \frac{3 \times 20}{4}$

$= 15$/year/sq km

Number of high-holding anchor drops associated with repair work on interfield lines existing at the commissioning date of the DONG A/S line $= L_E \times N_R \times P_{QD}$

$= 3 \times 10 \times 32 \times 10^{-2}$

$= 9.6$/year/sq km

Number of high-holding anchor drops associated with repair of subsequent interfield lines (as from the date of their installation) $= L_N \times N_R \times P_{QD}$

$= 3 \times 10 \times 32 \times 10^{-2}$

$= 9.6$/year/sq km

Total number of drops D =

initially:	9.6 + 0 = 9.6
during field development:	9.6 + 15 = 24.6
ultimately:	9.6 + 9.6 = 19.2

Number of such drops subject to negligence or position error $= D \times P_N$/year/sq km

Number of these 'random' drops which will interfere with the DONG A/S pipeline $= D \times P_N \times \frac{S_H}{1{,}000} \times f$ per year/km of DONG A/S pipeline

$= P$

All such events are assumed to lead to immediate or subsequent rupture.

Whence the probability of rupture of the DONG A/S 30-inch diameter gas transmission pipeline within the 1 kilometre stretch closest to the platform is:

$$\text{Initially:}\quad P_A = 9.6 \times 0.1 \times \frac{15}{1{,}000} \times 0.64 = 0.92 \times 10^{-2}\ \text{events/km/year}$$

$$\text{At peak:}\quad P_B = 24.6 \times 0.1 \times \frac{15}{1{,}000} \times 0.64 = 2.36 \times 10^{-2}\ \text{events/km/year}$$

$$\text{Ultimately:}\quad P_C = 19.2 \times 0.1 \times \frac{15}{1{,}000} \times 0.64 = 1.84 \times 10^{-2}\ \text{events/km/year}$$

(The assumed magnitude of P_{QD2} is therefore in the correct range of magnitude.)

It should be noted that while P_A and P_C are sensitive to changes in the assumptions concerning damage from routine maintenance and inspection work, P_B is not.

If we assume that field development (ie, the installation of the subsequent pipelines) takes place during the central period of the DONG A/S pipeline design life, then the average value of P is:

$$\left(\frac{13}{30} \times 0.92 + \frac{4}{30} \times 2.36 + \frac{13}{30} \times 1.84\right) \times 10^{-2}$$

$$= 1.51 \times 10^{-2}\ \text{events/km/year}$$

while a reduction of the assumed maintenance and inspection damage frequency (ie, P_{QD}) by one order of magnitude cuts the average value of P to:

$$\left(\frac{13}{30} \times 0.92 + \frac{4}{30} \times 1.53 + \frac{13}{30} \times 1.84\right) \times 10^{-2}$$

$$= 0.32 \times 10^{-2}\ \text{events/km/year}$$

The logarithmic average of these two alternative values is 0.70×10^{-2} events/km/year.

This calculation, in view of the many assumptions embodied in it, does not by itself prove anything. It demonstrates, however, that a theoretical analysis based on tenable assumptions regarding the actual projected situation in the south-east approach quadrant of the Tyra East platform complex gives no reason to doubt, let alone refute, the validity of the risk figures cited by DnV on the basis of historical experience.

APPENDIX H

Recommended Operation Precautions in the Platform Area

The following operational precautions are to be implemented and integrated with the overall platform complex safety philosophy of DUC.

(i) *Restricted area*

Vessel entry into a defined area including the high risk zone of the export pipeline in the Southeast quadrant should be strictly prohibited with the exception only of vessels with construction, repair, or maintenance errands in the area, and then only by express prior arrangement.

(ii) *Operator's supervisor on vessel*

Vessels should not enter the restricted zone until the operator's appointed supervisor has come on board, and the supervisor should remain on watch on board throughout the time that the vessel remains within the area.

(iii) *Use of operator's site map*

Vessels anchoring within the restricted zone should be required to plan their anchor spread in advance with the aid of the operator's own up-to-date site map. This site map should show all pipeline routes, platform locations, subsea obstacles, subsea debris, and the like.

(iv) *Contractor's surveyor on platform*

Contractors or others fielding construction, repair, maintenance, or other vessels in the restricted zone should, prior to moving in, transfer their own surveyor to the Tyra East platform complex. The surveyor should then use a tellurometer or other approved instruments to monitor all anchor running and other operations representing a potential hazard to the subsea pipelines. His sole duty should be to advise the vessel superintendent as to when it is safe to drop an anchor and inform him of the distance to any pipeline in the vicinity.

(v) *Daylight hours*

Anchor running and other potentially hazardous operations should only be permitted during daylight hours and under conditions of satisfactory visibility.

(vi) *Marking the site map*

The surveyor should be required to mark the precise position of each anchor drop and other potentially hazardous activity on the operator's site map.

(vii) *Copy of anchor pattern map*

As soon as the vessel's anchors have been set, a copy of the operator's site map, with the anchor pattern duly marked up, should be passed to the operator's supervisor for use in the control of any subsequent movement and for retention for later reference. Subsequent anchor patterns should likewise be recorded and copied to the supervisor.

(viii) *Operator's survey vessel*

As an additional precaution, in the event of major construction or repair works, the operator should, where possible, have a survey vessel on hand to provide an independent check of anchor positions prior to anchor running and again when the anchors have been set. In bad weather the survey vessel should perform a constant check to guard against dragging.

(ix) *Marker buoys*

The use of marker buoys to identify pipeline routes and obstacle locations is recommended in situations where it is feasible. The buoys should be deployed just prior to run-in. During anchor running some or all of the buoys may be run over and lost, but this does not negate their initial usefulness.

(x) *Heavy duty fixed anchor points*

To the extent that it is feasible within the context of the Tyra East platform complex development plan, one or more heavy-duty fixed anchor points should be installed in the seabed close to the TE-E platform in the south-east approach quadrant for use by construction and repair vessels. The configuration of fixed anchor points should be such as to obviate the necessity for dropping high-holding anchors in the vicinity of the export and interfield pipelines in the immediate platform area.

(xi) *Supply boat fixed anchor points*

To the extent that it is feasible within the context of the Tyra East platform complex development plan, one or more fixed anchor points should be provided in the north-west approach quadrant for use when appropriate by supply vessels, *inter alia*, in order to minimise any temptation for supply vessels to approach the platform complex from the south-east under adverse weather conditions.

(xii) *Diver assistance*

All anchoring at fixed anchor points or elsewhere within the restricted zone (other than the routine fixed point anchoring of supply vessels in the north-west quadrant) should be diver assisted.

CO-AUTHORS' BIOGRAPHIES

Michael B. Abbott, BSc, PhD, FICE, is Professor of Computational Hydraulics at the International Institute for Hydraulics and Environmental Engineering, Delft, The Netherlands. He is the designer of the *Dynamic Seapipe System* for the computer modelling of pipelines in the marine environment. Michael is a permanent advisor to the Danish Hydraulic Institute and a director with SOFT (Int) BV, The Hague. He is currently a consultant to VBB, Stockholm and Gränges Offshore, and was previously with Brown and Root, Rotterdam. He has published numerous papers about hydraulics and is author of the well-known book *Computational Hydraulics, Elements of the Theory of Free-surface Flow*, published by Pitman.

Frank Alexander, BSc, CEng, worked initially in the aero industry and then on rotating machinery for the oil and gas business. He joined Total Oil Marine in 1973, working in France and the UK before becoming Head of Engineering in Aberdeen in 1979.

Robert G. Bea, BS, MS, Registered Civil Engineer, is Senior International Consultant, Offshore Engineering Division, PMB Systems Engineering Inc, San Francisco. He has 28 years experience in engineering construction and research associated with offshore structures and pipelines and was previously employed by Shell for 17 years. Bob has published over 75 papers on offshore topics and has lectured widely on the design of foundations for platforms and pipelines. He received the James Croes Medal for his work on *Earthquake Response of Offshore Platforms* from the American Society of Civil Engineers in 1980.

Robert J. Brown, BS(CE), MS(CE), MASCE, (Registered Professional Engineer), is Chairman of R J Brown & Associates, a project management firm which has been responsible for the innovative engineering of submarine pipelines and offshore terminals on a worldwide basis, including, for example, installations in the Arctic and the Straits of Magellan.

Previously, Robert worked as an assistant chief engineer with Bechtel Corporation.

Mads B. Bryndum, MSc, MIngF (Denmark), is a senior structural and pipeline engineer with the Danish Hydraulic Institute. He has considerable experience in the design of the Danish North Sea oil and gas pipelines and was involved with the towed bundle flow-line installation in the Cormorant Field. Mads previously worked for Cowiconsultants, Copenhagen. His publications include, *inter alia*, the analysis of pipeline vibrations and pipeline lay analysis.

Robin S. Colquhoun, MA, CEng, FICE, FASCE, FPetInst, MIngF (Denmark), is a consultant to Flotec A/S, Oslo and the Danish Hydraulic Institute. He was the engineer in charge of the pipeline route surveys, pipeline safety analysis and protection concepts for the Danish natural gas transmission system marine pipelines in the North Sea and Danish Belts, and project engineer in charge of the detailed design of the 30-inch North Sea gasline. Robin has also been involved with several other pipeline projects including Chevron's Montanazo project, the Øresund pipeline between Denmark and Sweden and the Danbor flow-lines and pipelines.

Terry J. Corbishley, MSc, CEng, MRAes, MBIM, is a consultant and director of Maris (Oil Studies) Ltd. Terry's experience encompasses over 15 years in the offshore industry, embracing a very wide range of pipeline design and construction projects worldwide. These projects included major trunk pipelines, flow-line interfield networks, estuary crossings and shore approaches. For over five years he was Chief Engineer of one of the world's leading pipeline engineering consultants, where he was involved in the development of design philosophy and codes of practice. He has published technical papers on topics such as static and dynamic analysis of pipeline spans, and three-dimensional wave theory approach to pipeline stability.

Paul Davies, BSc, PhD, CEng, FICE, is Managing Director of J P Kenny & Partners Ltd, one of Europe's largest marine engineering consultants, involved with the design, engineering, procurement, construction and commissioning of trunk-lines, flow-lines, risers, pipeline barrier systems and underwater protection structures. Paul has published a number of technical papers and articles on the analysis of steel structures and has lectured on structural mechanics in Sweden and Poland.

Roger F. de la Mare, BSc, MBA, PhD, CEng, FIChemE, MBIM, is Chairman of the Postgraduate School of Technological Management at the University of Bradford and a visiting professor at the Norwegian Institute of Technology. His industrial career included 11 years experience with Courtaulds, Esso and Rio Tinto Zinc Corporation as a process engineer, technical superintendent, factory manager and corporate planner. At Bradford, he has played an important role in the re-design of all engineering programmes to include relevant courses in business studies. His research is mainly involved with engineering economics, maintenance management and reliability engineering, in which he is best known for his industrial reports: *Pipeline Reliability* and *Optimal Equipment Replacement Policies*, and his book *Manufacturing Systems Economics*, published by Holt, Rinehart and Winston. Roger has been employed regularly as a consultant by national and international companies.

David Geary, MA, PhD, MICorrSt, is Director of Corrosion and Protection Centre Industrial Services (CAPCIS), UMIST. He has been head of this university-based consultancy and R & D unit for the last 11 years, specialising in corrosion-associated problems of industry on an international basis. David's experience includes work on the selection of corrosion inhibitors and the internal and external corrosion coatings for submarine pipelines.

Robert M. Gibbons, BA(Sc), PhD, DIC, CEng, FIGS, FRCS, worked on the UK Gas Gathering Project 1979–1981 in the joint BGC-Mobil, and later, BGC-BP-Mobil teams. He manages the group in British Gas that develops and maintains computer simulation programs for the design of single and two-phase pipeline systems. He has published about 40 papers in the areas of thermodynamic properties and simulation of pipeline systems.

Helge Gravesen, BScE, is a senior engineer with Rambøll & Hanneman where he has been involved with the design of the Danish North Sea gas-line. Lately he has been responsible for the design work in connection with the large line from Norway, through Sweden and Denmark to West Germany, the Norske Hydro basic engineering for the Oseberg to shore line and the Danbor interfield lines.

Jørgen Juncher Jensen, PhD, is an associate professor in the Department of Ocean Engineering at the Technical University of Denmark. His research is mostly concerned with the buckling and collapse of pipes and pipelines.

Roger A. King, BSc, MSc, PhD, MICorrSt, is Deputy Manager – Long Term Projects, Corrosion and Protection Centre Industrial Services (CAPCIS), UMIST. Most of his work is involved with the cathodic protection of subsea pipelines and risers, including the evaluation of corrosion protection methods, microbiological corrosion, chemical dosage, vacuum drying and a/c interference with buried pipelines.

R. W. Luynenburg, Dip Ing, MRIE (Holland), works for R J Brown & Associates, The Hague. For the past six years, he has worked on problems involved with the design, construction and maintenance of many pipelines, tie-ins and risers. In particular, he has personally supervised pipeline installations by bottom-tow and tie-ins using, amongst other methods, the 'deflect-to-connect' technique. More recently he has been involved with pipeline spanning and retrieval problems in the North Sea.

Ian McKenzie, BSc, CEng, is Survey Systems Manager with UDI Group Ltd. He has been involved with pipeline acoustic inspection and the development of improved techniques since 1977.

Richard Nielsen, MSc, DTech, is Principal Consultant to Flotec A/S, Oslo, and has worked recently on projects for Statoil, Norske Hydro, Hamilton Brothers (UK) and Danbor (Mœrsk Offshore), as well as various joint industry research projects. Previously he was Senior Vice-President of Global Technical Services Ltd and then Global Engineering (Denmark) A/S, with responsibility for platform and pipeline design in New Zealand, Brunei and the Danish offshore fields. He was a joint venture board member and a chief consultant on the engineering of the Danish gas transmission system marine pipelines. He is the author of technical papers on a wide range of subjects.

Andrew C. Palmer, BA, PhD, CEng, FICE, is Vice-President, Engineering of R J Brown & Associates Corporation A/G, and was previously Professor of Civil Engineering, UMIST. He has been involved with the design and installation of several major pipeline projects including the Panartic Drake F-76 flowline system, which was the first arctic submarine pipeline, the Statfjord delta loading line, the trans-Alaskan pipeline system and several others in Canada and Abu Dhabi, most of which have involved considerable engineering innovation.

P. Ternrup Pedersen, PhD, is Professor of Ocean Engineering at the Technical University of Denmark. His research covers the mathematical modelling of pipelaying procedures and the buckling/collapse of pipelines.

Dudley P. Penick, BS, MS, is Manager of Gas Engineering for Mobil Exploration and Producing Services Inc. Dallas. He has 35 years experience with Mobil companies worldwide, specialising in natural gas. His experience includes gas production, treating, compression, processing, transportation, utilisation and oil/gas separation and recovery systems. Dudley manages an engineering group which provides consultant services in gas engineering to Mobil-affiliated companies which are engaged in exploration and production. He is also Mobil's representative on the Gas Processors Association Board of Directors.

René Quin was educated at the University of Strasbourg as a geophysicist. In 1957 he joined Total and took part in exploration ventures in Libya, Australia, Indonesia, Peru, North America and Greenland. Since 1979 he has been Technical Manager of Total Norske A/S in Stavanger, and Project Manager of the 'Cold Tapping' research and development projects.

Michael W. Rainbow, BSc, FGS, MIG, MHyd Soc, started his career as a seismologist and then moved into engineering geology and geophysics with successive companies including Hunting and Dames & Moore. By 1973 he was chiefly concerned with offshore site investigation by applied geophysical techniques. In 1980 he joined Gardline as Operations Manager of their Aberdeen base, and became involved in the Shell-backed joint venture with UDI to carry out the first inspections of North Sea pipelines employing the Colour Video Sonar Systems, and to develop the necessary interpretative and reporting techniques. In 1982 he joined Welldrill, a firm of petroleum exploration consultants, as a senior geologist/geophysicist and is now Manager of their Exploration Department.

A. R. Schultz, BS, MS, RegEngr, MASCE, MMarTech Soc, is Vice-President, Offshore Terminals and Floating Production Systems Division, R J Brown & Associates of America Inc. He is responsible for management and technical review of engineering and construction projects, and for development of deepwater systems using innovative concepts in the fields of SPM and marine technology. He has 14 years of international experience in project engineering and construction management of offshore terminals, pipelines, platforms and production systems. His experience is based on all phases of project development from feasibility and conceptual studies, model testing and design engineering through procurement, fabrication, installation and commissioning. He has worked overseas on projects in South East Asia and the Middle East and has been associated with projects for the North Sea, Gulf of Alaska, offshore West Africa, Gulf of Mexico and offshore Eastern Canada. Mr Schultz has presented and published papers on the use of SPM and floating production systems for the development of deepwater or marginal fields and on engineering and construction of offshore terminal pipeline systems.

Roe D. Strømmen, MSc, Dr Ing, is General Manager of CorrOcean A/S, Trondheim, Norway. His work has involved offshore cathodic protection and monitoring of North Sea pipelines and structures. He is also well-known for his pioneering work in computer modelling techniques.

William B. Thrasher, BS, retired in June 1984 after working for Mobil for 34 years. Most of this experience was involved with natural gas production and processing. Recently, he was Process Section Chief for Mobil Exploration and Producing Services Inc, Dallas.

Jens-Erik Thygesen, MSc, was a project leader at DONG, directing work on the land and marine pipelines of the Danish Natural Gas Transmission System (1979–81) and then Deputy Project Manager for the engineering and construction management of the total gas transmission system including the treatment plant, meter regulator stations, underground cavern storage and remote control and communications systems. Since 1983 he has been Head of DONG Marine and Technical Departments.

Nimal S. Wijesinghe, DipSurv, is an associate director with UDI Group Ltd and was previously Managing Director of COSAG (Hydrographic, Geophysical and Submarine Surveyors). His experience as a land and hydrographic surveyor covers 21 years to date. During this period he has worked for government departments, contractors, consultants and several oil companies, both in the UK, Sri Lanka and elsewhere. Nimal has six years experience as a chief surveyor and director in charge of planning, organising, directing and controlling survey activities with regard to the training of personnel, job pricing, setting up surveys, trouble shooting, plan preparation and the preparation of final reports.

SUBJECT INDEX

C

D

T

AUTHOR INDEX